Developments in Plant and Soil Sciences

VOLUME 81

The titles published in this series are listed at the end of this volume.

RESOURCE MANAGEMENT IN RICE SYSTEMS: NUTRIENTS

Resource Management in Rice Systems: Nutrients

Papers presented at the International Workshop on Natural Resource Management in Rice Systems: Technology Adaptation for Efficient Nutrient Use, Bogor, Indonesia, 2–5 December 1996

Edited by

V. BALASUBRAMANIAN

J.K. LADHA

and

G.L. DENNING

Partly reprinted from *Nutrient Cycling in Agroecosystems*, Volume 53, No. 1 (January 1999)

KLUWER ACADEMIC PUBLISHERS

DORDRECHT / BOSTON / LONDON

in cooperation with the
International Rice Research Institute

Library of Congress Cataloging-in-Publication Data

A C.I.P. Catalogue record for this book is available from the Library of Congress.

ISBN 0-7923-5199-1

Published by Kluwer Academic Publishers,
P.O. Box 17, 3300 AA Dordrecht, The Netherlands

Sold and distributed in North, Central and South America
by Kluwer Academic Publishers,
101 Philip Drive, Norwell, MA 02061, U.S.A.

In all other countries, sold and distributed
by Kluwer Academic Publishers,
P.O. Box 322, 3300 AH Dordrecht, The Netherlands

Printed on acid-free paper

Printed in the Netherlands

Contents

* Chapters indicated with an asterisk are reprinted from *Nutrient Cycling in Agroecosystems* Volume 53, No. 1 (January 1999)

PART II: FARMER ADOPTION OF NUTRIENT MANAGEMENT TECHNOLOGIES: CONSTRAINTS AND OPPORTUNITIES

PART III: APPROACHES AND METHODOLOGIES FOR EVALUATION AND PROMOTION OF CROP AND RESOURCE MANAGEMENT TECHNOLOGIES

Foreword

In the next three decades, we will require about 60% more rice than the 1994 production level of 535 million tons to feed the expanding population of rice consumers. This demand for more rice must be met from limited resources--land, water, energy, labor, and gene pool. To continuously increase food supplies with limited natural resources is a daunting challenge to the scientific community.

Most of the additional rice production must come from irrigated and favorable rainfed ricelands. We need to use more complex technologies and management practices to further intensify rice production systems and to conserve resource base from which all food is produced. The immense task before us is how to orient our research to generate innovative technologies and to organize/motivate different delivery agents to effectively disseminate them to rice producers. Farmers need much more education to receive, interpret, adapt, and use complex information and technologies. This is where IRRI's Crop and Resource Management Network (CREMNET) will make a significant contribution--by facilitating the on-farm evaluation and promotion of innovative, knowledge-intensive rice technologies and management practices to farmers.

As a part of the exchange of new research findings, innovations, and experiences, CREMNET/IRRI organized an international workshop on Natural Resource Management in Rice Systems: Technology Adaptation for Efficient Nutrient Use at Bogor, Indonesia, 2-5 Dec 1996, in collaboration with the Agency for Agricultural Research and Development (AARD), Indonesia. A total of 55 international scientists and development agents from public, cooperative, private, and NGO sectors participated in the workshop. They discussed and brainstormed the nutrient management issues involved in rice intensification and methods/approaches for dissemination of crop and resource management technologies including that of nutrients to rice farmers.

Papers presented in the workshop have been edited and published in this book for wider dissemination. The papers discuss knowledge-intensive nutrient management technologies for rice, farmer adoption constraints and opportunities for improvement, and methods and approaches for evaluation with and promotion of new technologies to farmers. I am confident that this book will be very useful to rice research and development workers all over the world.

Kenneth S. Fischer
Deputy Director General
for Research
International Rice Research
Institute

V. Balasubramanian, J. K. Ladha and G. L. Denning (eds.), Resource Management in Rice Systems: Nutrients, ix.

Preface

Rice is the major staple food in Asia; about 95% of the world's rice is produced and consumed here. Thus, the present and future food security of Asia will heavily depend on rice security for a foreseeable future.

Currently, Asian rice production increases at an annual rate of only 1.4%, which is below the population growth rate. This is rather disturbing to scientists, extension and development workers, policymakers, and consumers. By the year 2025, we need to produce about 60% more rice than what we produce today to meet the growing demand. About 70% of the additional production will have to come from the irrigated riceland and 20% from the favorable rainfed lowland which are already intensively cultivated. Further intensification of these ricelands must be pursued against a backdrop of shrinking land area and decreasing availability and increasing cost of production inputs: water, fertilizer, chemicals, labor, and energy.

Irrigated rice yield must be raised from the present level of 5 t ha^{-1} to 8 t ha^{-1} by the year 2020 to secure the targeted production. To reach this yield level, we have to increase the use of N fertilizers by 180% at 33% recovery efficiency, and by 87% at 50% recovery efficiency. We can clearly see the importance of promoting the efficient use of N fertilizers to produce the needed rice at minimum economic cost. The same is true for all other inputs.

We increasingly face ecological and environmental problems as a result of injudicious use of fertilizers and other chemicals (herbicides, pesticides) in intensive rice farming. The deteriorating soil and water quality and the rising agrochemical toxicity in farming are serious concerns. All these factors jeopardize efforts to sustain growth in food production. The big question before us is "Can we sustain high productivity with continued inefficient use of inputs and deterioration of the rice crop environment?".

A consultation sponsored by the Food and Agriculture Organization of the United Nations held at Bangkok in October 1996 pointed out the grave situation of declining productivity in intensive rice farms of Asia. The participants identified deterioration of soil quality as a major factor responsible for the downward yield trends. The other constraints, in the order of decreasing importance, are related to water, socioeconomic and policy aspects, buildup of biotic stresses, and varietal degeneration.

Given the above scenario, we should examine whether there are adequate technologies available to address the abovementioned rice production problems and challenges or there is a need to develop new technologies to remove some of these constraints. In addition, we may have to reorient and improve the technology delivery systems, especially in the dissemination of increasingly knowledge-intensive technologies to millions of small-scale rice farmers. We also have to devise attractive incentive systems to encourage farmers to efficiently produce more rice on less land.

V. Balasubramanian, J. K. Ladha and G. L. Denning (eds.), Resource Management in Rice Systems: Nutrients, xi–xii.

This book brings together the papers presented at the International Workshop on Natural Resource Management in Rice Systems: Technology Adaptation for Efficient Nutrient Use, held at Bogor, Indonesia, 2-5 December 1996. It deals with nutrient management issues that we face concomitant with intensification of rice farms. Part 1 is devoted to the discussion of emerging nutrient management strategies and options to optimize rice productivity and to sustain soil quality in intensive rice production systems. Farm-level constraints and opportunities for adoption by farmers of improved nutrient management technologies in rice cultivation are analyzed in Part II. The final section, Part III, explores the innovative approaches and methodologies used for evaluating and promoting crop and resource management technologies including that of nutrients.

V. Balasubramanian

Coordinator, Crop and Resource
Management Network (CREMNET) and
Chair of the Workshop Organizing
Committee

PART I

Nutrient management strategies and options

Strategies for nutrient management in irrigated and rainfed lowland rice systems

A. DOBERMANN[1] and P.F. WHITE[2]

[1]*International Rice Research Institute, P.O. Box 933, 1099 Manila, Philippines*

[2]*Cambodia-IRRI-Australia Project, P.O. Box 1, Phnom Penh, Cambodia*

Key words: irrigated rice, rainfed lowland rice, site-specific nutrient management, soil nutrient supply, fertilizer recommendations

Abstract

Attaining maximum economical yield is the priority of irrigated rice farmers. Current fertilizer management practices in intensive, irrigated rice systems are, however, not tailored to the large between-field differences in indigenous nutrient supply and crop demand. Substantial improvements in nutrient use efficiency and economic performance will require a site-specific nutrient management (SSNM) approach. The major components of a field-specific, knowledge-based strategy are quantification of crop nutrient requirements based on nutrient interactions and economic yield target; measurement of potential N, P, and K supply; and monitoring of plant N status to optimize N nutrition. Recent research has provided many of the practical tools needed, but they need further refinement and have to be applied in an integrated manner. Because of often poor soils, high degree of spatial variability of soils, spatial and temporal variability of water availability, high risk of crop failure, and fewer resources for capital expenditure, rainfed lowland rice farmers give priority to reducing risk. Farmers' ability to invest in cost-intensive innovative technologies is limited. As a consequence, the range of technologies likely to be adopted by rainfed lowland farmers is restricted. Farmers' knowledge and experience become vital for the efficient management of nutrients in these environments. We discuss the nutrient management strategy that is being developed for Cambodia. Technologies are generated and tested through research at selected representative sites. Technologies are extrapolated and applied through a new agronomic soil classification and probabilistic modeling that integrates farmer's knowledge and experience with empirical data generated from the research sites.

Challenges

Irrigated and rainfed lowland rice systems account for about 95% of the total rice production in Asia (Table 1). Associated with the Green Revolution in Asia, intensification of rice culture has occurred mainly in irrigated areas. At the same time, vast rainfed lowland and upland areas managed with low inputs had little progress during the past 20 or 30 yr. Undoubtedly, the intensive lowlands will remain the major source of rice production until, and probably beyond, 2025, but the agronomic and environmental pressure on highly productive irrigated lands is increasing. To keep pace with population growth, yields in both irrigated and favorable rainfed lowlands have to increase 60-70% within the next 30 yr (Table 1).

Along with this increase in grain yields, the demand for nutrients will increase substantially. For example, the current annual net nutrient removal with harvested grain from irrigated ricefields in Asia is in a range of 3.8-5.1 x 10^6 t N, 0.9-1.2 x 10^6 t K, 0.5-0.9 x 10^6 t P, and 0.2-0.5 x 10^6 t S (Table 2). Actual net nutrient removal from irrigated rice cropping systems is much greater because the amount of recycled straw is decreasing due to use of straw as forage or fuel or due to its removal to facilitate manual threshing

V. Balasubramanian, J. K. Ladha and G. L. Denning (eds.), Resource Management in Rice Systems: Nutrients, 1–26.

Table 1. Projected changes in rice area, yields, and total production required to sustain sufficient rice supply in Asia during the next 30 yr. The scenario shown assumes that the area under cultivation remains constant (*Source:* Cassman and Pingali 1995b).

Rice system	1991			2025		
	Area (10^6 ha)	Yield (t ha^{-1})	Production (10^6 t)	Area (10^6 ha)	Yield (t ha^{-1})	Production (10^6 t)
Irrigated	73.9	4.89	362	73.9	8.00	592
Rainfed lowland	38.7	2.30	89	38.7	3.50	136
Upland	10.5	1.07	11	10.5	1.07	11
Deepwater	10.0	1.53	15	10.0	1.53	15
Total	133.1	3.58	477	133.1	5.66	754

Table 2. Current and projected requirements of N,K,P, and S in irrigated rice systems of Asia. Estimates for 1991 are based on a harvest area of 74 million ha and an average yield of 4.9 t ha^{-1}. Estimates for 2025 assume a constant harvest area of 74 million ha and average yields of 8 t ha^{-1} to meet the projected rice demand of 592 million t from irrigated systems (Cassman and Pingali 1995b). All values are given on elemental basis.

Nutrient	Average nutrient content at harvest (%)		Uptake per t of grain yield[c] (kg t^{-1} grain)	Annual removal with grain (10^6 t yr^{-1})		Total annual uptake with grain and straw (10^6 t yr^{-1})	
	Grain	Straw		1991	2025	1991	2025
N[a]	1.05-1.40	0.50-0.80	15-22	3.8-5.1	6.2-8.3	5.4-8.0	8.9-13.0
Potassium[a]	0.25-0.33	1.30-2.00	15-25	0.9-1.2	1.5-2.0	5.4-9.0	8.9-14.8
Phosphorus[a]	0.15-0.25	0.05-0.10	2-4	0.5-0.9	0.9-1.5	0.7-1.4	1.2-2.4
Sulfur[b]	0.06-0.15	0.05-0.10	1.5-2.5	0.2-0.5	0.4-0.9	0.6-0.9	0.9-1.5

[a]N, P, and K concentrations in grain and straw measured as the interquartile range of 192 plots of long-term fertility experiments with rice at 11 sites in China, India, Indonesia, and Philippines (K. Cassman and A. Dobermann, unpubl. data).

[b]Sulfur concentrations in grain and straw based on literature data.

[c]Average total nutrient uptake in above ground biomass (grain + straw) per t of grain yield adjusted to 14% moisture content.

and rapid turnaround time between rice crops. To achieve the projected yield goals in irrigated rice systems of Asia, total annual crop demand will be about 9-13 x 10^6 t N, 9-15 x 10^6 t K, 1.2-2.4 x 10^6 t P, and 0.9-1.5 x 10^6 t S in 2025 (Table 2).

Breeding varieties with improved external nutrient acquisition capacity and high internal nutrient use efficiency remains an important strategy for all rice systems, the harsh rainfed environments in particular (Kirk et al 1997, Ladha et al 1997). Within the next 10 yr, new semidwarf rice with 13-15 t ha^{-1} yield potential in the tropics and resistance to major pests will be probably fully developed (Peng et al 1994) and adopted by farmers. It remains another challenge to also incorporate traits such as increased internal and external nutrient use efficiency. Because farmers can usually achieve yields up to 80% of the yield potential on an economically sustainable basis (Cassman and Pingali 1995a), the new plant types are likely to provide a quantum leap toward increased yields.

However, we do not know how sustainable this productivity increase will be, or, how soil and crop management should be changed to make it sustainable. Therefore, improving nutrient management (NM) is as equally important as breeding for high yield. At issue is whether the current NM strategies practiced by rice farmers are adequate to maintain the yield gains achieved over the long term and to support the future yield increases needed.

Research during the past 30 yr has led to many specific technologies and tools that are applicable across different rice-based cropping systems. Many of these technologies have been described elsewhere (Obcemea et al 1984, Youngdahl et al 1986, Buresh et al 1988, Takebe and Yoneyama 1989, Becker et al 1995, Dobermann et al 1995, George et al 1995, Kundu and Ladha 1995b, Singh et al 1995, Kundu et al 1996, Ladha et al 1996, Peng et al 1996) or are discussed in other papers in this proceedings. The technologies have different importance for different environments, but, in any case, economic rationality and adaptability determine farmers' choice of NM strategies (van Noordwijk and Scholten 1994).

The goals of NM differ in irrigated and rainfed rice systems, but for a given set of environmental and economic conditions, the amount of nutrients supplied to a crop should meet its demand to achieve the desired outcome targeted by a farmer. A delicate balance of nutrient supply and removal that satisfies the system and site-specific farming goals by also securing maintenance of soil and environmental quality over the long run has to be found. Differences in the biophysical and socioeconomic system characteristics have to be considered when options and strategies for NM are discussed.

Our paper will be limited to a general discussion of conceptual NM models for the two major rice production systems. We summarize some key issues related to strategies for NM, focusing on management targets, knowledge demand, and practical implementation. We discuss two recent examples of NM concepts proposed for irrigated and rainfed lowland rice environments.

Nutrient management recommendation domains

The success of any NM strategy depends on the spatial scale for which management decisions are made, and the issue of optimal recommendation and resource management domains has become a subject for debate in recent years (Byerlee 1987). Combined with high-yielding varieties, crop management strategies over the past 4 decades in Asia were driven by increased use of externally provided inputs and the so-called "package approach" based on blanket recommendations over wide areas (Byerlee 1996). Although a farmer's decisions are always related to his field, many of his deductions or the recommendations given to him are general . Farming philosophies in practically all rice systems are dominated by Type I or Type II recommendations (Byerlee 1987) [1]. Simple rules, use of little quantitative information, and large resource management domains predominate (Fig. 1).

Perhaps the most striking examples are blanket prescriptions for fertilizer use, which do not account for the variation in indigenous nutrient supply at different spatial scales.

---

[1] Type I - recommendations that are based on targeting only at the level of a recommendation domain; Type II - recommendations that are conditional on easily observable *qualitative* site and seasonal variables in a specific field; Type III - recommendations that are conditional on *quantitative* site and seasonal variables in a specific field, such as a soil test or pest count (Byerlee 1987).

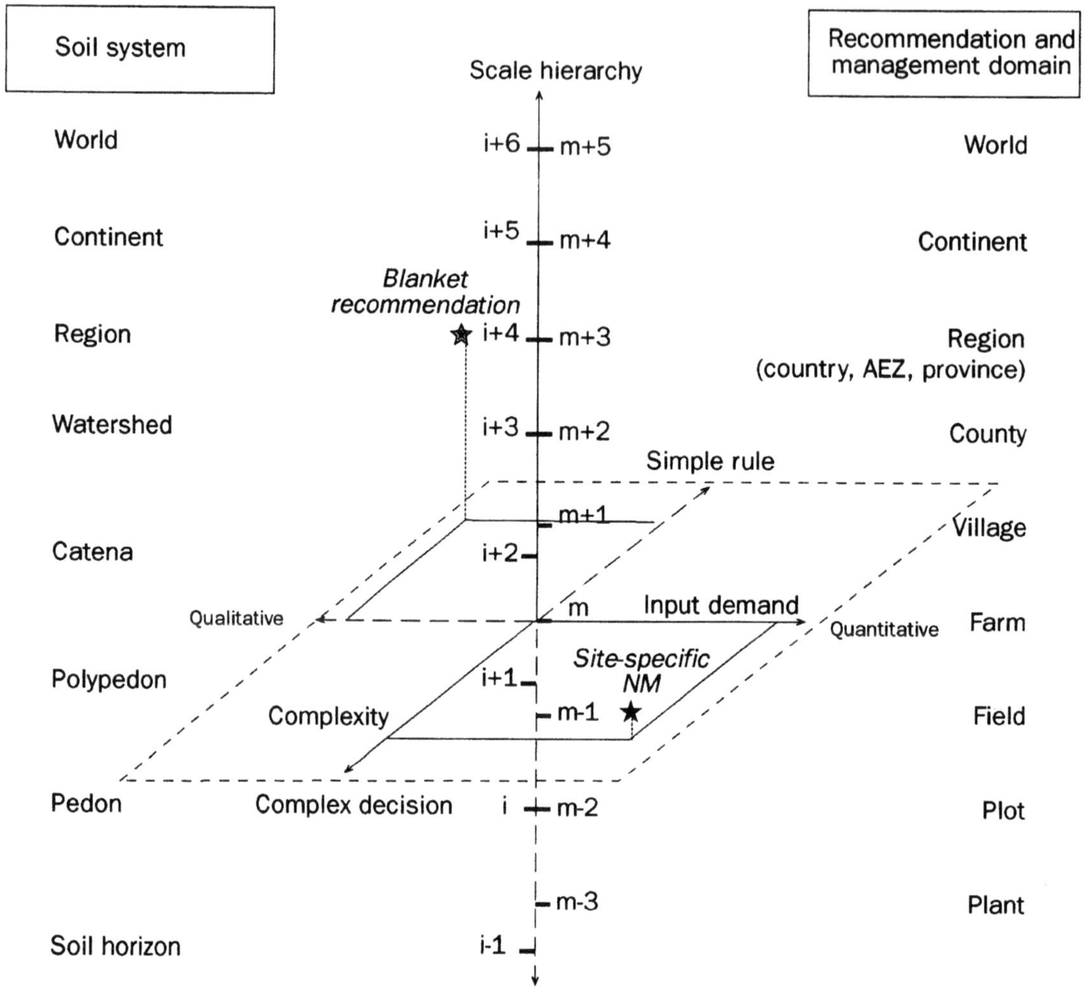

Figure 1. Scale diagram for three-dimensional classification of land resource management. The approximate positions of blanket and site-specific nutrient management are drawn as examples. Modified from the model classification diagram proposed by Hoosbeek and Bryant (1992).

Recent research has clearly demonstrated large between-farm variation in soil nutrient supply within rice-growing domains at scales of one or few villages. In small rice production domains of five Asian countries, the magnitude of difference in grain yield from unfertilized plots among farms within each domain ranged from 2.8 to 4.9 t ha^{-1} (Table 3). In two small rice production domains in the Mekong Delta in Vietnam and Central Luzon in the Philippines, resin adsorption quantities (as an index of soil nutrient supply) of N, P, and K after 7 d incubation had CVs of 42-85% among the farmers (Dobermann et al 1996a). Great between-farm variability in yield responses to P, K, Zn, and S was found in on-farm studies at another site in the Philippines (Angus et al 1990).

Table 3. Farm size and grain yield of unfertilized plots in farmers' fields in major rice-growing domains of Asia, 1994-95 (IRRI 1996).

Location	Fields/sites (no.)	Farm size of participating farmers in the project (ha)			Grain yield of unfertilized plots (kg ha^{-1})		
		Mean	Min	Max	Mean	Min	Max
Cuu Long Delta Rice Research Institute, Omon, Mekong Delta, Vietnam	32	0.92	0.23	3.60	4247	2490	5703
Philippine Rice Research Institute, Maligaya, Central Luzon, Philippines	48	1.91	0.35	5.00	4363	1835	6729
Research Institute for Rice, Sukamandi, Northwest Java, Indonesia	30	1.25	0.10	4.97	4887	3125	6260
Suphan Buri Rice Experiment Station, Suphan Buri, Thailand	30	2.04	0.96	3.52	3470	1964	4721
Tamil Nadu Rice Research Institute, Aduthurai, Tamil Nadu, India	30	2.47	0.29	6.40	4637	2488	5775

Although we do not have long-term on-farm monitoring studies covering the whole Green Revolution period, it is likely that this variation has resulted from different farming practices affecting nutrient balances over time frames of 20-30 yr. Farmers often do not follow the blanket recommendations of government agencies. In most rice-growing environments where farmers have no access to services such as soil testing, determining how much, what kind, when, and how to apply fertilizer remains a difficulty. Consequently, a large proportion of applied nutrients is not efficiently used, causing agronomic and environmental sustainability problems. Inefficient use of nutrients may not be simply because of unavailability of soil-testing facilities or differences in soil nutrient supply per se. Other important factors include variability in water management, general soil properties, application of organic matter, and pest and disease problems.

Little improvement in fertilizer use efficiency can be expected from the currently used blanket fertilizer recommendations, and we should endorse more site-specific approaches in the lowlands of Asia. Such approaches will succeed only if all essential nutrients and their interactions are integrated, if the additional labor required is minimum, and if the economic gain is great enough. It is also essential that scientific results are translated into simple and robust, nevertheless sound, recommendations. This specialized and difficult task requires that scientists and extension specialists work together.

Site-specific NM (SSNM) has become a reality in the rice-growing regions of Asia, but the actual size of a NM domain and its information demand will depend on the farming goals and available resources. In irrigated rice, field-specific NM, i.e., using single ricefields as the principal management and decision units, based on quantitative input data and relatively complex decisions will probably be necessary to achieve yields greater than 8 t ha^{-1} (Fig. 1). In rainfed lowland rice systems, the size of resource management domains may be more variable, depending on the actual spatial and temporal variation in key environmental factors. Coupled with less expensive information collection, soil-specific fertilizer recommendations may suffice the needs of a rainfed farmer.

More research is needed to quantify the agronomic, environmental, and economic benefits of NM based on different sizes of recommendation domains and different input demand. It is clear, however, that a move from regional scales and Type I recommendations to village, farm, field, or within-field scales of management combined with Type II and III recommendations (Fig. 1) will be a gradual process, with many transition forms and differences among crop management practices. Over the long term, NM, for example, will probably focus on single fields. Over the shorter term, other cost-effective methods, such as use of readily available soil information, local "soft" knowledge, or simple agronomic soil classification systems, may improve fertilizer recommendations at village or county scales, or, may be used to estimate a "window of opportunity" (Bouma 1995) for a specific soil type.

Nutrient management strategy for the intensive irrigated rice system

Biophysical and socioeconomic system characteristics

The two major irrigated rice cropping systems are double- and triple-crop monoculture rice, or rice - wheat. Together, they cover about 36 million ha in Asia and account for about 50% of the global rice supplies (Cassman and Pingali 1995b). Some important characteristics with regard to NM strategies are:

1. Irrigated rice areas are found on alluvial flood plains, terraces, inland valleys, and deltas in Asia. They have assured irrigation for one or more crops a year and soils are mostly fertile (Kawaguchi and Kyuma 1977).
2. Before the Green Revolution, traditional varieties with long growth periods of 130-210 d were grown in single-crop systems, the harvest index and yields were low, and much of the straw remained in the field (Uexkuell and Beaton 1992). Natural components of the nutrient balance (sedimentation, organic residues, biological N_2 fixation [BNF]) played a major role in securing the sustainability of traditional rice systems at, however, low yield levels.
3. The early-maturing, high-yielding semidwarf indica varieties introduced in the 1960s responded well to fertilizer N additions. The higher yield potential promoted the construction of new irrigation facilities. Farmers were then able to grow two or even three rice crops per year and the average grain yields of irrigated rice rose to about 4.9 t ha^{-1} in 1991 (Cassman and Pingali 1995b). To facilitate land preparation for the next crop, farmers cut the entire crop and removed or burned the straw (Uexkuell and Beaton 1992). The nutrient demand, both in terms of total amounts taken up by the rice crops and in terms of rates of supply, increased rapidly. Even though nutrient inputs from natural sources such as BNF still play an important role, mineral fertilizer inputs are dominant factors of the overall nutrient balance in intensive rice systems.
4. Intensification has led to subtle changes in soil quality, including supply of resources for crop growth. Symptoms such as declining rice productivity in fully fertilized plots or nutrient omission plots (Flinn and De Datta 1984, Nambiar 1994, Cassman and Pingali 1995a), declining soil N-supplying capacity (Cassman et al 1995, Cassman et al 1996a), changing soil organic matter quality (Olk et al 1996a), diminishing returns to fertilizer N application

(Cassman and Pingali 1995a), exhaustion of soil P or K reserves (Tiwari 1985, Uexkuell 1985, De Datta et al 1988, Chen et al 1992, Dobermann et al 1996b,c, Dobermann and Oberthuer 1997), soil physical and microbial changes due to continuous flooding (Sharma and De Datta 1986, Kundu and Ladha 1995a, Roger 1996), and large between-farm variation in indigenous soil nutrient supply (Angus et al 1990, Dobermann et al 1996a, Oberthuer et al 1996, Olk et al 1996b) provide evidence that the soil resource base is changing. Most of these changes appear related to factors such as increasing length of submergence and negative nutrient balances.

5. N is deficient in practically all intensive irrigated rice systems. Due to secured water supply, good yield response to fertilizer N is usually observed and N management is the key factor toward achieving high productivity (Savant and De Datta 1982, De Datta and Patrick 1986, Cassman et al 1994, 1997).
6. Both negative and positive P balances can be found in irrigated rice areas (Dobermann et al 1996b), but, in the past, sufficient K nutrition and maintenance of soil K reserves have been neglected in many fertilizer recommendations. Most farmers do not apply much K fertilizer, and straw is increasingly removed from the field. For example, in 63 farms in Central Luzon, Philippines, the K balance was negative in most farms and average

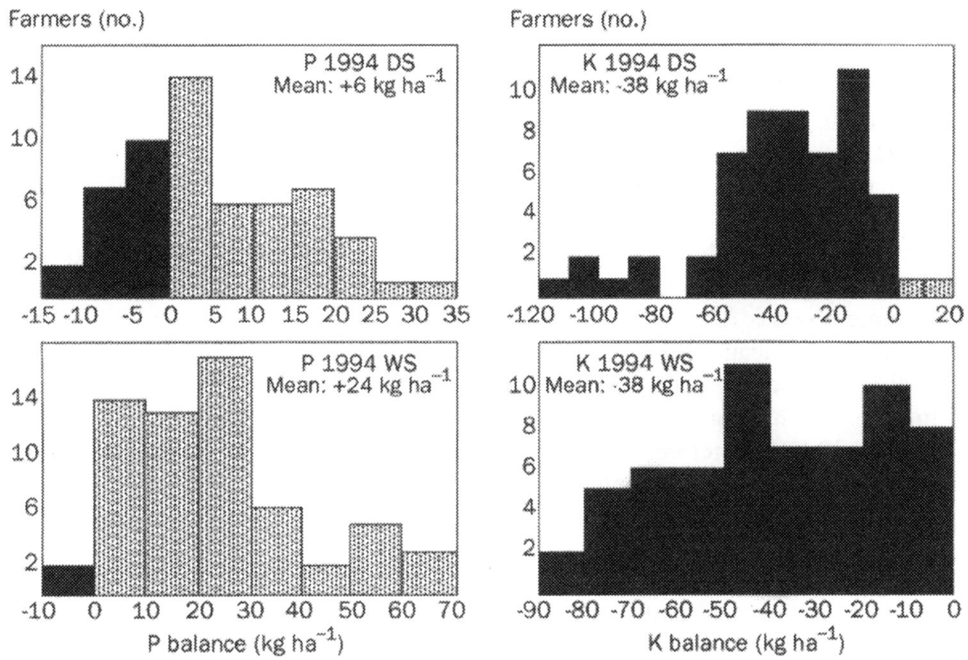

Figure 2. **P and K balance in farmers' fields with rice - rice cropping system. Data shown are based on measurements in 63 farmers' fields of Nueva Ecija, Philippines, during the 1994 dry and wet seasons. Negative balance shown as dark bars.**

losses of 38 kg K ha^{-1} $crop^{-1}$ were estimated in the 1994 dry and the wet seasons (Fig. 2). Potassium deficiency has become a constraint in soils that were previously not considered as K-limiting (De Datta and Mikkelsen 1985, Tiwari 1985, Uexkuell 1985, Mohanty and Mandal 1989, Chen et al 1992, Oberthuer et al 1995, Dobermann et al 1996c).

7. Due to relatively safe water supply, the risk of crop failure is usually less than in rainfed rice areas. Therefore, attaining maximum economic yield is the priority of irrigated rice farmers and many of them have more resources at their disposal than rainfed rice farmers. Nevertheless, investments in fertilizers or other nutrient inputs such as organic manure or green manure depend heavily on socioeconomic factors, including market availability of fertilizers, prices, timely availability of money, other commitments, or availability and cost of labor.

Requirements for nutrient management strategies

Have research and technology transfer kept pace with the demand for improved NM technologies in irrigated rice? Perhaps not. Despite intensive research on mechanisms causing N losses from applied fertilizer, average N use efficiency in farmers' fields has remained unchanged at about 30-40% over the past 25 yr (Cassman and Harwood 1995). Agronomic N use efficiency varies widely and, at the farm level, is often suboptimal due to less appropriate timing and rate of N applications (Peng et al 1996) or insufficient supply of other nutrients (Dobermann et al 1996a). Although researchers started to raise concern about the danger of negative K balances and soil K depletion many years ago (Kemmler 1980, De Datta and Mikkelsen 1985, Uexkuell 1985), that has not led to a significant improvement of K management. Other nutritional disorders such as micronutrient deficiencies have become more important management factors (Randhawa and Katyal 1982).

Past experience has shown that subtle and more complicated forms of soil and environmental degradation occur in the intensive rice system and present bigger problems than the more obvious forms such as salinity, erosion, or acidity. Because of long-term flooding and intensification of rice cropping, problems such as soil P or K depletion (De Datta et al 1988) or changing soil organic matter quality (Olk et al 1996a) develop slowly over time. New technologies often mask the disturbing effects on soil quality (Cassman and Harwood 1995). Therefore, future farming strategies will need to change to securing short-term farming gains and maintenance of land quality and profitability over the long run. The NM requirements differ for each essential nutrient:

1. *Dynamic soil- and plant-based management is needed for N.* The ability to adjust the quantity of applied N in relation to variation in indigenous N supply is as important as the timing, placement, and source of applied N (Cassman et al 1996b, Peng et al 1996). N management should include

- estimating crop demand, potential indigenous nutrient supply, and recovery from applied inorganic and organic sources over time to predict the total amount of N needed;
- estimating soil N release during early growth stages to identify the need for a basal N application; and
- monitoring plant N status to optimize timing of split applications in relation to crop demand and soil N supply.

2. *Management of P and K.* P and K management requires a long-term strategy because P and K are not easily lost or added to the root zone by the biological and chemical processes. Therefore, the issue of maximizing the recovery efficiency of fertilizer P and K is less important than predicting the need for applied nutrient and the amount to apply. Management must emphasize the maintenance of available soil nutrient supply to ensure that soil P and K supply does not limit crop growth and thus reduce N use efficiency. Changes in potential indigenous nutrient supply can be predicted as a function of the partial nutrient balance. Key components of P and K management should be

- estimation of crop demand, potential indigenous nutrient supply, and recovery from applied inorganic and organic sources over time to predict the P and K inputs required to maintain a targeted yield level;
- decision about timing of K applications depending on soil K buffering characteristics and relationships to pest management; and
- knowledge of the relationship between the P and K balance and changes in potential indigenous nutrient supply over time.

Research has provided much information on P and K balances and crop requirements in irrigated rice, but little is known about the effects of fertilizer and crop management practices on recovery efficiencies and residual effects.

3. *Diagnosis of potential deficiencies.* Diagnosis is the key management tool for nutrients such as Mg, Zn, and S. Once identified as a problem, deficiencies can be alleviated by regular or irregular (single) measures as part of a general recommendation. Similarly, diagnostic criteria can be used to identify other nutritional disorders such as salinity, Fe toxicity, or B toxicity, to make adjustments in the N, P, and K management which accounts for these limitations, or to alter soil management practices to reduce the severity of these toxicities.

Farm management should also become more based on a cropping systems perspective, i.e., management of a sequence of the same or different crops over the long run. This particularly relates to maintenance of an adequate overall nutrient balance by integrated measures such as fertilizer application, organic manure, or crop residue management.

Framework for soil-specific nutrient management in irrigated rice

Site-specific crop management or precision farming

The concept of changing spatial and temporal scales of land resource management forms the core of site-specific crop management (SSCM), which is gaining popularity in developed countries. SSCM tries to find a profitable and sustainable balance between agricultural food production and quality of land and water resources (Robert et al 1995a). A SSCM system is based on matching resource application and agronomic practices with soil attributes and crop requirements as they vary across a site.[2] Economically, the precise calculation and placement of inputs suggest a more efficient and profitable use of enterprise resources. Ecologically, this precision offers the prospect of reducing environmental risks associated with blanket field treatments and provides the ability to work with the natural diversity within a domain (McBratney and Whelan 1995).

By definition, SSCM practiced in some developed countries focuses on managing spatial and temporal variability within a single and usually large field. Varying the depth of soil tillage according to microrelief and soil texture, changing variety and/or seed rate according to soil type, adjusting the rate of fertilizer application according to variation in soil test values, adjusting N rates based on plant N status, selective liming of certain field parts only, varying the rate of pesticide sprays based on actual crop stress or damage, or managing spatially variable weed populations are typical examples of SSCM operations in developed countries (Robert et al 1995b).

During the last 10 yr, research on SSCM has been very much technology driven, resulting in a wide range of tools now available. The accuracy of real-time global positioning systems (GPS, land- or satellite-based) has been improved significantly, and GPS provide the georeferencing needed for varying operations according to the specific location within a single field (ASAE 1991). Other important tools include geostatistical sampling and mapping procedures; sensors for continuous measurement of crop yield (yield mapping); "on-the-go sensors" for continuous measurement of some soil properties, real-time weed and pest damage recognition systems; farm-oriented geographic information systems (GIS) for processing and storing digital maps; expert systems or models as decision aids; and variable-rate technology (VRT) for precise continuous adjustment of application rates (McBratney and Whelan 1995, Robert et al 1995b).

In a broader sense, however, SSCM should be viewed as a farming philosophy and a system to implement that philosophy, aiming at replacing prophylactic application of agricultural inputs by demand-based application (Searcy 1995). Because of this, SSCM is also relevant to rice farming in Asia. Implementing SSCM in the intensive rice system with (whole) small fields as the basic production units is feasible. Conceptually, adjusting tillage, sowing, and fertilizer or pesticide rate separately for small fields or farms (< 1 ha .... <10 ha) in an Asian domain is similar to adjusting according to soil variation within a large field (>10 ha .... >100 ha) in North America. An "Asian variant" of SSCM in the

---

[2] "Site-specific crop management is an information and technology based agricultural management system to identify, analyze, and mange site-soil spatial and temporal variability within fields for optimum profitability, sustainability, and protection of the environment"(Robert et al 1995a).

intensive rice system would probably include operations at different spatial scales and with very different information demand. Generally, SSCM in Asia can be built around much less sophisticated technology than implementing SSCM in large fields, where GPS, mapping systems, computer technology, and VRT are minimum requirements.

Integrated pest (IPM) and nutrient management (INM) is probably a priority issue to get SSCM in irrigated rice started, but its concepts and information requirements differ. The IPM concept focuses on improving farmers' knowledge so that they develop a better "feel" for the real pest situation in a specific field. Training plays a vital role in IPM, and field-specific decisions may be based on both qualitative and quantitative information. In any case, most of this information is "visible" or easily obtainable. Precision NM in rice, on the other hand, requires more quantitative information such as soil tests, leaf N monitoring, and accurate measurements of yields and externally provided nutrient inputs (Dobermann et al 1996a). Therefore, during the next 2 decades, NM will probably focus on managing between-field spatial variability, temporal variability in plant N status occurring within one growing season, and temporal variation in soil N, P, and K status from season to season. Other SSCM measures in irrigated rice may include selection of cropping systems, crop varieties, or sowing dates according to climatic characteristics and soil properties; or field-specific soil tillage and irrigation based on knowledge of soil texture, soil morphology, and rooting depth.

Site-specific nutrient management in irrigated rice

We propose a new framework for SSNM based on the requirements for groups of different nutrients defined earlier (Dobermann et al 1996a), as a major component of SSCM in irrigated rice. A key assumption for this is that significant increases in nutrient use efficiency require quantitative information about indigenous nutrient supply and turnover of externally applied nutrients on a field- or farm-specific basis.

An iterative research and extension approach to achieve this capability is illustrated in Figure 3 through a succession of rice crops. This approach provides a nutrient decision support system (NuDSS) that integrates the various data inputs and provides the user with more cost-effective fertilizer recommendations. The potential (or indigenous) supply of N (S_N), P (S_P), and K (S_K, all in kg ha^{-1}) is a pivotal parameter needed to predict the applied rates of N, P, and K fertilizer required for a given yield level (Fig. 4). For practical purposes, potential indigenous nutrient supply can be defined as the amount of nutrient a crop takes up from indigenous resources under optimum conditions in the field, i.e., when all other nutrients are amply supplied and other limitations to growth are absent (Janssen et al 1990).

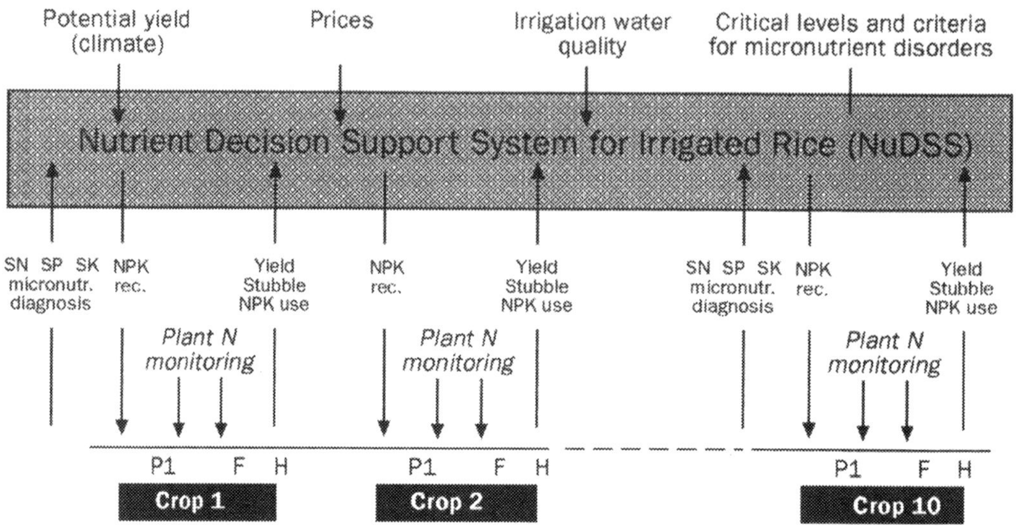

Figure 3. Iterative long-term, site-specific nutrient management strategy for intensive, irrigated rice systems (Dobermann et al 1996a).

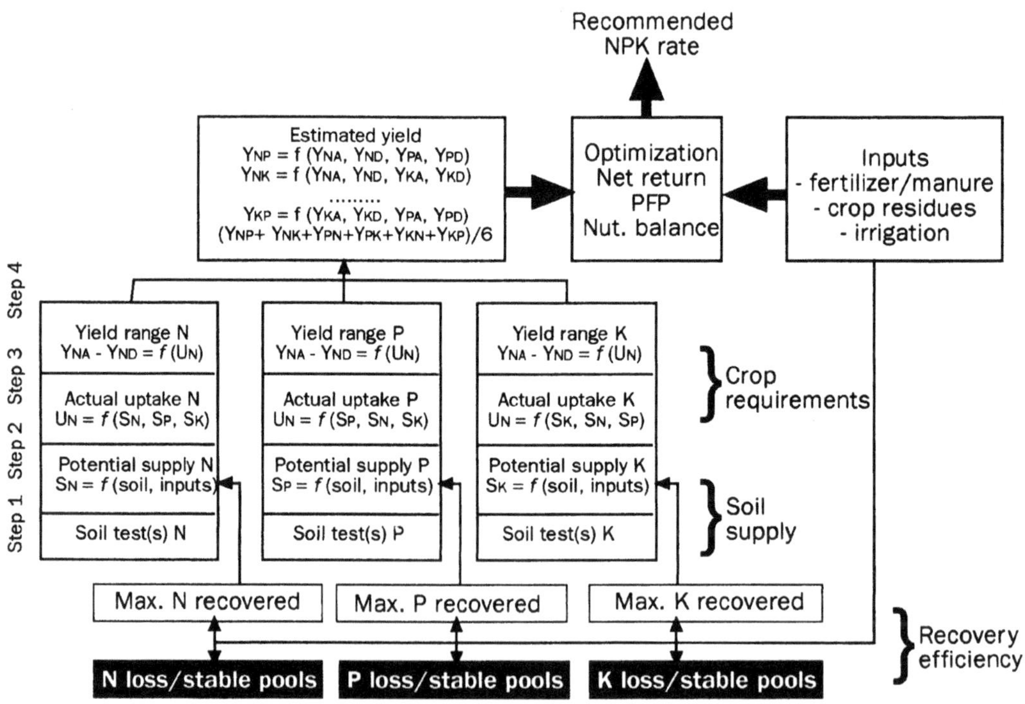

Figure 4. Proposed fertilizer recommendation model for irrigated rice based on the QUEFTS model (Dobermann et al 1996a, Janssen et al 1990).

The following steps are required:

1. measurement of S_N, S_P, and S_K and diagnosis of other nutritional disorders in year one;
2. recommendation for NPK use and alleviation of other nutritional problems based on the NuDSS;
3. optimization of timing and amount of applied N; and
4. estimation of actual grain yield, stubble (straw) returned to the field, and amount of fertilizer used. The latter is then used to predict the change in S_N, S_P, and S_K during the first crop cycle, based on the estimated nutrient balance (Fig. 5). For nutrients such as P and K, reasonable estimates of the nutrient balance can be obtained by estimating nutrient inputs from manure, fertilizers, and crop residues, and nutrient removal with grain and straw (Dobermann et al 1996b,c). The predicted S_N, S_P, and S_K values are then used to develop fertilizer recommendations in the second crop cycle. This procedure is followed for a succession of crops.

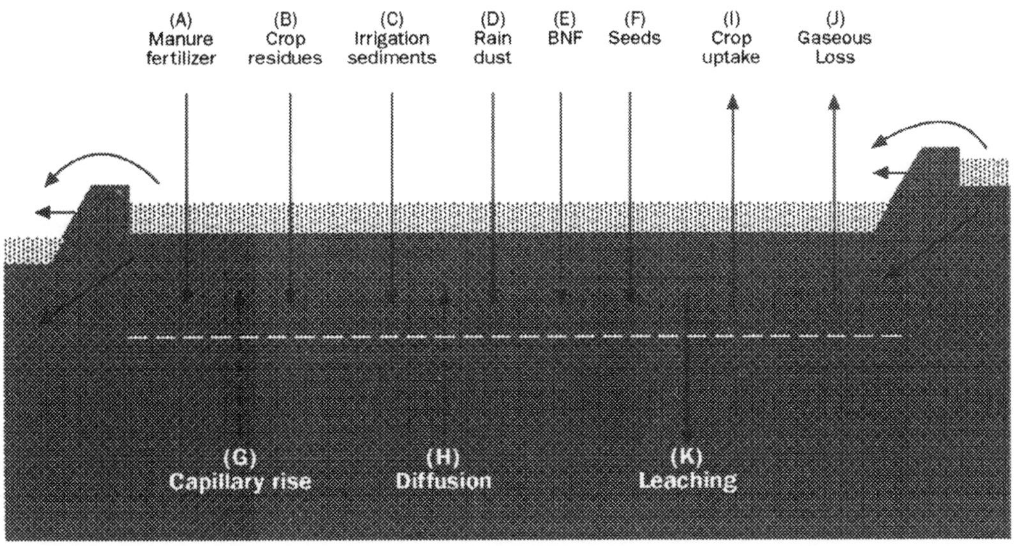

Figure 5. Components of the nutrient balance in the plow layer of an irrigated ricefield.

Within each cropping season, timing and splitting of N applications are crucial decisions determining synchrony of N supply and crop N demand. The decisions can be based on (1) 3-5 split applications following agronomic principles, (2) regular monitoring of plant N status up to flowering stage, using tools such as the chlorophyll meter (perhaps done by a village technician) or green leaf color charts (Peng et al 1996, Takebe and Yoneyama 1989), or (3) prediction of N split applications using simplified simulation models such as Manage-N (ten Berge et al 1997a,b). Many of these approaches have been developed independently and their performance has not been compared widely. Future research has to clarify where which method is most suited.

The entire iterative process may include several years during which soil testing or other methods for estimating potential indigenous nutrient supply will not be required, but plant N monitoring within each cropping season is recommended. The inputs needed for each crop (grain yield, stubble left, fertilizer use of previous crop) can be easily obtained from the farmers, if they keep good records. We speculate the process may require about 10 rice cropping periods and could evolve to account for changes in crop management practices such as crop establishment method, tillage, pest control, and water management, which would affect yield levels and soil nutrient supply.

The top row in Figure 3 shows general site information required for decision making. Climatic yield potential is needed as a boundary condition for estimating crop nutrient requirements. Potential yields can be default values, experts' estimates, or predictions from crop models. In most lowland sites in the humid tropics of Asia, yield levels of 6 t ha^{-1} in the wet season and 9-10 t ha^{-1} in the dry season can be obtained with the current short-duration varieties (Kropff et al 1993). Local commodity prices are needed as input for the net return optimization module in the NuDSS. If available, data about the nutrient content in irrigation water and leaching losses can be incorporated to improve the estimated nutrient balance. Critical levels and criteria for identifying nutrient disorders other than N, P, and K can be default values in the NuDSS, or more site-specific information could be used.

The core of the NuDSS will be a module for estimating the amounts of fertilizer N, P, and K needed to achieve a certain yield target (Fig. 4). Its construction follows the approach used in the QUEFTS model developed for maize (Janssen et al 1990, Smaling and Janssen 1993). QUEFTS is based on equations describing (i) S_N, S_P, and S_K as a function of chemical soil test values (step 1), (ii) actual NPK uptake (U_N, U_P, U_K) as a function of S_N, S_P, and S_K (step 2), and (iii) grain yield as a function of U_N, U_P, and U_K (steps 3 and 4), acknowledging interactions among the three macronutrients (Fig. 4). Steps 2 and 4 are based mainly on theoretical considerations, and details can be found elsewhere (Janssen et al 1990, Smaling and Janssen 1993).

In an unfertilized plot, the srepwise calculation procedure provides an estimate of grain yield that can be achievea by utilizing indigenous nutrient resources only. In a fertilized plot or field, nutrients added from inorganic and organic fertilizers contribute directly to potential supply for the current crop and to losses from the system or nutrient transfer to stable pools that become available only over longer time periods. Thus, transfer functions for estimating this recovery efficiency are used to estimate this increase in potential supply, and the calculation results in a yield estimate for each nutrient input combination tested (Fig. 4). For routine extension and advisory work, the target will be to find the best combination of nutrient inputs for maximizing net return by maintaining the resource base, i.e., ensuring high partial factor productivity for N and maintaining adequate levels of S_N, S_P and S_K.

Implementation considerations

Key implementation issues include the following:

- The scope for using qualitative soil information (Type II recommendations for fertilizer use) for SSNM in irrigated rice is probably limited. Most irrigated riceland is found on large floodplains or coastal plains. Due to lack of easily visible soil-landscape relationships, it is often difficult to judge the spatial distribution of soil types and estimate S_N, S_P, or S_K (Oberthuer et al 1996). Understanding of the relationships between general (stable) soil properties and dynamic soil properties related to nutrients and microbial activities is insufficient for fine-tuning NM at high yield levels.
- The overall soil testing and fertilizer recommendation paradigm has changed very little since the 1950s and may restrict adoption and benefits of SSNM in the irrigated rice system. Rhizospheric effects (Kirk et al 1997), often unknown N supply from BNF (Roger and Ladha 1990, Kundu and Ladha 1995a), and differences in soil depth (Kundu et al 1996) affect the overall nutrient supply to a growing rice plant so that classical "static" soil tests may not be suitable predictors of S_N, S_P, or S_K. We have started to look for alternatives such as on-farm nutrient omission plots (Cassman et al 1996a, Dobermann et al 1996a) or resin capsules for in situ soil testing (Dobermann et al 1997). They seem to be promising tools for estimating soil nutrient-supplying capacity in irrigated ricefields, but rigorous on-farm testing remains to be done.
- The high cropping intensity in tropical rice and the huge number of single management units (field, parcel) that must be handled may set limits to implementing SSNM. Two or more crops are grown per year and the number of observations and decisions to be made would be large. Currently, it seems difficult to conduct certain SSNM measures such as soil testing or regular plant monitoring on a field-specific basis or, if done, their costs per hectare may be too high. In most regions, the demand for service would easily exceed current facilities.
- Because of differences in climate, yield potential, and nutrient use efficiency, NM and information demand for wet and dry season rice differ. Therefore, season-specific calibration data sets for the major relationships between potential nutrient supply, nutrient uptake, and grain yield have to be generated first for the major rice-growing regions.
- Managing the organic phase in a SSNM approach remains uncertain. Organic matter quality seems to affect soil N supply more than organic matter content (Cassman et al 1996a). Decomposition and humus formation pathways are

sensitive to the timing of incorporating organic materials; we need better means of accounting for this in a nutrient balance-oriented SSNM.

- Availability of accurate but simple and user-friendly decision aids will determine adoption of a knowledge-intensive technology such as field-specific NM. At IRRI, we have initiated the development of a NuDSS for irrigated rice. This software will become a tool for research and extension. It will have all facilities for practical farm management over the long run. Farmers themselves will need better means of accurately measuring nutrient inputs and outputs (yield) on a per field basis. That would include improvement of their recording practices toward establishing long-term farm databases.
- In many Asian countries, facilities for more sophisticated farmer support need to be built. These include soil testing laboratories and a soil testing program, perhaps with the involvement of the private sector; fertilizer recommendation services; objective information about new fertilizer products; and use of mass media (radio, TV, newspapers) for extension of new technologies.

There are probably more unique problems. The important point is that many of the approaches and tools needed for SSNM in irrigated ricelands are already well known or in an advanced development stage. Too often, they have been applied independently, or technologies proposed by researchers have not found their way into the routine farm operations.

Nutrient management strategy for rainfed lowland rice systems

Biophysical and socioeconomic system characteristics

The rainfed lowland rice ecosystem has many unique characteristics which have important impacts on strategies for NM and which are in contrast to the type of management suitable for the irrigated systems. There are many subecosystems within the rainfed lowland rice ecosystem so that it is difficult to specify general characteristics with relevance to NM. Important considerations include

1. A high degree of spatial and temporal variability in both soil type and water availability (Zeigler and Puckridge 1995). Predicting the magnitude of a response to nutrient input for any given situation is, therefore, difficult unless prior knowledge exists for each individual circumstance.
2. The soils often have poorer chemical and physical properties than soils in irrigated systems. For example, sandy Ultisols and Alfisols predominate in the rainfed lowlands of Cambodia, Thailand, and Laos. These soils have very low native fertility, organic matter, and cation exchange capacity (CEC) levels (Ragland and Boonpuckdee 1987, Pheav et al 1996). The poor soil quality increases the difficulty of NM and restricts realistic potential yields obtainable on these soils (Willet and Intrawech 1988).

3. Farmers in the rainfed lowlands generally have less resources for capital expenditure and limited access to credit compared with farmers in irrigated systems (Zeigler and Puckridge 1995). Farmers' ability to invest in techniques to monitor the nutrient status of their soils or crops or, indeed, invest in any innovative technologies is, therefore, limited. The options for improved NM are consequently few.
4. There is a greater chance of crop failure in the rainfed systems than in the irrigated systems, and if crops fail, rainfed farmers often have fewer options for supplementary food or income. Avoidance of risk is likely to be more important in the decision making of rainfed farmers than of irrigated farmers (Zeigler and Puckridge 1995). Consequently, again the range of technologies willing to be adopted by rainfed lowland farmers is further restricted.

Requirements for nutrient management strategies

These characteristics have implications on the development of NM strategies: The first is the unsuitability of blanket technology recommendations. Strategies need to be aimed at specific targets which have relatively small recommendation domains (Pingali et al 1996). Because many such recommendation domains exist, developing these strategies using normal empirical experimentation is expensive and time consuming. The high cost contrasts with the low yields and few resources of most farmers and is usually prohibitive. NM strategies must, therefore, have a mechanism for accurate extrapolating from general recommendation to the specific circumstances faced by a farmer.

Second, strategies must be flexible and can be varied by farmers depending on the progress of the season and/or the outcome of the previous season. Farmers need to maximize the advantage gained from a better-than-average season, while also being able to minimize the risk and losses associated with a poor season. Farmers must necessarily make informed decisions about changing strategies to suit the seasonal progress and other changing circumstances.

Third, management strategies must also be able to accommodate the aims of the farmer who, because of particular preferences or social needs, may favor management strategies or fertilizer types that are not aimed at maximizing yields or economic returns. For example, in Cambodia, some farmers are reluctant to apply inorganic fertilizer to traditional varieties grown for their own consumption because they believe it affects the rice flavor. Farmers also sometimes apply sea salt as a fertilizer to rice because of its very low cost.

Finally and most important, the strategy must provide a structure for the transfer of knowledge and experience between farmers and allow incremental improvements of the technologies with lessons learned. In this way, the NM strategies can be assimilated into the knowledge base of the farming family (Pingali et al 1996). The strategies must also facilitate the entry of new technologies as they are developed.

Within this context, recent research in Cambodia has led to a new NM strategy tailored to the needs of resource-poor rice farmers operating on soils of different quality. We describe this framework below as one example of how information about crucial factors affecting NM decisions in rainfed lowland rice areas can be obtained and processed.

Framework for nutrient management in rainfed rice in Cambodia

The NM strategy that is being developed for Cambodia has two major stages (Fig. 6). First, new technologies are generated and adapted from elsewhere in the region and then tested through research at selected representative sites at research stations and farmers' fields. The research follows the normal empirical approach, with attention given to site characterization (i.e., soil type) and other factors affecting NM. Second, technologies are extrapolated and applied through the use of an agronomic soil classification and probabilistic modeling that integrates farmers' knowledge and experience with empirical data generated from the research sites. The strategy involves a high component of individual farmer and extension officer adaptation so that technologies appropriate for their individual situations are generated. The aim is to create a system that structures and integrates knowledge on NM from a varied range of information sources but is flexible enough for updating and developing this knowledge. We attempt to provide a means for the flexible application of technologies in farmers' fields using simple terms easily understood by a range of people.

The approach is still being tested but has so far been very well received. The two key components of the second stage of the strategy are described.

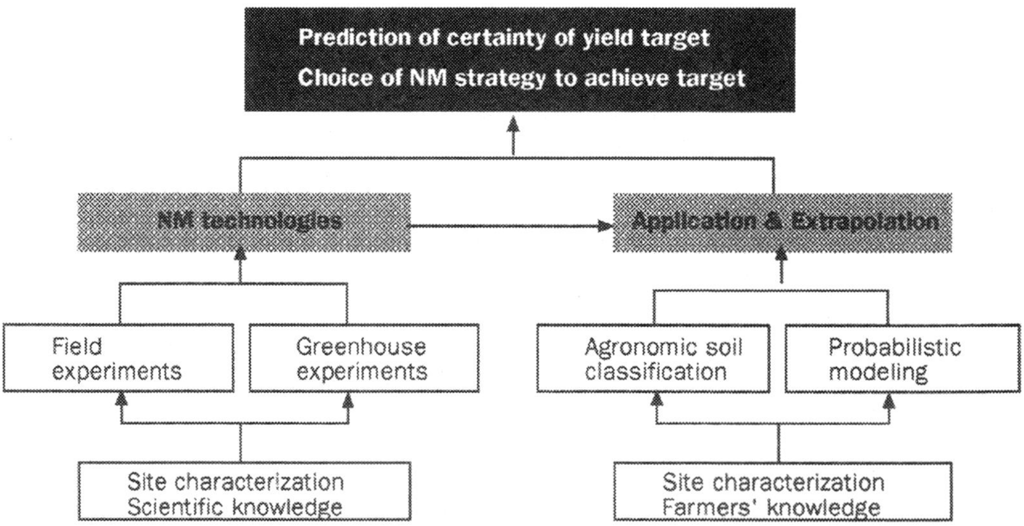

Figure 6. Development and major components of a nutrient management strategy proposed for rainfed lowland rice in Cambodia.

Agronomic soil classification system for Cambodia

An agronomic soil classification (White et al 1996) was developed for Cambodia to provide an effective means of communication about soils for researchers, extension workers, and farmers. The soil classification systems already available in the country were not used by this group of people because they were regarded as too complicated. As a result, on a national or provincial basis, soils were discussed in vague terms or were ignored, leading to difficulties in interpreting and extending research results (CIAP 1994).

The classification system consists of three levels (Table 4): soil group, soil phase, and a modified version of the Fertility Capability Classification (FCC) (Buol et al 1973, Sanchez et al 1982, Sanchez and Buol 1985). Soil groups are based on pedogenic criteria using mainly morphological characteristics of the soil to separate phases (Table 5). The soil phase level is agronomically based and uses soil characteristics that affect crop production as criteria for separating phases. The FCC is a quantitative assessment of the soil fertility constraints and provides guidelines for management.

The classification is based on only two diagnostic layers of the soil, which generally correspond to the top 20 cm of soil (representing the topsoil) and the 20-50 cm layer (representing the subsoil). Emphasis is placed on features easily detectable in the field, such as texture, color, depth of horizons, presence or absence of cracks, gravel content, and compacted layers. Soil analytical laboratory data are used only to support the classification. The FCC enables more accurate quantitative classification at specific locations.

Table 4. The three components of the agronomic soil classification used in Cambodia (*Source:* White et al 1996).

Soil group	Soil phase	FCC code
Prey Khmer	Coarse sandy phase	S d h e k o p
	Fine sandy phase	S d h e k op
Prateah Lang	Shallow phase	SC or R+ d g h e k o p
	Loamy subsoil phase	SL d g h e k o p (i +)[a]
	Clayey subsoil phase	SC d g h e k o p
Labansiek	Nonpetroferric phase	C d a e i+ k
	Petroferric phase	CR d a e i+ k
Orung		LS g d k e p o
Krakor	Noncracking phase	LC g or C g
	Cracking phase	C g (g+) v
Bakan		LC g e k p o
Kbal Po	Thionic phase	C g (g+) c (a)
	Nonthionic phase	C g (g+)
Kein Svay		L d g (p)
Toul Samroung	Brown phase	C g d (v) (o) i
	Gray phase	C g d (v) (o) i
Koktrap	Fertile phase	LC or C g (g+) d k i
	Nonfertile phase	LC (R+) g (g+)d h (a) k v i or C (R+) g (g+) d h (a) k v i
Kampong Siem	Gravelly phase	C g (d) p v
	Nongravelly phase	C g (d b) p v

[a]Modifiers in parentheses indicate uncertainty in the application of the modifier.

Soil groups, more specifically, are defined as a unit of morphologically similar soils that occur at the same position on the landscape (Table 5). Hence, a black cracking clay occurring on the old alluvial terraces is classified separately from a black cracking clay occurring on the expansive floodplain. Similarly, all soils with a deep sandy profile occurring on the old alluvial terraces are grouped together. This broad criterion assumes a link between topographic location and morphology, and pedogenic processes.

Local experience and expertise are important in determining the number and type of phases subdividing a group. The definition of soil phase is flexible; much of the information contained within local farmer soil classifications can be incorporated into the soil phase level where needed (Talawar 1996). Class limits for the phases, nevertheless, are based on data gained from the scientific literature.

The FCC is a technical system for grouping soils according to the kinds of problems they present for agronomic management. The FCC provides a checklist of identifiable properties that previous research and experience have found to influence the effectiveness of crop management. For any particular soil, the classification is presented as a code consisting of topsoil and subsoil textural classes and condition modifiers which represent chemical or physical soil properties that affect soil management (e.g., Lehk). The interpretation of the code then provides information to guide the user in choosing the right practices for the classified soil (Sanchez and Buol 1985). Some additional soil modifiers were necessary when the FCC was applied to Cambodian soils.

Soil classification at the soil group and phase levels is sufficient for extension workers and farmers; the more experienced user may use the FCC. The classification is open-ended, allowing more groups and phases to be added as more information becomes available.

The strength of the classification system is its ease of use--the soil can be classified at several locations quickly and easily. The reliance on subjective criteria, however, can cause some difficulty in classification, particularly if it occurs at the border between two groups.

Probabilistic modeling

Probabilistic modeling is used to integrate empirically derived data with experts' opinion. Hence, the expected response of rice to a NM strategy will be modified based on the farmers' (in this case, the expert) local knowledge and experience. The appropriateness of the strategy can then be more accurately defined and adjusted to individual situations.

The outputs of the model are the following: 1) the prediction of the certainty of a predefined yield outcome resulting from a given NM strategy for the particular set of circumstances faced by the farmer and 2) the prediction of the NM strategy that will achieve a predefined yield target with a given certainty for the conditions facing the farmer (Fig. 6).

Farmers' and other experts' knowledge is obtained from sample interviews in the target area. The aim of the interviews is to obtain information on the biophysical factors controlling yield, as well as the socioeconomic factors influencing farmers' decisions. For example, these questions will be asked: 'Do you think that applying N 2 wk after trans-

Table 5. Examples of group concept and phase description used in the agronomic soil classification system for Cambodia (*Source*: White et al 1996).

Group 0 - Prey Khmer[a]

A soil occurring on the old alluvial terraces or the alluvial/colluvial plains which has a sandy textured profile extending deeper than 50 cm (10-12% of total rice area)

Fine sandy phase: the sand is predominantly fine sand (<0.5mm).

Coarse sandy phase: the sand is predominantly coarse sand (>0.5mm).

Group 1 - Prateah Lang

A soil occurring on the old alluvial terraces or the alluvial/colluvial plains which has a sandy topsoil less than 40 cm thick over a subsoil which has a loamy or clayey texture (25-30% of total rice area)

Shallow phase: the depth of the topsoil is 20 cm or less and the texture of the subsoil is clayey or a very hard layer occurs within 20 cm of the surface. The very hard layer is a plow pan or a consolidated iron pan.

Clayey subsoil phase: the depth of the topsoil is 20 cm or greater, a very hard layer is not present within the top 20 cm of the profile and the texture of the subsoil clayey.

Loamy subsoil phase: the subsoil has a loamy texture and a very hard layer is not present within 20 cm of the surface.

Group 4 - Krakor

A soil within a gray to brown but not dark gray, very dark brown or black, loamy or clayey textured topsoil over a sandy, loamy or clayey subsoil occurring on the active floodplains. (15% of total rice area).

Cracking phase: moderate to large cracks form in the soil surface when it dries. The cracks penetrate deeper that 5 cm into the soil when it dries.

Noncracking phase: cracks are not formed in the soil surface when it dries or only surface cracks which do not penetrate more than 5 cm into the soil are formed.

Group 6 - Kbal Po

A soil with a dark gray, very dark brown to black, clayey topsoil which forms large deep cracks over a clayey subsoil, occurring on the active floodplains. Sandy strata in the subsoil may occur (15% of total rice area).

Thionic phase: the subsoil pH is below 4.5.

Nonthionic phase: the subsoil pH is above 4.5.

Group 9 - Koktrap

A soil occurring on the old alluvial terraces which has a dark gray, very dark brown to black topsoil with a clayey or loamy texture over a light gray or light brown, loamy or clayey subsoil (5% of total rice area).

Infertile phase: unfertilized yields on farmer's fields are less than 1,200 kg rice ha^{-1}*, generally around 500 to 800 kg* ha^{-1}*. Slight severe bronzing of the leaves of rice grown on these soils frequently occurs.*

Fertile phase: unfertilized yields on farmer's fields are more than 1,200 kg rice ha^{-1}*, generally more than 1,500 kg* ha^{-1} *and rice bronzing does not occur.*

[a]It should be noted that the groups have been named arbitrarily. The occurrence of the soils is not restricted to the area indicated by the name. All soils can be found in many provinces.

planting will improve rice growth?' or: 'Does the availability of urea at the start of the season affect the amount of fertilizer you apply?' The farmers' response (yes, no, I don't know, maybe, definitely yes) is scaled from -5 to +5. (definitely not =-5, definitely yes = +5, I don't know =0). These responses are then converted into probability statistics and combined with the empirical data following the procedure of Katz (1991). To combine the empirical data and farmers' responses, the data must be converted into a probability statistic using Bayesian theory (Hart et al 1978) and then weighted for the different types of information in terms of its necessity to achieve the model (likelihood necessity) and its effect on the model outcome (likelihood sufficiency).

Implementation considerations

Combining the agronomic soil classification and the probabilistic modeling provides a unique system of NM suitable for the rainfed lowlands which requires NM strategies to be widely modified to suit local conditions. The agronomic soil classification allows a range of people with varied expertise and backgrounds to communicate about soils in simple, easily understood terms and, more important, to recognize the soils in the field. We have used the new system to work out preliminary fertilizer recommendations for modern rice varieties in Cambodia. These recommendations are now being tested in Cambodia (Table 6).

The probabilistic modeling will allow more accurate extrapolation from the general to the specific by utilizing farmers' unique knowledge of their situation. The system will fit within the normal scientific research structure, which will generate the empirical data and the model (Fig. 6). The computer program (probabilistic model) will be based in provincial offices and run by extension personnel, in collaboration with district officers and farmers. By using quantitative and subjective data, and technical and nontechnical terms, the system can act as a reservoir of knowledge and a means by which experience can be shared and applied.

Outlook

The next 20 yr will be a period of major changes in the rice systems of Asia. It will probably mark the introduction and adoption of a new generation of modern rice varieties with a yield potential of 13 t ha^{-1} or even more. Parallel to this, we hope to see a major

Table 6. Preliminary soil- and environment-specific fertilizer recommendations (kg NPK ha^{-1}) for modern rice varieties in Cambodia.

	Soil							
	Bakan and Orung	Prateah Lang	Prey Khmer	Kampong Siem	Kbal Po and Krakor (dry season)	Koktrap (infertile phase)	Koktrap (fertile phase)	Toul Samro-ung
Favorable environment[a]	73:40:30 If no K is applied: 62:40:0	60:29:30 If no K is applied: 42:23:0	52:15:20 If no K is applied: 40:12:0	110:35:0	122:46:0	94:63:22 If no K is applied: 70:46:0	100:52:0	100:520
Unfavorable environment	45:23:20 If no K is applied: 36:23:0	40:23:20 If no K is applied: 30:15:0	28:12:10 If no K is applied: 28:12:0	66:23:0	-	48:46:15 If no K is applied: 42:35:0	62:40:0	62:40:0

[a]A favorable environment is where there is sufficient water and only short periods of minor water deficiency occur. An unfavorable environment is where the crop suffers one or more periods of drought. If farmers cannot apply K fertilizer, recommended rates of N and P fertilizer are decreased. If not stated otherwise, all rates given refer to wet season rainfed rice.

change in the farming philosophy toward farming by soil and crop needs. Long-term sustainability of high productivity and ecological functions of the resource will hopefully become major targets of human activities in the intensive rice system. Rainfed lowland rice farmers will hopefully have better NM strategies to reduce risk of crop failure and increase yields and income in their environment. Both new rice varieties and improved management have the potential to create another Green Revolution in Asia. Our research must ensure that this second Green Revolution has impact on favorable and less favorable rice environments.

We hope to be able to reverse many of the negative effects of rice farming on the soil resource base, which have occurred during the past 30 yr, by a more resource-oriented approach such as SSCM. Associated with this, we need better procedures to ascertain subtle changes in soil quality or long-term consequences of present-day management. This also includes issues such as biodiversity in the soil-floodwater system, dynamics of soil organic matter, and contamination of soil and water by agrochemicals.

To reverse the negative effects of rice farming on the soil resource base, we need a major revision of our traditional field experimentation techniques. Useful as they are, classical crop response trials or long-term experiments conducted in small plots in research stations provide only snapshots of the much more diverse rice environment. Agronomic and environmental research has to be integrated at scales that are relevant for assessing both on-farm and off-farm effects of land use (Nielsen et al 1995). Thus, generating the knowledge needed for practical farm management has to include primary research in farmers' fields beyond simple demonstration trials. This process is more complicated, associated with more risk, and often more expensive, but it is the only way to capture the vast variety of biophysical and socioeconomic conditions. If our primary databases are limited, the recommendations we give to farmers will be limited as well.

The two examples of NM strategies presented here may look complicated, but will be indeed rather simple in terms of inputs needed and overall usability. They are not perfect, but they propose a more holistic means to manage nutrients for rice. They are models we can work with, to be modified and improved as we apply them. We plan to develop the complete SSNM package for irrigated rice presented here during the next 5 yr, and test it in on-farm studies with more than 200 rice farmers in China, India, Indonesia, Thailand, Vietnam, and the Philippines, within the IRRI-NARS project on Reversing Trends of Declining Productivity. Similarly, the NM system proposed for rainfed rice in Cambodia will undergo intensive on-farm validation.

References

Angus J F, St.-Groth C Fd, Tasic R C (1990) Between-farm variability in yield responses to inputs of fertilizers and herbicide applied to rainfed lowland rice in the Philippines. Agric. Ecosyst. Environ. 30: 219-234.

ASAE--American Society of Agricultural Engineers (1991) Automated agriculture for the 21st century. American Society of Agricultural Engineers, St. Joseph, MI.

Becker M, Ladha J K, Ali M (1995) Green manure technology--potential, usage, and limitations--a case study for lowland rice. Plant Soil 174: 181-194.

Bouma J (1995) Methods to characterize soil resource variability in space and time. Pages 201-207 in Robert P C, Rust R H, Larson W E (Eds) Site-specific management for agricultural systems. ASA, CSSA, and SSSA, Madison, WI.

Buol S W, Sanchez P A, Cate R B Jr., Granger M A (1973) Soil fertility capability classification. Pages 126-141 in Bornemisza E, Alvarado A (Eds.) Soil management in tropical America. North Carolina State University, NC.

Buresh R J, De Datta S K, Padilla J L, Samson M J (1988) Effect of two urease inhibitors on floodwater ammonia following urea application to lowland rice. Soil Sci. Soc. Am. J. 52: 856-861.
Byerlee D (1987) From adaptive research to farmer recommendations and extension advice. Agric. Admin. 27: 231-244.
Byerlee D (1996) Knowledge-intensive crop management technologies: concepts, impacts and prospects in Asian agriculture. Paper presented at the International Rice Research Conference, 3-5 Jun 1996, Bangkok, Thailand.
Cassman K G, De Datta S K, Olk D C, Alcantara J, Samson M, Descalsota J P, Dizon M (1995) Yield decline and the N economy of long-term experiments on continuous, irrigated rice systems in the tropics. Pages 181-222 in Lal R, Stewart B A (Eds.) Sustainable management of soils. Lewiston Publishers, CRC Press Inc., Michigan.
Cassman K G, Dobermann A, Sta.Cruz P C, Gines G C, Samson M I, Descalsota J P, Alcantara J M, Dizon M A, Olk D,C (1996a) Soil organic matter and the indigenous N supply of intensive irrigated rice systems in the tropics. Plant Soil 182: 267-278.
Cassman K G, Gines G C, Dizon M, Samson M I, Alcantara J M (1996b) N use efficiency in tropical lowland rice systems: contributions from indigenous and applied N. Field Crops Res 47: 1-12.
Cassman K G, Harwood R R (1995) The nature of agricultural systems--food security and environmental balance. Food Policy 20: 439-454.
Cassman K G, Kropff M J Yan Zhen De (1994) A conceptual framework for N management of irrigated rice in high-yield environments. Pages 81-96. in Virmani S S (Ed.) Hybrid rice technology: new developments and future prospects. International Rice Research Institute, P.O. Box 933, Manila, Philippines.
Cassman K G, Peng S, Olk D C, Ladha J K, Reichardt W, Dobermann A, Singh U (1997) Opportunities for increased nitrogen-use efficiency from improved resource management in irrigated rice systems. Field Crops Res. (in review)
Cassman KG, Pingali P L (1995a) Extrapolating trends from long-term experiments to farmers fields: the case of irrigated rice systems in Asia. Pages 64-84 in Barnett V, Payne R, Steiner R (Eds.) Agricultural sustainability in economic, environmental, and statistical terms. J. Wiley & Sons, London.
Cassman K G, Pingali P L (1995b) Intensification of irrigated rice systems: learning from the past to meet future challenges. GeoJournal 35: 299-305.
Chen F X, Chen Y A, Zou C M (1992) Potassium balance in red earth paddy soil of southern Hunan and its application. Pages 193-194 in Proceedings of the International Symposium on Paddy Soils. Chinese Academy of Sciences, Nanjing.
CIAP--Cambodia-IRRI-Australia Project (1994) Annual research report for 1994. Cambodia-IRRI-Australia Project, Phnom Penh.
De Datta S K, Gomez K A, Descalsota J P (1988) Changes in yield response to major nutrients and in soil fertility under intensive rice cropping. Soil Sci. 146: 350-358.
De Datta S K, Mikkelsen D S (1985) Potassium nutrition of rice. Pages 665-699 in Munson R D, Sumner M E, Bishop W D (Eds.) Potassium in agriculture. ASA, CSSA, and SSSA, Madison, WI.
De Datta S K, Patrick W H Jr. (1986) N economy of flooded rice soils. Development in Plant and Soil Sciences (26). Martinus Nijhoff Publishers, The Hague.
Dobermann A, Cassman K G, Peng S, Pham Sy Tan, Cao Van Phung, Sta.Cruz P C, Bajita J B, Adviento M A A, Olk D C (1996a) Precision NM in intensive irrigated rice systems. Pages 133-154 in Proceedings of the International Symposium on Maximizing Sustainable Rice Yields through Improved Soil and Environmental Management, 11-17 Nov 1996, Khon Kaen, Thailand. Department of Agriculture, Soil and Fertilizer Society of Thailand, Department of Land Development, and International Society of Soil Science, Bangkok, Thailand.
Dobermann A, Cassman K G, Sta.Cruz P C, Adviento M A A, Pampolino M F (1996b) Fertilizer inputs, nutrient balance, and soil nutrient-supplying power in intensive, irrigated rice systems. III. Phosphorus. Nutr, Cycling Agroecosyst. 46: 111-125.
Dobermann A, Cassman K G, Sta.Cruz P C, Neue H U, Skogley E O, Pampolino M F Adviento MAA (1995) Dynamic soil tests for rice. Pages 343-365 in Fragile lives in fragile ecosystems. Proceedings of the International Rice Research Conference, 13-17 Feb 1995. International Rice Research Institute, P.O. Box 933, Manila, Philippines.
Dobermann A, Oberthuer T (1997) Fuzzy mapping of soil fertility--a case study on irrigated riceland in the Philippines. Geoderma (in press)
Dobermann A, Pampolino M F, Adviento M A A (1997) Resin capsules for on-site assessment of soil nutrient supply in lowland rice fields. Soil Sci. Soc. Am. J. (in press)
Dobermann A, Sta.Cruz P C, Cassman K G (1996c) Fertilizer inputs, nutrient balance, and soil nutrient-supplying power in intensive, irrigated rice systems. I. Potassium uptake and K balance. Nutr. Cycling Agroecosyst. 46: 1-10.
Flinn J C, De Datta S K (1984) Trends in irrigated-rice yields under intensive cropping at Philippine research stations. Field Crops Res. 9: 1-15.
George T, Ladha J K, Garrity DP, Torres R O (1995) N dynamics of grain legume weedy fallow flooded rice sequences in the tropics. Agron. J. 87: 1-6.
Hart P E, Duda R O, Einaudi M T (1978) PROSPECTOR--a computer-based consultation system for mineral exploration. Math. Geol. 10: 589-610.

Hoosbeek M R, Bryant R B (1992) Towards the quantitative modeling of pedogenesis--a review. Geoderma 55: 181-210.

IRRI--International Rice Research Institute (1996) Mega project on reversing trends of declining productivity in intensive irrigated rice systems, 1994-1996. Progress report, May 1995-Jun 1996. International Rice Research Institute, P.O. Box 933, 1099 Manila, Philippines.

Janssen B H, Guiking F C T, Van der Eijk D, Smaling E M A, Wolf J, Van Reuler H (1990) A system for Quantitative Evaluation of the Fertility of Tropical Soils (QUEFTS). Geoderma 46: 299-318.

Katz S S (1991) Emulating the prospector expert system with a raster GIS. Computers Geosci. 17: 1033-1050

Kawaguchi K, Kyuma K (1977) Paddy soils in tropical Asia: their material nature and fertility. The University Press of Hawaii, Honolulu.

Kemmler G (1980) Potassium deficiency in soils of the tropics as a constraint to food production. Pages 253-275 in Priorities for alleviating soil-related constraints to food production in the tropics. International Rice Research Institute, P.O. Box 933, 1099 Manila, Philippines.

Kirk G J D, Courtois B, George T, Senadhira D (1997) Opportunities to improve phosphorus efficiency and soil fertility in rainfed lowland and upland rice ecosystems. Field Crops Res. (in review)

Kropff M J, Cassman K G, van Laar H H, Peng S (1993) N and yield potential of irrigated rice. Plant Soil 155/156: 391-394.

Kundu D K, Ladha J K (1995a) Efficient management of soil and biologically fixed N_2 in intensively cultivated rice fields. Soil Biol. Biochem. 27: 431-439.

Kundu D K, Ladha J K (1995b) Enhancing soil N use and BNF in wetland rice. Exp. Agric. 31: 261-277.

Kundu D K, Ladha J K, Lapitan-de Guzman E (1996) Tillage depth influence on soil N distribution and availability in a rice lowland. Soil Sci. Soc. Am. J. 60: 1153-1159.

Ladha J K, Kirk G J D, Bennett J, Peng S, Reddy C K, Reddy P M, Singh U (1997) Opportunities for increased N use efficiency from improved lowland rice germplasm. Field Crops Res. (in review)

Ladha J K, Kundu D K, Coppenolle M G A, Peoples M B, Carangal V R, Dart P J (1996) Legume productivity and soil N dynamics in lowland rice-based cropping systems. Soil Sci. Soc. Am. J. 60: 183-192.

McBratney A B, Whelan B M (1995) The potential for site-specific management of cotton farming systems. CRC for Sustainable Cotton Production, Narrabi, NSW.

Mohanty S K, Mandal L N (1989) Transformation and budgeting of N, P and K in soils for rice cultivation. Oryza 26: 213-231.

Nambiar K K M (1994) Soil fertility and crop productivity under long-term fertilizer use in India. Indian Council for Agricultural Research, New Delhi.

Nielsen D R, Wendroth O, Parlange M B (1995) Opportunities for examining on-farm soil variability. Pages 95-132 in Robert P C, Rust R H. Larson W E (Eds.) Site-specific management for agricultural systems. ASA, CSSA, and SSSA, Madison, WI.

Obcemea W N, De Datta S K, Broadbent F E (1984) Movement and distribution of fertilizer N as affected by depth of placement in wetland rice. Fert. Res. 5: 125-148.

Oberthuer T, Dobermann A, Neue H U (1996) How good is a reconnaissance soil map for agronomic purposes? Soil Use Manag.e 12: 33-43.

Oberthuer T, Dobermann A, Neue H U (1995) Spatial modeling of soil fertility--a case study in Nueva Ecija, Philippines. Pages 689-705 in Fragile lives in fragile ecosystems. Proceedings of the International Rice Research Conference, 13-17 Feb 1995. International Rice Research Institute, P.O. Box 933, Manila, Philippines.

Olk D C, Cassman K G, Randall E W, Kinchesh P, Sanger L J, Anderson J M (1996a) Changes in chemical properties of soil organic matter with intensified rice cropping in tropical lowland soils. Eur. J. Soil Sci. 47: 293-303.

Olk D C, Cassman K G, Simbahan G, Sta.Cruz P C, Abdulrachman S, Nagarajan R, Pham Sy Tan, Satawathananont S (1996b) Congruence of N fertilizer management by farmers and soil N supply in tropical irrigated lowland rice systems. Pages 29-38 in Proceedings of the International Symposium on Maximizing Sustainable Rice Yields through Improved Soil and Environmental Management, 11-17 Nov 1996, Khon Kaen, Thailand. Department of Agriculture, Soil and Fertilizer Society of Thailand, Department of Land Development, and International Soil Science Society, Bangkok,Thailand.

Peng S, Garcia F V, Laza R C, Sanico A L, Visperas R M, Cassman K G (1996) Increased N-use efficiency using a chlorophyll meter on high-yielding irrigated rice. Field Crops Res. 47: 243-252.

Peng S, Khush G S, Cassman K G (1994) Evolution of the new plant ideotype for increased yield potential. Pages 5-20 in: Cassman K G (Ed.) Breaking the yield barrier. International Rice Research Institute, P.O. Box 933, Manila, Philippines.

Pheav S, White P F, Oberthuer T, Reyes R T (1996) Management constraints of the major rice-growing soils of Cambodia. Pages 537-549 in Proceedings of the International Symposium on Maximizing Sustainable Rice Yields through Improved Soil and Environmental Management, 11-17 Nov 1996, Khon Kaen, Thailand. Department of Agriculture, Soil and Fertilizer Society of Thailand, Department of Land Development, and International Soil Science Society, Bangkok, Thailand.

Pingali P L, Hossain M, Pandey S, Price L (1996) Economics of nutrient management in Asian rice systems: towards increasing knowledge intensity. Paper presented at a workshop on nutrient use efficiency of cropping system(s), 13-15 Dec 1995, International Rice Research Institute, Los Banos, Laguna, Philippines.

Ragland J, Boonpuckdee L (1987) Fertilizer responses in Northeast Thailand. 1. Literature review and rationale. Thai. J. Soils Fert. 9: 65-79.

Randhawa N S, Katyal J C (1982) Micronutrient management for submerged rice soils. Pages 192-211 in Transactions of the 12th International Congress. of Soil Science. Symposium papers II: Vertisols and rice soils of the tropics. International Soil Science Society, New Delhi.

Robert P C, Rust R H, Larson W E (1995a) Preface. Pages xiii-xiv in Robert P C, Rust R H, Larson W E (Eds.) Site-specific management for agricultural systems. ASA, CSSA, SSSA Madison, WI.

Robert P C, Rust R H, Larson W E (1995b) Site-specific management for agricultural systems. ASA, CSSA, and SSSA, Madison,WI.

Roger P A (1996) Biology and management of the floodwater ecosystem in ricefields. International Rice Research Institute, P.O. Box 933, 1099 Manila, Philippines and L' institut francais de recherche scientifique pour le developpement en cooperation, France. 250 p.

Roger P A, Ladha J K (1990) Estimation of biological N_2 fixation and its contribution to N balance in wetland rice fields. Pages 128-133 in Transactions of the 14th International Congress of Soil Science, Kyoto, Japan, Aug 1990. International Soil Science Society, Kyoto.

Sanchez P A, Buol S W (1985) Agronomic taxonomy for wetlands. Pages 207-228 in Wetland soils: characterization, classification, and utilization. International Rice Research Institute, P.O. Box 933, Manila 1099, Philippines.

Sanchez P A, Couto W, Buol S W (1982) The fertility capability classification system: interpretation, applicability and modification. Geoderma 27: 283-309.

Savant N K, De Datta S K (1982) N transformations in wetland rice soils. Adv Agron. 35: 241-302.

Searcy S W (1995) Engineering systems for site-specific management: opportunities and limits. Pages 603-612 in Robert P C, Rust R H, Larson W E (Eds.) Site-specific management for agricultural systems. ASA, CSSA, and SSSA, Madison, WI.

Sharma P K, De Datta S K (1986) Physical properties and processes of puddled rice soils. Adv. Soil Sci. 5:139-178.

Singh U, Cassman K G, Ladha J K, Bronson K F (1995) Innovative N management strategies for lowland rice systems. Pages 229-254 in Fragile lives in fragile ecosystems. Proceedings of the International Rice Research Conference, 13-17 Feb 1995. International Rice Research Institute, P.O. Box 933, Manila, Philippines.

Smaling E M A, Janssen B H (1993) Calibration of QUEFTS, a model predicting nutrient uptake and yields from chemical soil fertility indices. Geoderma 59: 21-44.

Takebe M, Yoneyama T (1989) Measurement of leaf color scores and its implication to N nutrition of rice plants. Jpn. Agric. Res. Q. 23: 86-93.

Talawar S (1996) Local soil classification and management practices: bibliographic review. Research Paper No. 2. Department of Anthropology, University of Georgia, Athens, USA.

ten Berge H F M, Quinghua Shi, Zhiming Zheng, Rao K S, Riethoven J J M, Xuhua Zhong (1997a) Numerical optimization of N application to rice. Part II. Field evaluations. Field Crops Res. 51(1/2): 43-54.

ten Berge H F M, Thiyagarajan T M, Quinghua Shi, Wopereis MCS, Drenth H and Jansen M J W (1997b) Numerical optimization of N application to rice. Part I. Description of MANAGE-N. Field Crops Res. 51(1/2):29-42.

Tiwari K N (1985) Changes in potassium status of alluvial soils under intensive cropping. Fert. News 30: (9): 17-24.

Uexkuell H R V (1985) Availability and management of potassium in wetland rice soils. Pages 293-305 in Wetland soils: characterization, classification and utilization. International Rice Research Institute, P.O. Box 933, 1099 Manila, Philippines.

Uexkuell H R V, Beaton J D (1992) A review of fertility management of rice soils. Pages 288-300 in Kimble J M (Ed.) Proceedings of the 8th International Soil Correlation Meeting (VIII. ISCOM): Characterization, classification, and utilization of wet soils. United States Department of Agriculture, Soil Conservation Service, Lincoln.

van Noordwijk M, Scholten J H M (1994) Effects of fertilizer price on feasibility of efficiency improvement: case study for a urea injector for lowland rice. Fert. Res. 39: 1-9.

White P F, Oberthuer T, Pheav Sovuthy (1996) The soils used for rice production in Cambodia: a manual for their recognition and management. International Rice Research Institute, P.O. Box 933, 1099 Manila, Philippines.

Willett I R, Intrawech A (1988) Preliminary studies of chemical dynamics of sandy paddy soils of Tung Kula Ronghai, Northeast Thailand. CSIRO Divisional Report No. 95. Commonwealth Scientific and Industrial Research Organisation, Melbourne.

Youngdahl L J, Lupin M S, Craswell E T (1986) New developments in N fertilizers for rice. Fert. Res. 9: 149-160.

Zeigler R S, Puckridge D W (1995) Improving sustainable productivity in rice-based rainfed lowland systems of south and southeast Asia. GeoJournal 35: 307-324.

Sustaining productivity of lowland rice soils: issues and options related to N availability

D.K. KUNDU[1] and J.K. LADHA[2]

[1]*Directorate of Rice Research, Rajendranagar, Hyderabad 500030, India*

[2]*International Rice Research Institute, P.O. Box 933, Manila, Philippines*

Key words: N fertilization, N availability, prerice nitrate accumulation, subsoil N supply, hardpan, deep tillage

Abstract

The subsistence rice farming of the prechemical era efficiently sustained the N status of soils by maintaining a balance between N loss and N gain from biological N_2 fixation. Intensive monocropping with excessive use of chemical N fertilizers has raised many concerns about N sustainability in lowland rice soils. Long-term fertilizer experiments in India and the Philippines recorded declining N availability to lowland rice over the years, regardless of the changes in total N content of the soils. While total N content was maintained in soils continuously flooded or saturated, there are indications showing its decline in soils subjected to drying and flooding cycles. Obviously, N supply in the lowlands is dependent not only on total soil N content but also on N transformations, root access, accumulation of reducing substances, and associated factors that govern N availability to plants. Recent research has identified several factors responsible for the declining N content in lowland soils: nitrate accumulation in soils during the dry phase followed by its loss on reflooding, small amounts of crop residues returned to the soils, little use of organic manures and symbiotic N_2 fixing systems, and heavy loss of applied fertilizer N. Continuous flooding and intense puddling greatly reduce permeability, allow accumulation of reducing substances in the rhizosphere, and retard mineralization and N uptake in some soils. Further, shallow tillage practiced in intensively cropped lowlands often reduces plow layer thickness (to <15 cm) and nutrient availability from subsoil layers. This paper discusses these issues and suggests various options to help sustain the N fertility and productivity of lowland rice soils.

Introduction

Rice (*Oryza sativa*) is the world's most important food crop. More than 40% of the world's population depends on rice as the major source of calories. Most of the rice in tropical countries is produced in lowland areas. To feed the increasing global population, the world's annual rice production must increase from the present 528 million t to 760 million t by the year 2020. Nitrogen is a key input limiting rice production worldwide. To produce a ton of grain, the rice crop takes up an average of 20 kg N ha^{-1} from the soil over 3-5 mo. For rice productivity to be sustained at present levels, let alone improved in the future, the N removed in harvested produce or lost from the system must be replaced by N fertilizers or must be obtained through biological N_2 fixation (BNF).

Most of the world's rice production is in Asia where, for centuries, farmers have practiced a cultural system that ensured modest but stable yields, yet maintained a degree of N fertility in the soil. Additions of N through BNF balanced the losses of N through crop harvest and other mechanisms, creating a dynamic equilibrium (Ladha 1997). This equilibrium was disturbed by the need to increase rice production and high-yielding rice varieties (HYVs) which can use additional N were thus developed. Compared with traditional varieties, the HYVs remove larger amounts of N from the system. With irrigation, cultivation was intensified to two and even three crops per year. Another trend detrimental to soil N fertility is the increase in area being cropped only to cereals.

V. Balasubramanian, J. K. Ladha and G. L. Denning (eds.), Resource Management in Rice Systems: Nutrients, 27–44.

Yield trends from long-term continuous cropping experiments conducted in the Philippines, India, Indonesia, Thailand, and Bangladesh indicate that, even with the best available cultivars and scientific management, rice yields (holding input levels constant), has declined over time since the early 1980s (Flinn and De Datta 1984, Cassman and Pingali 1995, Nambiar and Ghosh 1984 [as cited in Pingali et al 1997]). Farm monitoring data from the Philippines showed that average wet season rice yields were 4.2 t ha^{-1} in Central Luzon and 4.7 t ha^{-1} in Laguna in the early 1980s. Since then, yields have gradually declined to such an extent that they were 0.5 t ha^{-1} lower in 1990 in both domains. In Ludhiana, Punjab, India, where an intensive rice - wheat double-cropped system is practiced, average rice yield attained by 1980 was 4.0 t ha^{-1} and has remained relatively constant thereafter (Cassman and Pingali 1995). Such declining or stagnant yields have raised concerns about the long-term sustainability of intensive rice production systems.

The declining yields, holding input levels and crop management practices constant, could indicate one or both of the following: (a) the genetic potential of later generation varieties is lower than that of earlier varieties, and (b) the productive potential of the rice resource base itself has been declining over time. Substantial evidence at present indicates that the genetic potential of later generation varieties is not different from that of early-generation varieties (Khush 1990 and Cassman et al 1994 [as cited in Pingali et al 1997]). Current data seem to show a long-term degradation of the rice resource base resulting from continuous flooding and puddling associated with intensive rice cultivation (Kundu and Ladha 1995b). Cassman and Pingali (1995) hypothesized that the yield declines are associated with decreased availability of native soil N, citing increasing doses of fertilizer N used to maintain the original yield levels. In long-term fertilizer experiments in India, N availability to lowland rice declined over the cropping years even when recommended levels of inorganic fertilizers were used (Table 1). These observations further suggest a close relationship between rice yields and N availability.

In this paper, we discuss different aspects of N fertility of lowland rice soils, examine impacts of various soil and crop management practices on N fertility, and suggest some strategies to sustain N fertility over the long term.

Determinants of N fertility of lowland rice soils

Soil N fertility is the status of a soil with respect to its ability to provide plants with N in the form, amount, and rate needed for optimum growth. The N fertility of a rice soil primarily depends on (1) total N content in the soil, (2) N transformations in the soil, and (3) various factors that influence N absorption by rice roots. Total N concentration in the plow layer of 410 lowland rice soils in tropical Asia ranged from 0.02 to 0.92% on air-dry weight basis, and averaged 0.13% (Table 2). About 78% of these soils contained less than 0.15% N. Assuming a plow layer of 15 cm and a bulk density of flooded soil in this layer of 1.0 g cm^{-3}, these soils contain less than 2250 kg N ha^{-1}. If there is no replenishment, growing rice for about 30 successive seasons will exhaust the N reserve of the soils (assuming a grain yield level of 3.5-4.0 t ha^{-1}). Besides N removal by plants, losses of N

Table 1. Changes in N uptake of wet season rice over the years at four long-term experimental sites in India (*Sources*: IARI 1989, Nambiar 1994).

Site and soil group	Fertilizer/manure treatment	N uptake by rice[a] (kg N $ha^{-1}crop^{-1}$)	
		1971-85	1985-87
Bhubaneswar, *Haplaquept*	NPK[b]	67	57
	NPK. + FYM[c]	79	75
Hyderabad, *Tropaquept*	NPK	78	59
	NPK + FYM	96	92
Pantnagar, *Hapludoll*	NPK	124	92
	NPK + FYM	139	128
Barrackpore, *Eutrochrept*	NPK	70	59
	NPK + FYM	73	70[d]

[a]Total N harvested in grain plus straw at maturity.
[b]Recommended doses of NPK fertilizers applied every crop season.
[c]Farmyard manure, field-fresh sample with 20-25% moisture, @ 10-15 t ha^{-1} applied once every year.
[d]Over 1986-87.

Table 2. Total nitrogen concentration in the plow layer of rice soils in tropical Asia (*Source*: Kawaguchi and Kyuma 1977).

Country	Soil samples (n)	Average total N[a] (%)
Bangladesh	53	0.13 (0.05-0.52)
Cambodia	16	0.10 (0.03-0.29)
India	73	0.08 (0.02-0.20)
Indonesia	44	0.12 (0.05-0.59)
Malaysia (West)	41	0.28 (0.05-0.92)
Myanmar	16	0.10 (0.04-0.18)
Philippines	54	0.15 (0.06-0.35)
Sri Lanka	33	0.13 (0.03-0.64)
Thailand	80	0.09 (0.02-0.27)
Tropical Asia	410	0.13 (0.02-0.92)

[a]Figures in parentheses indicate range.

from the soils take place through ammonia volatilization, leaching, denitrification, and runoff.

Generally only the N in the top 15-20 cm soil layer is considered in evaluating N availability. Recent studies, however, have shown that the soil below the top 15 cm layer contains considerable amount of N (Table 3), and that under favorable conditions, more than 40% of total N taken up by a rice crop from the 0-45 cm soil profile comes from the subsoil layers (Table 4). The amount of N mineralized during a rice-growing season

Table 3. Depth distribution of total N content in some lowland rice soils of tropical Asia.

Country and soil series/group	Soil depth (cm)	Total N (%)	Reference
Bangladesh, *Lokdeo*	0-15	0.058	Brammer (1978)
	15-38	0.050	
	38-64	0.040	
India, *Gangetic alluvium*	0-11	0.080	Dolui and Guhathakurta (1988)
	11-33	0.070	
	33-50	0.060	
Indonesia, *Gley soils*	0-15	0.100	Soepraptohardjo and Suhardjo (1978)
	> 15	0.070	
Malaysia, *Kampong Tepus*	0-15	0.240	Paramananthan (1978)
	15-35	0.190	
Myanmar, *Meadow soils*	0-15	0.160	Goung et al (1978)
	20-35	0.100	
Philippines, *Maahas clay*	0-15	0.124	Kundu et al (1996)
	15-30	0.071	
	30-45	0.037	

Table 4. Apparent N removal by rice (kg N ha^{-1}) from different layers of a submerged rice soil as influenced by tillage depth[a]. Los Baños. Philippines, 1993 dry season (*Source:* Kundu et al 1996).

Tillage depth, T(cm)	Soil layer, S			Total
	0-15 cm	15-30 cm	30-45 cm	
15	35.5	4.7	0.6	40.8
25	29.0	23.2	1.2	53.4
40	37.6	28.1	2.2	67.9

LSD (P=0.05) to compare two T means at each S: 8.5 and two S means at each T:8.7

[a]Estimated from the difference in mineral N between unplanted and planted soils, at 10 wk after planting. Mineral N refers to the 2 *M* KCl-extractable NH_4-plus NO_3-N.

commonly ranges from 3 to 8% of the total N present in the soils (Kundu and Ladha 1995a). Any soil, water, or crop management practice that influences the thickness of the plow layer, the mineralization rate of organic N, the absorption of N by rice plants, or the recovery of applied fertilizer N may have significant effects on soil N fertility. To sustain N fertility, it is necessary to maintain total N content and to improve other soil properties so that the required plant-available N is supplied to rice in the long term.

Intensive lowland rice culture and soil N fertility

Lowland soils remain flooded or saturated during most of the rice-growing season. The number of rice crops grown annually in tropical lowlands varies with the duration of the flooded periods. Tropical lowland rice cultures may be grouped into three major classes, depending on duration of flood and nonflooded periods. These are a) year-round lowland rice culture, b) lowland rice culture with short nonflooded season, and c) lowland rice culture with long nonflooded season. In the first class, three rice crops are cultivated in a year, usually with irrigation, and the fields remain constantly flooded or saturated. Some of the irrigated double-cropped ricelands also remain flooded or saturated throughout the year. The second type of rice culture is found in much of Asia's irrigated double-cropped riceland where there is a 1-2-mo dry period between rice crops. Some rainfed areas with long wet seasons support two successive rice crops, leaving a 2-3-mo dry period. Lowland rice culture with long nonflooded season is dominant in the tropics because most riceland is rainfed or can be irrigated only in the wet season. The soil remains nonflooded 4-8 mo of the year and a single crop of rice is grown in the wet season.

Some features of intensive rice cropping in the lowlands that have adverse impact on soil N fertility are:

- Flooding and drying cycles of varying duration which occur in the soils between and during rice crops.
- Neglect in the recycling of crop residues and in the use of bulky organic manures.
- Superficial tillage used to prepare the lands for planting crops (shallow tillage commonly practiced in intensively cropped rice lowlands and continuous flooding and repeated puddling of the soil without thorough dry tillage under the year-round lowland rice culture).
- Large-scale abandonment of legume-cereal crop rotations in favor of cereal monocropping as a result of government subsidy and support for cereal crops in many countries.
- Huge losses and very low recovery of applied fertilizer N by the crops, making them heavily dependent on the native soil N pool to meet total N requirement.

Major issues and options related to the sustenance of N fertility

To sustain the N fertility of lowland rice soils, several issues require consideration. We discuss the major ones in the following sections.

Conservation of nitrate N accumulated in the soil during nonflooded periods

Tropical lowland rice soils typically undergo alternate saturation and drying. They remain flooded or saturated during most of the rice-growing season, but between rice crops, the soil usually dries and becomes aerobic. At this time, the soil is either left fallow or some dryland crops are grown, depending on moisture availability. When flooded or saturated soils dry and become aerobic, NH_4 formed from mineralization of organic N or from N fertilizer is oxidized to NO_3 which may accumulate in the soil or be used by the plants. Most of the NO_3 that is not used by the plants may be lost through leaching and denitrification when the soils are subsequently flooded for rice cultivation.

Ponnamperuma (1985) measured NO_3 N levels in 28 Philippine lowland soils prior to flooding. Nitrate N concentrations ranged from 5 to 39 mg kg^{-1} soil (about 15-117 kg N ha^{-1}) and averaged 13 mg kg^{-1} (about 39 kg N ha^{-1}). Buresh et al (1989) measured 39-91 kg NO_3 N ha^{-1} in the top 60 cm of the soil at the end of the dry season at three Philippine sites. As measured by George et al (1993, 1994) in two successive years, maximum level of mineral N (i.e. NH_4 N plus NO_3 N), accumulated in a Philippine lowland before flooding the soil for wet season rice, ranged from 117 to 164 kg ha^{-1}. On another lowland site in the Philippines, maximum levels of mineral N measured by Ladha et al (1996) in the top 60 cm soil layer during the dry season ranged from 36 to 63 kg ha^{-1} in one year and 113 kg ha^{-1} in the following year.

In intensively cropped rice - wheat systems, large quantities of mineral N may remain in the soil after wheat harvest. In a sandy loam soil in India, soil NO_3 after wheat harvest averaged 82 kg N ha^{-1} (Sharma et al 1985). Nitrate accumulation in the soil during a mungbean crop following wheat was substantial. Before the start of the wet season, the soil contained 223 kg NO_3 N ha^{-1}. More recent studies conducted in rainfed lowlands in the Philippines (where rice is rotated with diverse dry season cash crops receiving heavy fertilization, tillage, and irrigation) have shown accumulation and loss of more than 500 kg NO_3 N ha^{-1} yr^{-1} from the 100 cm soil profile (Tripathi et al 1997; Shrestha and Ladha 1998). The N loss from the system averaged 270 kg N ha^{-1} and ranged from 240 kg N ha^{-1} (when dry season tobacco was grown) to 575 kg N ha^{-1} (when sweet pepper preceded rice). Much of the lost N found its way into the groundwater, causing pollution (Shrestha and Ladha 1998).

The negative impacts of NO_3 loss on the long-term N fertility of lowland soil have so far been ignored. Crop productivity and the N fertility of such lowlands are not likely to be sustained unless measures are taken to limit NO_3 buildup and to conserve it. If crops can be grown during the dry period, they can assimilate this NO_3 and prevent its loss. The assimilated N can then be recycled to the subsequent rice crop by incorporating plant residues into the soil. Singh (1984) speculated that legumes grown in rotation with lowland rice can scavenge soil mineral N. Buresh et al (1993) showed that *Sesbania rostrata* grown for 45-60 d and incorporated into the soil before planting lowland rice reduced the loss of soil N. George et al (1994) grew mungbean and *S. rostrata* during the dry-to-wet transition period and conserved 66-125 kg N ha^{-1} (Table 5). In another study, George et al (1995) conserved 57-75 kg soil N ha^{-1} by growing grain legumes such as soybean, cowpea, and mungbean during dry season. The maximum level of mineral N in the top 60 cm soil layer of a rice lowland during a dry season, as measured by Ladha et al (1996), was 113 kg N ha^{-1}, and forage legume *Crotalaria juncea* grown through the season recovered 80 kg soil N ha^{-1} from the site.

Recycling of crop residues and use of organic manure

Because the N-supplying capacity of rice soils originates from organic matter, continued addition of organic matter is important to sustain soil N fertility. Organic manure applied to the soil increases not only its total N content but its cation exchange capacity as well. It also improves the soil physical properties and increases the amount of CO_2 available to

Table 5. Soil N uptake (kg ha^{-1}) by legumes and weeds during the dry-to-wet season transition (May to July) in a Philippine rice lowland[a], 1990-91 (*Adapted from* George et al 1994).

Fallow management during the dry-to-wet season transition	Maximum mineral N measured in the soil prior to flooding the field for wet season rice		Soil N uptake by aboveground parts of the dry-to-wet transition crops	
	1990	1991	1990	1991
Sesbania rostrata	118	164	97	125
Mungbean	118	125	66	80
Weeds	118	117	35	46
Weed-free	118	155	–	–
LSD (0.05)	–	28	25	18

[a]The field was kept weed-free and subjected to periodic tillage and water application during the preceding dry season (February to May).

plants and microorganisms. The application of organic materials through incorporation of straw, stubble, animal manure, or green manure adds N to rice soils. Of the farm organic substances, rice straw is the most suited to lowland rice. In farmers' fields in the tropics, straw yield corresponding to 1 t of sun-dried rice is about 1.5 t, and it contains about 9 kg N. Straw removed from the field for industrial uses or burned represents a total loss of its N content. Incorporating straw is a simple way of increasing the N content of lowland ricefields. Incorporating rice straw at 5 t ha^{-1} each season can add about 30 kg N ha^{-1} to the soil. Water regime determines the long-term effects of straw application on the soil's organic matter and N content. Under dryland conditions, application of organic materials over the years may not bring about an appreciable increase in organic matter or N content over the equilibrium value for the environment and cropping system. But anaerobic conditions during the flooded phase of lowland rice cultivation retard organic matter decomposition and cause an increase in both organic matter and N. If two rice crops are grown per year in a tropical lowland where none or one crop was grown before, accretion of C and N is detectable in 2-3 yr. Flooded fallow, compared with the normal practice of dry fallow between crops, enhances the contribution of straw to N fertility of lowland rice soils. At IRRI, incorporating the straw produced in situ twice a year caused an average increase of 48 kg N ha^{-1} $season^{-1}$ in two field experiments lasting 7 yr each. In a drum study with three soils over a 5-yr period, the increase due to straw incorporation was 40 kg ha^{-1} $season^{-1}$, about 10 kg more than the straw's N content (Table 6). The additional 10

Table 6. Nitrogen gains in lowland rice soils due to straw incorporation in 3 experiments over 7 yr (*Source*: Ponnamperuma 1984).

Type of study	No. of rice crops grown	N (%) Straw removed	N (%) Straw incorporated	Yearly N gains (kg ha^{-1})
Field	14	0.1880	0.2150	77
Field	14	0.1787	0.2183	113
Drum	10	0.1415	0.1615	80

kg of N probably came from BNF stimulated by straw acting as an energy source for heterotrophs and as a CO_2 supplement to surface phototrophs. Anaerobic soil conditions, compared with dryland soils, not only preserve more straw N and guard against nitrification and subsequent leaching but also favor N conservation and BNF (Roper and Ladha 1995).

Depending on the N concentration and the C-N ratio of applied organic materials, there is a net mineralization or immobilization of N, and a greater or lesser increase in soil organic matter content. The mineralized N is available for uptake by the rice crop and some may be lost. Available research data suggest that 10-20% of the N freshly supplied through organic materials with a high C-N ratio such as rice straw and stubble is assimilated by the rice crop, 10-20% is lost through various pathways, and 60-80% is immobilized or stored in the soil (Koyama 1981). Of the N supplied through fresh organic materials with a low C-N ratio such as green manure, compost, or animal manure, 20-30% is assimilated by the current rice crop, 20-30% is lost, and 40-60% is immobilized or stored in the soil (Shi et al 1980, Koyama 1981, Zhu et al 1983). Incorporation of organic materials with low N concentrations or a high C-N ratio will thus lead to a greater increase in soil organic matter. If a constant amount of organic material (for example, 10 t ha^{-1} of compost) is applied every year, the soil organic matter content will gradually increase, with a concomitant decrease in the rate of decomposition of the applied material. Continuous application of organic material over a long period increases the annual output of decomposed organic matter (Suzuki et al 1990) which ultimately reaches that of the annual addition of organic materials. Thus, an equilibrium of organic matter is reached such that if 10 t ha^{-1} of compost is successively added to a ricefield every year, an equal amount of N (that is, about 50 kg N ha^{-1}) contained in the compost is released annually. This amount is independent of the rate of organic matter decomposition in the soil. Significant positive impacts of farmyard manure (FYM) application on sustenance of N fertility in rice lowlands have been recorded in long-term fertilizer experiments in India (Tables 1, 7, and 8).

Table 7. Total N and organic C status of a rice soil (0-22 cm) after 14 yr of cropping[a] with different fertilizer/manure treatments in Barrackpore, West Bengal, India. Soil analyzed in 1985 (*Source*: Mandal et al 1991).

Fertilizer/manure treatment	Total N content in soil (kg N ha^{-1})	Organic C concentration in soil (%)
No fertilizer (control)	1348	0.42
50% of recommended NPK[b]	1473	0.43
100% of recommended NPK	1496	0.45
100% of recommended NPK+FYM[c]	1758	0.52
Initial control (as in 1972)	1655	0.52

[a]Annual cropping sequence: jute - rice - wheat.
[b]Recommended NPK doses were 120-26-50 kg ha^{-1} for rice (wet season), 120-26-50 kg ha^{-1} for wheat (dry season), and 60-13-50 kg ha^{-1} for jute (prerice wet season).
[c]Farmyard manure @ 10 t ha^{-1} applied once every year before sowing of jute.

Table 8. Mean annual N balance[a] and change in available N[b] in a Tropaquept intensively grown to rice for 14 yr (1972-86) using different fertilizer/manure treatments. Hyderabad, Andhra Pradesh, India (*Source*: IARI 1989).

Fertilizer/manure treatment	N applied (kg N ha^{-1} yr^{-1})	Crop N removal (kg N ha^{-1} yr^{-1})[c]	N balance (kg N ha^{-1} yr^{-1})	Change in soil N (kg N ha^{-1})[d]
No fert. (control)	0	63	-63	-54
50% NPK	115	105	10	-31
100% NPK[e]	230	151	79	-22
100% NPK + FYM[f]	268	190	78	-18
150% NPK	345	207	138	-44

[a]N applied through fertilizer/manures minus N removal by crops.
[b]Alkaline permanganate ($KMnO_4$) extractable N.
[c]N harvested in grains plus straw at maturity.
[d]Over the 14-yr period.
[e]Recommended NPK dose: 115-20-30 kg ha^{-1} for each rice crop.
[f]Farmyard manure @ 10 t ha^{-1} applied once a year before planting of wet season rice.

Although bulky organic manures like crop residues, FYM, and compost have been the mainstay in the soil fertility management program in many farms, the rapid increase in the use of chemical fertilizers has relegated manures to an inferior position. Wide use of crop residues and organic manures as sources of plant nutrients, however, is very often constrained by their competitive use as sources of fuel and feed for cattle in rural areas. Tanaka (1973), in his studies of various methods of utilization of rice straw in different countries, observed that the straw is largely burned in the Philippines, Thailand, Indonesia, and Malaysia, and used as animal feed or bedding in India, Bangladesh, Sri Lanka, Pakistan, Myanmar, and Egypt. Other means of straw disposal include use as fuel, mulch, roofing, packing material, and in mushroom culture and paper manufacture. By using the available technology for biogas production, the two apparently conflicting aims of getting better fuels and good quality manures from cattle dung and crop residues can be achieved to a large extent.

Enhancement of BNF

In subsistence farming systems in many parts of tropical Asia, a single crop of wetland rice has been grown each year continuously for centuries with no N fertilizer and with no apparent decline in soil N content. Among the processes which contribute to the natural supply of N to wetland rice, BNF appears to be crucial for the maintenance of soil N fertility. The N balance data of some long-term field experiments recently reviewed by Kundu and Ladha (1995b) indicate that the contribution of BNF to the soil during a rice-growing season ranged from 19 to 38 kg N ha^{-1} in Japan, from 28 to 33 kg N ha^{-1} in Thailand, and from 30 to 52 kg N ha^{-1} in the Philippines. This N contribution came largely from indigenous associative and free-living N_2-fixing microorganisms, which include heterotrophic and phototrophic bacteria and cyanobacteria (blue-green algae) present in the soil-plant-floodwater system of ricefields. Roger and Ladha (1992) estimated the contribution of indigenous BNF to lowland rice at 14-50 kg N ha^{-1} $crop^{-1}$. Kundu and Ladha (1995b) suggested that excessive reduction of soil created under continuous flooding and intense puddling adversely affect indigenous BNF in lowland rice. Apparent contributions of indigenous BNF to Maahas clay at Los Banos, Philippines, decreased from 103 kg N ha^{-1} yr^{-1} in the first 12 yr to 59 kg N ha^{-1} yr^{-1} during the next 5 yr. Shifting from continuous wetland rice cropping to multiple-cropping systems focusing on wet-dry rotation increased the population of N_2-fixing microorganisms in the soil of some Chinese farms (Xie 1992).

When exogenous BNF systems, i.e., symbiotic N_2-fixing systems are used in rice lowlands, the N contribution can be much larger (Table 9). Leguminous green manures and grain legumes played an important role in maintaining soil N fertility under traditional rice production systems of the tropics. However, government subsidies and support for growing more rice and the availability of chemical fertilizers have virtually dislodged legumes from the rice systems of many developing countries. Promoting symbiotic N_2-fixing systems can greatly help in sustaining N fertility in lowland rice soils (Ladha and Kundu 1997). The 45-60-d fallow period between two rice crops can be used to grow fast-growing leguminous green-manure crops such as *Sesbania* and *Crotalaria*, or short-duration grain legumes such as cowpea and mungbean. They usually accumulate 80-120 kg N ha^{-1}, and at least 50% of this amount comes from symbiotic N_2 fixation (Ladha et al 1992). In an irrigated double-cropped rice lowland of the Philippines, two green manure legumes or N_2-fixing aquatic fern *Azolla* each of 45-60-d duration could be grown and incorporated. Each of these legumes accumulated 60-90 kg N ha^{-1} (Table 10) of which 50% came from BNF. The green-manure legumes or residues of the grain-legume crops may then be incorporated into the field to supply the fixed N_2 to soil. The

Table 9. N_2 fixed by different species of grain, forage, and green manure legumes in rice lowlands.

Location and legume species	Growth duration (d)	%Ndfa[a]	Amount of N_2 fixed (kg N ha^{-1})	Reference
Los Baños, Philippines				
Crotalaria	190-195	72-81	199-223	Ladha et al (1996)
Clitoria	190-195	78-79	200-240	
Desmanthus	190-195	78-80	196-226	
Siratro	190-195	69-74	91-132	
Pigeon pea	190-195	71-72	111-167	
Sesbania rostrata	45-65	68-94	70-458	Ladha et al (1992)
Aeschynomene afraspera	56	68-76	105-145	
Mungbean	125	64	37	Peoples et al (1995)
Indigofera tinctoria	225	70	79	
Sesbania cannabina	45-55	93	119-188	Pareek et al (1990)
'Clark' soybean	70-84	69	149	George et al (1995)
'UPSY2' soybean	70-84	74	176	
'IT82D-889' cowpea	70-84	48	60	
'Pelungthay' cowpea	70-84	56	78	
'Pagasa 1' mungbean	70-84	45	61	
'Pagasa 2' mungbean	70-84	61	90	
Chiang Mai, Thailand				
Green gram	57-64	89-90	64-66	Peoples et al (1995)
Black gram	66	95-98	119-140	
Soybean	64-73	78-87	26-57	

[a]Percentage of plant N derived from N_2 fixation.

Table 10. Dry matter (DM) yield and N accumulation of two green manure (GM) legumes[a] grown annually in an intensively cultivated rice lowland in Philippines, 1985-89 (*Adapted from* Ventura and Watanabe 1993).

Year	First rice season	First GM legume: Duration (d)	DM yield ($t\ ha^{-1}$)	N accumulation ($kg\ ha^{-1}$)	Second rice season	Second GM legume: Duration (d)	DM yield ($t\ ha^{-1}$)	N accumulation ($kg\ ha^{-1}$)
1985-86	May-Aug	46	3.2	74	Oct-Feb	60	2.7	64
1986-87	Jun-Sep	53	4.0	79	Dec-Apr	61	3.2	55
1987-88	Jul-Oct	53	5.0	90	Jan-Apr	57	3.6	65
1988-89	Jul-Oct	67	6.0	82	Feb-May	62	3.4	85
1989-90	Jul-Nov	50	2.7	61				

[a]Flood-tolerant and stem-nodulating *Sesbania rostrata.*

Azolla can be grown before transplanting rice or as a cover crop with rice, which can subsequently be incorporated into the soil. The fresh weight for a mat of *Azolla* thus grown is commonly 10-15 t ha^{-1}, accumulating about 20-30 kg N. At least 70% of this N is derived from the atmosphere through symbiotic N_2 fixation (Roger and Ladha 1992). In a continuous cropping experiment at IRRI, two rice crops were grown alternately with either a fallow or a green manure crop of *Sesbania* or *Azolla.* Much higher contributions of BNF were recorded in *Sesbania* and *Azolla* than in urea-treated plots. The N balance in the 0-50 cm soil profile after 8 yr showed annual gains of 56 kg N ha^{-1} under urea, 84 kg N ha^{-1} under *Sesbania,* and 148 kg N ha^{-1} under *Azolla* treatment. Kundu and Pillai (1992) suggested that in many double-rice cropped areas of India, a short-duration grain legume could be accommodated during the fallow period to increase the annual foodgrain productivity by 1 t ha^{-1} and make a net saving of 30 kg N ha^{-1}. In rainfed lowlands--where only one crop of rice is grown in wet season leaving a long dry-season fallow due to lack of water--growing of grain or forage legumes helps in improving the soil N balance (Table 11). However, the beneficial role of grain or forage legumes in the sustenance of soil N status depends on their N harvest index (NHI), which is the fraction of total N removed in the harvested produce. To play a positive role in the maintenance of soil N fertility, they must leave behind more N from N_2 fixation than the amount of soil N they remove. To achieve a positive soil N balance after legume harvest, the proportion of total N derived from N_2 fixation (%Ndfa) must exceed the NHI.

A recent survey indicated that green manure is currently used by a small proportion of farmers in India, China, and the Philippines. Where there is sufficient water for growing green manure and when it does not directly prevent growth of more remunerative crops, the principal limitations to green manure use are the labor requirement and the difficulties in propagating the legumes. Seed availability is often a problem with *Sesbania.* Although the plants produce a large number of flowers and seeds, a large proportion is often damaged by insects. The use of stem cuttings has been suggested as an alternative to seed propagation but is likely to be too labor-intensive to be practical. Biomass production of *Sesbania* can be stimulated by ratooning the plants after they have established, and this is an additional way to reduce the amount of seed required initially. It is unlikely for green manure to be adopted solely for its beneficial effects on soil fertility. But where other benefits like suppression of weeds, reduction of incidence of pests, or control of soil erosion are also found, farmers may be encouraged to use them.

Table 11. Contributions of one grain and four forage legumes[a] (kg N ha^{-1}) grown during dry season to soil N balance in a Philippine rice lowland, 1991 (*Adapted from* Ladha et al 1996).

Legume	Aboveground N[b] in the legumes			N returned to the soil from aboveground residues	Apparent N balance in soil[c]	Apparent N contribution of legume roots[d]
	Total	Ndfs	Ndfa			
Pigeon pea	154	43	111	113	+70	26
Crotalaria	277	54	223	118	+64	19
Clitoria	256	56	200	114	+58	33
Desmanthus	251	55	196	137	+82	20
Siratro	132	41	91	81	+40	17

[a]Legume seeds sown in December/January, forage or pods harvested periodically, and the aboveground legume residues cut and incorporated into soil in June.
[b]Ndfs=N derived from soil; Ndfa=N derived from N_2 fixation.
[c]N returned to soil through legume residue minus soil N removed by the legume.
[d]Represents the difference in N uptake by rice from legume (with aboveground residue removed) and fallow plots.

A major problem for adoption is that the green manure legumes often occupy land at a time when some more remunerative crops could be grown. A possible compromise is the use of grain or forage legumes which can provide young pods/grains as food/ vegetables or periodic cuttings of foliage as fodder while still leaving sufficient green leafy material to be used as green manure.

Reducing losses of applied fertilizer N

Inorganic fertilizer N applied to ricefields is partly assimilated by the crop, partly immobilized in the soil, and partly lost through ammonia volatilization, denitrification, leaching, and runoff. Most Asian farmers broadcast fertilizer N as urea into the floodwater 2-4 wk after transplanting rice. Craswell et al (1981) found that broadcasting urea into floodwater resulted in an average 30% recovery of fertilizer N by the rice crop in the dry and wet seasons. Broadcasting and incorporating two-thirds of the urea before transplanting and the remainder at panicle initiation increased the recovery to 40%. The contribution of fertilizer N to the labile soil N pool is small, through the recycling of the crop roots and stubble left in the field after harvest. The fraction of applied fertilizer N that contributes to the labile soil N pool may not exceed 10% (Wada et al 1971, Koyama et al 1977). Higher values generally reported in the literature (Yoshida and Padre 1975, Koyama 1981) are probably the result of exchange between labeled fertilizer N and nonlabeled soil N. Maeda and Shiga (1978) observed that the ratio between additional mineralization of soil organic N due to fertilizer N application and the immobilization of added fertilizer N is about 1, thus Koyama (1981) concluded that the net effect of fertilizer N application on the availability of soil N is small or absent.

Much of the applied fertilizer N is thus lost from the soil-crop system. Total N recovery of ^{15}N-labeled urea in grain, straw, stubble, and soil at harvest was 65-67% in two Philippine rice lowlands where prilled urea was applied in recommended splits, indicating a 33-35% loss of the applied fertilizer N (De Datta 1987). Any means of improving the recovery of applied fertilizer N, either in crop or in soil, will greatly help in sustaining the N fertility. Several ^{15}N balance studies conducted in the Philippines have shown that the loss of fertilizer N from the soil-crop system can be substantially brought down to less than 5% by placing urea at a 5-10 cm depth rather than by broadcasting and incorporating it into the surface soil (Table 12). With improved water control at the time of its application, loss of fertilizer N can be considerably reduced. In IRRI studies, fertilizer application and incorporation into the mud without standing water resulted in only 13% of the applied N being detected in the subsequently applied floodwater, compared with the farmers' practice of broadcasting fertilizer into standing water, which showed 59% of the applied N remaining in floodwater. Grain yields of rice from plots where N was applied to the soil without standing water were higher (by 0.5-0.9 t ha^{-1}) than those obtained by applying fertilizer N into 5 cm of standing water, suggesting a net saving of 10-18 kg N ha^{-1}.

Maintenance of optimum soil physical conditions

The deterioration of soil physical properties create unfavorable chemical and microbiological environments in intensively cultivated ricelands and thereby reduce the availability of soil and biologically fixed N_2 (Kundu and Ladha 1995b).

a) Permeability: Continuous flooding and repeated puddling of soils without dry tillage drastically reduce percolation rate, sometimes to about zero. Permeability is important in maintaining the long-term fertility of rice lowlands because the percolating water carries dissolved oxygen to the active root zone and dilutes or eliminates toxic substances, renewing the soil environment in the process. The percolating water, however, may carry some nutrients past the rooting zone as well. Permeability of fertile ricefields thus are neither high nor too low; the actual rates depend on local conditions. The water percolation rates of most high-yielding ricefields have been found to range from 20 to 30 mm d^{-1} in Japan (Takai and Wada 1977) to 9-15 mm d^{-1} in the Jiangsu and Shanghai provinces of China (Yang and Chen 1961, Anonymous 1978). In Malaysia, Sugimoto (1969) obtained higher rice yields with the 10 mm d^{-1} percolation rate than with no percolation. Continued incorporation of rice stubble into fields with inadequate percolation builds up the levels of readily decomposable organic substances and creates excessively reduced soil conditions, thus favoring accumulation of toxic substances in the plow layer. In one such ricefield in China, permeability declined from 7.8 to 1.5 mm d^{-1}, and average redox potential of the soil in the plow layer decreased from 260 to 173 mV (Hseung et al 1980). Very low rates of soil organic matter decomposition and N mineralization in poorly drained ricefields (with water percolation rates <5 mm d^{-1}) were observed due to excessively reduced conditions (Takai and Wada 1977, Kanke and Kanazawa 1986). Large accumulation of substances like ferrous iron, nitrites, sulfides, alcohols, aldehydes, ketones, reducing sugars, aliphatic hydroxy and unsaturated acids, mercaptans, organic sulfides, ferulic acid, sinapic acid, etc, in the soils, may strongly inhibit the absorption of N by rice plants (Table 13).

Soil N availability in continuously saturated ricefields may be considerably improved by soil drying and adoption of suitable cropping systems. In triple-cropped areas, one rice

Table 12. Recovery of ^{15}N-labeled urea by transplanted rice as affected by time and method of N fertilizer application[a] (*Source*: Schnier et al 1990).

Treatment	Recovery of ^{15}N-labeled fertilizer (% of applied N)				
	Straw	Root	Grain	Soil	Unaccounted for
First application					
Broadcast and incorporated	11	2	16	26	45
Band-placed, 14 DAT	19	3	43	25	10
USG point-placed, 1 DAT	21	2	52	21	4
Second application					
Broadcast, 5-7 DBPI	14	3	47	19	17
Band-placed, 5-7 DBPI	22	2	56	17	3

[a]87 kg N ha^{-1} was applied in two splits: (2/3 basal + 1/3 at 5-7 DBPI). DAT=days after transplanting, USG=urea supergranule, DBPI=days before panicle initiation.

Table 13. Comparative N uptake by rice in gley and normal soils developed from the same parent materials in Hunan, China (*Source*: Pan et al 1986).

Parent material	Soil type	Total N (%)	Total reducing substances (meq $100g^{-1}$)	N uptake by rice (mg pot^{-1})
Red clay	Gley	0.28	5.0	294
	Normal	0.15	2.3	786
Slate	Gley	0.23	6.0	432
	Normal	0.19	2.3	959

crop may be replaced by a suitable upland crop. A continuous system of rice - rice - barley has maintained a high and stable grain yield level over a long period in the southern coastal plain of Zhejiang, China (Li 1993). In double-rice cropping areas, a dry fallow period of at least 60 d yr^{-1} can be achieved by planting appropriate rice varieties. However, the desired degree of soil drying may not be achieved with continuous double-rice cropping, and in such areas, practicing a thorough wet-dry land rotation system using only one rice and an upland crop after every 2 yr or so might be beneficial (Kundu and Ladha 1995b). To overcome the stagnation in productivity in continuously cropped wetland ricefields, similar multiple-cropping systems employing flooded rice and upland crop rotations were introduced in China in the 1980s. After 5-6 yr of practicing such systems, soil N fertility improved significantly through improvement of physical properties, prevention of formation of secondary gley layers in the soil, and promotion of the growth of N_2-fixing microorganisms (Xie 1992).

b) Thickness of the plow layer: Intensively cropped ricefields often are not thoroughly plowed due to poor trafficability. Fields are prepared every season by shallow rototilling and puddling. Continuous preparation of land in this way makes the plow layer thinner and helps develop a hard impervious soil layer, often at 15 cm depth, which may prevent rice roots from accessing N in subsoil layers. The importance of plow layer thickness in lowland rice production was shown in Korea (Table 14). A rice - wheat - cowpea cropping sequence practiced for 20 yr caused a hardpan to develop at 15 cm in a Hapludoll soil in India (Rawat et al 1996). Lowland rice can derive a significant part of

Table 14. Differences in some soil physical properties between high-yielding and neighboring irrigated rice fields of Korea[a] (*Source*: Jo and Um 1990).

Ricefield	Year of survey	Thickness of plow layer (cm)	Depth of root zone (cm)	Penetration resistance[b] ($kg\ cm^{-2}$)		Grain yield ($t\ ha^{-1}$)
				T	S	
High-yielding	1983	20.1	39.4	4.8	6.5	8.1
	1984	18.4	32.9	3.8	5.1	8.2
Mean		19.3	36.2	4.2	5.8	8.1
Neighboring	1983	13.5	28.1	6.1	9.9	4.4
	1984	14.8	25.0	4.9	8.2	4.6
Mean		14.2	26.6	5.6	9.0	4.5

[a]Average of 34 fields in both years. [b]T=topsoil, S=subsoil.

Table 15. Yield and N uptake of lowland rice in a Typic Hapludalf in India as influenced by different tillage treatments. Palampur, Himachal Pradesh, 1983-84 (*Source:* Acharya and Sood 1992).

Tillage treatment	Grain yield ($t\ ha^{-1}$)		Total N uptake ($kg\ ha^{-1}$)	
	1983	1984	1983	1984
Zero tillage	2.3	3.7	48.6	71.2
15 cm tillage	3.8	5.0	68.8	103.9
30 cm tillage	4.4	5.9	91.9	123.8

its N requirements from the subsoil. In a microplot experiment, about 40% of total N absorbed by the unfertilized rice plants came from below 20 cm (Ventura and Watanabe 1984). Similarly, in the Tai Lake region of China, Chen and Zhu (1986) found that the subsoil contributed 16-49% to the total soil N supply to rice plants. To sustain N fertility of rice lowlands, it is necessary to alleviate constraints to subsoil N availability. The important role of tillage practices on N availability to rice was observed in an Indian lowland (Table 15). Recent field experiments in the Philippines have shown that increasing the depth of primary tillage from 15-25 cm to 40 cm can increase the availability of N to rice in farms having a hard impervious soil layer at 15-17 cm depth (Kundu et al 1996). The benefit of a deep tillage was realized in three successive rice cropping seasons without any adverse impact on the balance of total N in the soil (Table 16).

Conclusions

The N-supplying capacity of lowland rice soils has declined over the years of cropping, even when the total N content of the soils did not decrease. Retarded mineralization of soil organic N, accumulation of toxic substances in the root zone, and reduced thickness of the plow layer have been identified as major factors responsible for the declining N availability to rice. Further research is needed to determine the optimum thickness of the plow layer, the optimum water percolation rate, and the critical concentration of reducing

Table 16. Rice yield and N uptake in a Philippine lowland as influenced by tillage depth.[a] Los Baños, Laguna, 1993-94 (*Source*: Kundu et al 1996).

Tillage depth (cm)	Grain yield ($t\ ha^{-1}$)	N uptake in grain plus straw ($kg\ ha^{-1}$)	Total N in 0-45 cm soil after rice harvest ($kg\ ha^{-1}$)[b]
15	14.1	220.3	4322 ± 287
25	15.2	238.9	4408 ± 396
40	16.5	264.3	4506 ± 206

[a]Three crops of rice were successively grown after imposition of primary tillage treatments. [b]Before the rice crops, the soil had 4394 ± 463 kg N ha^{-1}.

substances in the rice soils of different regions and situations. To sustain total N content in the soils, it is necessary to maintain or increase N input and reduce N losses. Further increase in N input from BNF will require promotion of symbiotic N_2-fixing systems through integration of appropriate legumes in rice lowlands. Reduction of N losses will have to come from improved recovery of fertilizer N and conservation of NO_3 formed in the soils during the nonflooded period.

References

Acharya C L, Sood M C (1992) Effect of tillage methods on soil physical properties and water expense of rice on an acidic alfisol. J. Indian Soc. Soil Sci. 40: 409-414.

Anonymous (1978) Characteristics of the soil fertility and its improvement of the high-yielding paddy soil in the Shanghai suburbs. Sci. Agric. Sin. 2: 66-72.

Brammer H (1978) Rice soils of Bangladesh. Pages 35-55 in Soils and rice. International Rice Research Institute, P.O. Box 933, Manila, Philippines.

Buresh R J, Chua T T, Castillo E G, Liboon S P, Garrity D P (1993) Fallow and *Sesbania* effects on soil nitrogen dynamics in lowland rice-based cropping systems. Agron. J. 85: 316-321.

Buresh R J, Woodhead T, Shepherd K D, Flordelis E, Cabangon R C (1989) Nitrate accumulation and loss in a mungbean/lowland rice cropping system. Soil Sci. Soc. Am. J. 53: 477-482.

Cassman K G, Pingali P L (1995) Extrapolating trends from long-term experiments to farmers' fields: the case of irrigated rice systems in Asia. Pages 63-84 in Barnett V, Payne R, Steiner R (Eds.) Agricultural sustainability: economic, environmental and statistical considerations. John Wiley & Sons Ltd.

Chen D L, Zhu Z L (1986) Nitrogen supply of subsoil in flooded rice field. Soils (China) 18: 33-34.

Craswell E T, De Datta S K, Hartantyo M, Obcemea WN (1981) Time and mode of nitrogen fertilizer application to tropical wetland rice. Fert. Res. 2: 247-259.

De Datta S K (1987) Advances in soil fertility research and nitrogen fertilizer management for lowland rice. Pages 27-41 in Efficiency of nitrogen fertilizers for rice. International Rice Research Institute, P.O. Box 933, Manila, Philippines.

Dolui A K, Guhathakurta D (1988) Sulphur fractions and carbon, nitrogen and sulphur relationships in some soils of West Bengal. J. Indian Soc. Soil Sci. 36:53-58.

Flinn J C, De Datta S K (1984) Trends in irrigated rice yields under intensive cropping at Philippine research stations. Field Crops Res. 9: 1-15.

George T, Ladha J K, Buresh R J, Garrity D P (1993) Nitrate dynamics during the aerobic soil phase in lowland rice-based cropping systems. Soil Sci. Soc. Am. J. 57: 1526-1532.

George T, Ladha J K, Garrity D P, Buresh R J (1994) Legumes as nitrate catch crops during the dry-to-wet transition in lowland rice cropping systems. Agron. J. 86: 267-273.

George T, Ladha J K, Garrity D P, Torres R O (1995) Nitrogen dynamics of grain legume-weedy fallow-flooded rice sequences in the tropics. Agron. J. 87:1-6.

Goung Y E, Win K, Htin W (1978) Rice soils of Burma. Pages 57-71 in Soils and rice. International Rice Research Institute, P.O. Box 933, Manila, Philippines.

Hseung Y, Xu Q, Yao X, Zhu Z (1980) Effect of cropping system on the fertility of paddy soil. Acta Pedol. Sin. 17: 101-119.

IARI—Indian Agricultural Research Institute (1989) Annual report 1985-86/1986-87, All India Coordinated Research Project on Long Term Fertilizer Experiments, Indian Agricultural Research Institute, New Delhi.
Jo I S, Um K T (1990) Field drainage and rice growth. Pages 152-157.in Transactions of the 14th International Congress of Soil Science, Kyoto, Vol 1.
Kanke B, Kanazawa S (1986) Effect of drainage on soil saccharides and microbial activities in poorly drained paddy fields. Pages 594-595 in Transactions of the 13th International Congress of Soil Science, Hamburg. Vol 2.
Kawaguchi K, Kyuma K (1977) Paddy soils in tropical Asia: their material nature and Fertility. University of Hawaii, HI.
Koyama T (1981) The transformations and balance of nitrogen in Japanese paddy fields. Fert. Res. 2: 261-278.
Koyama T, Shibuya M, Tokuyasu M, Shimomura T, Ide T, Ide K (1977) Balance sheet and residual effects of fertilizer nitrogen in Saga paddy field. Pages 289-296 in Proceedings of the International Seminar on Soil Environment and Fertility Management in Intensive Agriculture, Society of the Science of Soil and Manure, Tokyo, Japan.
Kundu D K, Ladha J K (1995a) Enhancing soil nitrogen use and biological nitrogen fixation in wetland rice. Exp. Agric. 31: 261-277.
Kundu D K, Ladha J K (1995b) Efficient management of soil and biologically fixed N_2 in intensively cultivated rice fields. Soil Biol. Biochem. 27: 431-439.
Kundu D K, Ladha J K, Lapitan-de Guzman E (1996) Tillage depth influence on soil nitrogen distribution and availability in a rice lowland. Soil Sci. Soc. Am. J. 60: 1153-1159.
Kundu D K, Pillai K G (1992) Integrated nutrient supply system in rice and rice-based cropping systems. Fert. News 37(4): 35-41.
Ladha J K (1997) Role of biological nitrogen fixation in replenishing soil nitrogen pool in cropping systems. In Elmerich C et al (Eds.) Proceedings of the International Congress on Nitrogen Fixation. Kluwer Academic Press, The Netherlands (in press)
Ladha J K, Kundu D K (1997) Towards sustaining the nitrogen fertility of lowland rice soils: issues and options. In Rupela O P (Ed.) Managing legumes nitrogen fixation in cropping systems of Asia. International Crops Research Institute for the Semi-arid Tropics, Hyderabad. (in press)
Ladha J K, Kundu D K, Angelo-Van Coppenolle M G, Peoples M B, Carangal V R, Dart P J (1996) Legume productivity and soil nitrogen dynamics in lowland rice-based cropping systems. Soil Sci. Soc. Am. J. 60: 183-192.
Ladha J K, Pareek R P, Becker M (1992) Stem-nodulating legume-rhizobium symbiosis and its agronomic use in lowland rice. Adv. Soil Sci. 20: 147-192.
Li S Y (1993) Yield stability and fertilizer efficiency of long-term triple cereal cropping in paddy fields of China. Biol. Fert. Soils 16: 151-153.
Maeda K, Shiga H (1978) Relationship between mineralization and nitrogen influenced by various conditions of submerged soils. Soil Sci. Plant Nutr. 24: 515-524.
Mandal B C, Saha M N, Adhikari M (1991) Yield of rice and nutrient status of soil due to continuous cropping and manuring in a jute-rice-wheat sequence on Indo-Gangetic alluvial soil. J. Indian Soc. Soil Sci. 39: 689-692.
Nambiar K K M (1994) Soil fertility and crop productivity under long-term fertilizer use in India. Indian Council of Agricultural Research, New Delhi.
Pan S, Chen J, Yue Z (1986) Nutrient supply of gleyed paddy soils. Pages 365-376 in Current progress in soil research in People's Republic of China. Jiangsu Science and Technology Publishing House, Nanjing.
Paramananthan S (1978) Rice soils of Malaysia. Pages 87-98 in Soils and rice. International Rice Research Institute, P.O. Box 933, Manila, Philippines.
Pareek R P, Ladha J K, Watanabe I (1990) Estimating N_2 fixation by *Sesbania rostrata* and *S. cannabina* (syn. *S. aculeata*) in lowland rice soil by the ^{15}N dilution method. Biol. Fert. Soils 10: 77-88.
Peoples M B, Herridge D F, Ladha J K (1995) Biological nitrogen fixation: an efficient source of nitrogen for sustainable agricultural production? Plant Soil 174: 3-28.
Pingali P L, Hossain M, Gerpacio RV (1997) Asian rice bowls: the returning crisis? CAB International, Wallingford, Oxon, UK. p.341.
Ponnamperuma F N (1985) Chemical kinetics of wetland rice soils relative to soil fertility. Pages 71-80 in Wetland soils: characterization, classification and utilization. International Rice Research Institute, P.O. Box 933, Manila, Philippines.
Ponnamperuma F N (1984) Straw as a source of nutrients for wetland rice. Pages 117-136 in Organic matter and rice. International Rice Research Institute, P.O. Box 933, Manila, Philippines.
Rawat M S, Tripathi R P, Nand Ram (1996) Long-term effect of puddling and fertilizer use in rice-wheat-cowpea sequence on structural properties of soils. J. Indian Soc. Soil Sci. 44: 364-368.
Roger P A, Ladha J K (1992) Biological nitrogen fixation in wetland rice fields: estimation and contribution to nitrogen balance. Plant Soil 141: 41-55.
Roper M B, Ladha J K (1995) Biological nitrogen fixation by heterotrophic and phototrophic bacteria in association with straw. Plant Soil 174: 211-224.
Schnier H F, Dingkuhn M, De Datta S K, Marqueses E P, Faronilo J E (1990) Nitrogen-15 balance in transplanted and direct-seeded flooded rice as affected by different methods of urea application. Biol. Fert. Soils 10: 89-96.

Sharma K N, Bhandari A L, Kapur M L, Rana D S (1985) Influence of growing various crops in five different fixed rotations on the changes in nitrate and total nitrogen content of soils. J. Agric. Sci. Camb 104: 609-613.
Shi S, Wen Q, Lian H (1980) The availability of nitrogen of green manures in relation to their chemical composition. Acta Pedol. Sin. 17: 240-246.
Shrestha R K, Ladha J K (1998) Nitrate in groundwater and integration of N-catch crop in intensive rice-based cropping system to reduce nitrate leaching. Soil Sci. Soc. Am. J. (in press)
Singh N T (1984) Green manures as sources of nutrients in rice production. Pages 217-228 in Organic matter and rice. International Rice Research Institute, P.O. Box 933, Manila, Philippines.
Soepraptohardjo M, Suhardjo H (1978) Rice soils of Indonesia. Pages 99-113 in Soils and rice. International Rice Research Institute, P.O. Box 933, Manila, Philippines.
Sugimoto K (1969) Studies on plant-water relationship of paddy in Muda river irrigation project area of West Malaysia. Department of Agriculture, Malaysia. p.145.
Suzuki M, Kamekawa M, Sekiya S, Shiga H (1990) Effect of continuous application of organic or inorganic fertilizer for sixty years on soil fertility and rice yield in paddy field. Pages 14-19 in Transactions of the 14th International Congress of Soil Science, Kyoto, Japan. Vol. 4.
Takai Y, Wada H (1977) Effects of water percolation on fertility of paddy soils. Pages 216-222 in Proceedings of the International Seminar on Soil Environment and Fertility Management in Intensive Agriculture. Society of the Science of Soil and Manure, Tokyo, Japan.
Tanaka A (1973) Methods of handling the rice straw in various countries. Int. Rice Com. Newsl. 22: 1-20.
Tripathi B P, Ladha J K, Timsina J, Pascua S R. (1997) Nitrogen dynamics and balance in intensified rainfed lowland rice-based cropping systems. Soil Sci. Soc. Am. J. 61:812-821.
Ventura W, Watanabe I (1984) Dynamics of nitrogen availability in lowland rice soils: the role of soil below plow layer and effect of moisture regimes. Phillipp. J. Crop Sci. 9: 135-142.
Ventura W, Watanabe I (1993) Green manure production of *Azolla microphylla* and *Sesbania rostrata* and their long-term effects on rice yields and soil fertility. Biol. Fert. Soils 15: 241-248.
Wada G, Shoji S, Takahashi J, Saito K, Shinbo I (1971) The fate of fertilizer nitrogen applied to the paddy field and its absorption by rice plant. Proc. Crop Sci. Soc. Jpn. 40: 293.
Xie W (1992) Crop rotation in paddy field and soil fertility. Proceedings of the International Symposium on Paddy Soils. Institute of Soil Science, Academia Sinica, Nanjing. p. 467.
Yang G, Chen J (1961) On the significance of constant renewal of soil condition as affected by the permeability of paddy soil. Acta Pedol. Sin. 9: 65-71.
Yoshida T, Padre B C Jr (1975) Effect of organic matter application and water regimes on the transformation of fertilizer nitrogen in a Philippine soil. Soil Sci. Plant Nutr. 21: 281-292.
Zhu Z, Liao X, Cai G, Chen R, Wang Z (1983) On the improvement of the efficiency of nitrogen of chemical fertilizers and organic manures in rice production. Soil Sci.135:35-39.

Interpreting fertilizer use efficiency in relation to soil nutrient-supplying capacity, factor productivity, and agronomic efficiency

D.C. OLK[1], K.G. CASSMAN[2], G. SIMBAHAN[1], P.C. STA. CRUZ[1,3], S. ABDULRACHMAN[4], R. NAGARAJAN[5], PHAM SY TAN[6], and S. SATAWATHANANONT[7]

[1]*International Rice Research Institute, P.O. Box 933, 1099 Manila, Philippines*

[2]*Department of Agronomy; University of Nebraska, P.O. Box 830915, Lincoln, Nebraska, 68583-0915; USA*

[3]*Philippine Rice Research Institute, Maligaya, Muñoz, Nueva Ecija, Philippines*

[4]*Research Institute for Rice, Jl. Raya No. 9, Sukamandi 41256, Subang, West Java, Indonesia*

[5]*Tamil Nadu Rice Research Institute, Aduthurai 612101, Tamil Nadu, India*

[6]*Division of Agronomy, Cuu Long Delta Rice Research Institute, Omon, Can Tho, Vietnam*

[7]*Pathum Thani Rice Research Center, Thanyaburi, Pathum Thani 12110, Thailand*

Key words: soil N, fertilizer N, agronomic efficiency, productivity, irrigated rice

Abstract

Although efficient use of N remains a critical constraint to productivity in irrigated lowland rice, a comprehensive database does not exist for the efficiency of on-farm management of N and other nutrients. In 1994, IRRI initiated its Mega Project on Reversing Trends of Declining Productivity in Intensive Irrigated Rice Systems in selected rice production domains of five tropical Asian nations to improve on-farm fertilizer use efficiency and to monitor long-term productivity trends as related to fertilizers and other inputs. Data are reported here for the first crop cycle, the 1994-95 dry season. The indigenous soil N supply (INS) was estimated by aboveground crop N uptake and grain yield (GY) in plots without applied N established in farmers' fields under otherwise favorable growth conditions. The fertilizer N rate each farmer applied to his/her field surrounding these plots was recorded; GY was also measured in that area. In each domain, GY in unfertilized plots varied considerably among farms, as the range between maximum and minimum values within each domain was at least 2.8 t ha^{-1}, thus of comparable magnitude to mean GY for these plots. Fertilizer N rates varied from 36 to 246 kg ha^{-1} across all domains, but their lack of relationship to INS contributed to relatively low fertilizer N efficiency and high variability in efficiency among farms. Mean agronomic efficiency (ΔGY/applied N rate) for each domain was only 6-15 kg grain kg^{-1} N, while values for individual farmers ranged from 0 to 59 kg grain kg^{-1} N. Initial data on P and K fertilizer management also suggest highly variable applications at suboptimal efficiency. These results indicate the potential for greater fertilizer efficiency from improved congruence between the indigenous soil supply and applied fertilizer and emphasize the need for field-specific nutrient management. Although agronomic efficiency and partial factor productivity (GY/applied N rate) can each be used to describe the efficiency of fertilizer applications, a complete analysis of nutrient management should include both terms, grain yield, fertilizer rates, and native soil fertility.

Introduction

Asia's food supply has kept pace with rapid population growth during recent decades, primarily because of widespread use of modern rice varieties with high yield potential. The shorter growth duration of these semidwarf varieties has allowed multiple cropping each year in the irrigated lowland systems. Long-term field experiments, begun at IRRI field stations shortly after the introduction of modern rice varieties indicate, however, that after several consecutive years of such intensified cropping with constant and high fertilizer inputs, grain yields begin declining (Cassman and Pingali 1995). Although several factors may contribute to this trend, a primary cause appears to be declining soil N availability (Cassman et al 1995). IRRI is investigating the changes in soil N cycling that

V. Balasubramanian, J. K. Ladha and G. L. Denning (eds.), Resource Management in Rice Systems: Nutrients, 45–55.

occur under intensified irrigated rice cropping (Olk et al 1996a), as well as crop management options for mitigating the declining yield trend.

To properly assess how Asian rice production may be affected by declining yields with constant inputs or declining productivity with varying input levels, similar studies are needed under on-farm conditions. The first step toward this assessment is the development of a long-term database for on-farm productivity trends. A comprehensive database does not yet exist, however.

To address this deficiency, IRRI in 1994 began its Mega Project on Reversing Trends of Declining Productivity in Intensive Irrigated Rice Systems. A primary objective was to establish a database on farmers' inputs, grain yield and nutrient removal, and selected soil properties in double-cropped irrigated rice systems in major rice-growing domains of tropical Asia. Another objective was to improve on-farm fertilizer use efficiency while maintaining or increasing grain yields through a better balance between applied fertilizer rates and actual soil nutrient supply.

A central assumption to both objectives is that soil has an inherent and distinct capacity to supply nutrients to the crop, independent of fertilizer (Cassman et al 1996b, Appel 1994). From this perspective, declining productivity may result from either a decline in the soil nutrient-supplying capacity (SNSC), declining fertilizer use efficiency, or both. Maximum fertilizer use efficiency would be achieved by optimal utilization of both the SNSC and applied fertilizers under proper crop management.

In the following sections we describe the structure of the Mega Project, report estimates of the SNSC and fertilizer N use efficiency measured in the first crop cycle of project activities (1994-95 dry season), and provide initial interpretations of the results.

Materials and methods

Structure of the Mega Project on Declining Productivity

Details of the project structure were provided by Olk et al (1996b). Project activities were conducted at five sites representing domains of double-cropped irrigated lowland rice in each of five tropical Asian nations (Table 1). Each domain was centered around a national or provincial rice research institute or research station. The project staff at each facility managed a long-term fertility experiment located at the research station to establish yield trends when recommended rates of fertilizer inputs are applied during 10-25-yr periods, and to identify potential nutrient limitations that develop from continuous rice cropping at high yield levels.

On-farm research activities were conducted in 24-36 farmers' fields near villages in a 15-20 km radius of the research station. Each field was managed by the farmer following his/her own decisions, except for treatment plots within these fields that were established and managed by the project team. Neither manure nor N, P, or K fertilizer was applied to the -F treatment plot. The N-supplying capacity of the soil in each field, hereafter called indigenous N supply (INS), was defined as aboveground crop N uptake in this treatment, except for calculations of fertilizer use efficiency. Because crop N uptake was closely correlated with grain yield (GY) at all sites (Olk et al 1996b), GY was also used as an index of INS. In an adjacent treatment plot, P and K fertilizers without any N were ap-

Table 1. Collaborating institutions in the IRRI Mega Project on Reversing Trends of Declining Productivity in intensive, irrigated rice systems.

	Cuu Long Delta Rice Research Institute (Omon)[a]
Region	Mekong Delta, Vietnam
Latitude, longitude	10^0 34' 80" N, 100^0 7' 50" E
Soil type (USDA)	Fluvaquentic Humuaquept
	Philippine Rice Research Institute (Maligaya)
Region	Central Luzon, Philippines
Latitude, longitude	15^0 40' 21" N, 120^0 53' 26" E
Soil type (USDA)	Entic Pellustert
	Research Institute for Rice (Sukamandi)
Region	Northwest Java, Indonesia
Latitude, longitude	06^0 18' 02" S, 107^0 37' 46" E
Soil type (USDA)	Aeric Tropaqualf
	Suphan Buri Rice Experiment Station (Suphan Buri)
Region	Central Floodplains, Thailand
Latitude, longitude	14^0 28' 80" N, 100^0 05' 23" E
Soil type (USDA)	Aeric Tropaquept
	Tamil Nadu Rice Research Institute (Aduthurai)
Region	Tamil Nadu, India
Latitude, longitude[b]	10^0 52' N, 79^0 20' E
Soil type (USDA)	Udorthentic Pellustert or Udorthentic Chromustert

[a]Name in parentheses is the primary local community where on-station and farmer field trials are conducted.
[b]Latitude and longitude available only to the minute.

plied at rates recommended by project scientists to ensure adequate P and K supply (+PK plot). Higher aboveground crop N uptake and GY in the +PK plots than the -F plots would indicate that P and/or K deficiency limited plant growth in the -F plots. In these cases, INS would be based on the +PK plots. Adjoining these two plots, an identical area under farmer's fertilizer practice was also demarcated (FP plot). Three sets of -F, +PK, and FP plots, or replications within fields, were established at different locations in each farmer's field.

Plant and soil measurements

Grain yield was measured in the center of each treatment plot for either 120 hills (transplanted) or a 5-m^2 area (direct-seeded rice). Sampled plants were threshed and grains were allowed to sun-dry for 1-2 d to about 10% moisture content. Grain yields were reported on a basis of 14% moisture content. Nitrogen concentration of plant samples was measured on Kjeldahl digests, although the exact methodology varied among sites.

The framework of Cassman et al (1996b) was used to evaluate N use efficiency achieved by farmers in each of the five domains. This framework defines partial factor productivity from applied N as

$$PFP(N) = Y/N_r$$

where Y is grain yield and N_r is the amount of N fertilizer applied. Because grain yield at a given N_r represents the sum of yield without N input (Y_0) plus the incremental increase that results from N application (ΔY), PFP(N) can be expressed as:

$$PFP(N) = (Y_o + \Delta Y)/N_r$$

or

$$PFP(N) = (Y_o/N_r) + (\Delta Y/N_r)$$

The incremental efficiency from applied N, $\Delta Y/N_r$, is the conventional definition of the agronomic efficiency of applied N [AE(N)]. For all calculations involving AE(N) or PFP(N), Y_o was defined as grain yield in the +PK treatment.

Efforts in the initial stage of the project focused on N efficiency. The high real cost of N fertilizer input and the low N fertilizer use efficiency in irrigated lowland rice (De Datta and Buresh 1989; Crasswell and Vlek 1979) provided the rationale for this emphasis. In addition, decreasing soil N availability was associated with the yield declines noted in the IRRI long-term experiments (Cassman et al 1995).

The partial factor productivity for NPK fertilizers was defined as the ratio of GY to the summed inputs of N, P, and K fertilizers in units of kg ha^{-1} (elemental basis). For all parameters measured at the on-farm sites within each domain, a mixed model analysis of variance was used to determine the significance of fixed effects from fertilizer treatments on GY, and the random effects from farmers' fields on GY and N efficiency parameters (Cassman et al 1996b).

Only two sites was GY in the -F plots significantly correlated ($P<0.05$) with either soil organic C or total soil N, and the highest correlation coefficient was 0.50 (data not shown). These parameters are often considered as indices of soil N availability. However, this general lack of correlation in irrigated lowland ricefields agrees with previous results of Cassman et al (1996a) in the Philippines.

Results

Variability in INS within recommendation domains

In the 1994-95 dry season ,GY in -F treatment plots varied significantly ($P<0.001$, data not shown) among farmers' fields within each domain or site; the range between maximum and minimum values within each domain was at least 2.8 t ha^{-1}, and is thus of comparable magnitude to mean GY for these plots (Table 2). Mean GY in the FP plots ranged between 4.6 and 6.3 t ha^{-1} at the different sites.

Relationship between INS and farmer practice

Fertilizer N rates of the farmers were not related to the INS in their fields. The fertilizer N rate applied to the FP plots was not significantly correlated ($P>0.17$) with total N uptake in the -F treatment at any site nor with GY at four of the five sites ($P>0.38$). The increase

Table 2. Mean grain yields (kg ha^{-1}) in on-farm treatment plots fertilized with (i) no N, P, or K (-F), (ii) P and K (+PK), or (iii) normal farmer's fertilizer practice (FP), 1994-95 dry season.

Treatment plot		Omon (32)[a]	Maligaya (43)	Sukamandi (30)	Suphan Buri (26)	Aduthurai (28)
-F	Mean	4247	4387	4887	3335	4637
	Std dev	(874)[b]	(1251)	(841)	(801)	(825)
+PK	Mean	4635	4519	5265	3470	4755
	Std .dev	(842)	(1339)	(796)	(748)	(836)
FP	Mean	5353	6163	5926	4624	6285
	Std dev	(732)	(1265)	(589)	(672)	(1066)

[a]Number of farmers' fields sampled. Each farm value is the mean of three replicate plots.
[b]Standard deviation of farm means.

in GY associated with N input was also not correlated with fertilizer N rates. Correlation coefficients between increase in GY associated with fertilizer N input and INS were significant and negative for the sites. They varied from -0.28 to -0.70, indicating a relatively high degree of scatter (e.g., Fig. 1). Fertilizer N rates applied by farmers in the FP plots varied widely within each domain (Table 3), which indicates that farmers did not follow the blanket recommendations provided by extension agencies in these domains.

Because of the significant variation in INS among farmers' fields at each site and the lack of correlation between farmers' N fertilizer rates and INS1, PFP(N) and AE(N) also varied widely among fields within each domain (Table 3). These parameters were not significantly correlated with each other at any site (data not shown). In four domains, AE(N) decreased significantly with increasing N fertilizer input although the degree of correlation was generally low (e.g., Fig. 2). At RIR, there was no correlation between AE(N) and N fertilizer rate because AE(N) was extremely low in most fields. At all sites, mean AE(N) was much lower than the values of 24-33 kg grain kg^{-1} of applied N that can be achieved at high yield levels with proper crop and fertilizer management (Peng et al 1996).

Mean yield response to N fertilization by site was 0.7-1.7 t ha^{-1}, which was equivalent to 16-40% of the mean GY in -F plots (Table 2). This implies that most of the crop N uptake in the FP plots was derived from INS, AE(N) values were low, and PFP(N) was positively correlated to Y_0/N_r at all sites (r = 0.67-0.90, P< 0.0001). Because of widely varying N fertilizer rates among farmers, however, Y_0 (i.e., GY in +PK treatment) by itself predicted less than 4% of the variation in PFP(N) at most sites (e.g., Fig. 3). No soil characteristic predicted more than 16% of PFP(N) at any site (data not shown).

PFP(NPK) also varied greatly within the domain of each site (Table 4). Lack of GY response to applied P and K at Aduthurai both on-farm (Table 2) and in the long-term fertility experiment at the research station (data not shown) suggests that P and K rates applied by farmers may be in excess of actual requirements.

The potential for improved on-farm nutrient management is illustrated by the yield gap between the NPK treatment in the on-station long-term experiment (Table 5) and the on-farm FP treatment (Table 2). At four of the five sites, GY in the NPK treatment on-station was 1.2-2.5 t ha^{-1} higher than that attained by farmers in FP plots. One reason for this gap is that on-station N fertilizer rates were higher than those applied by most farmers. For

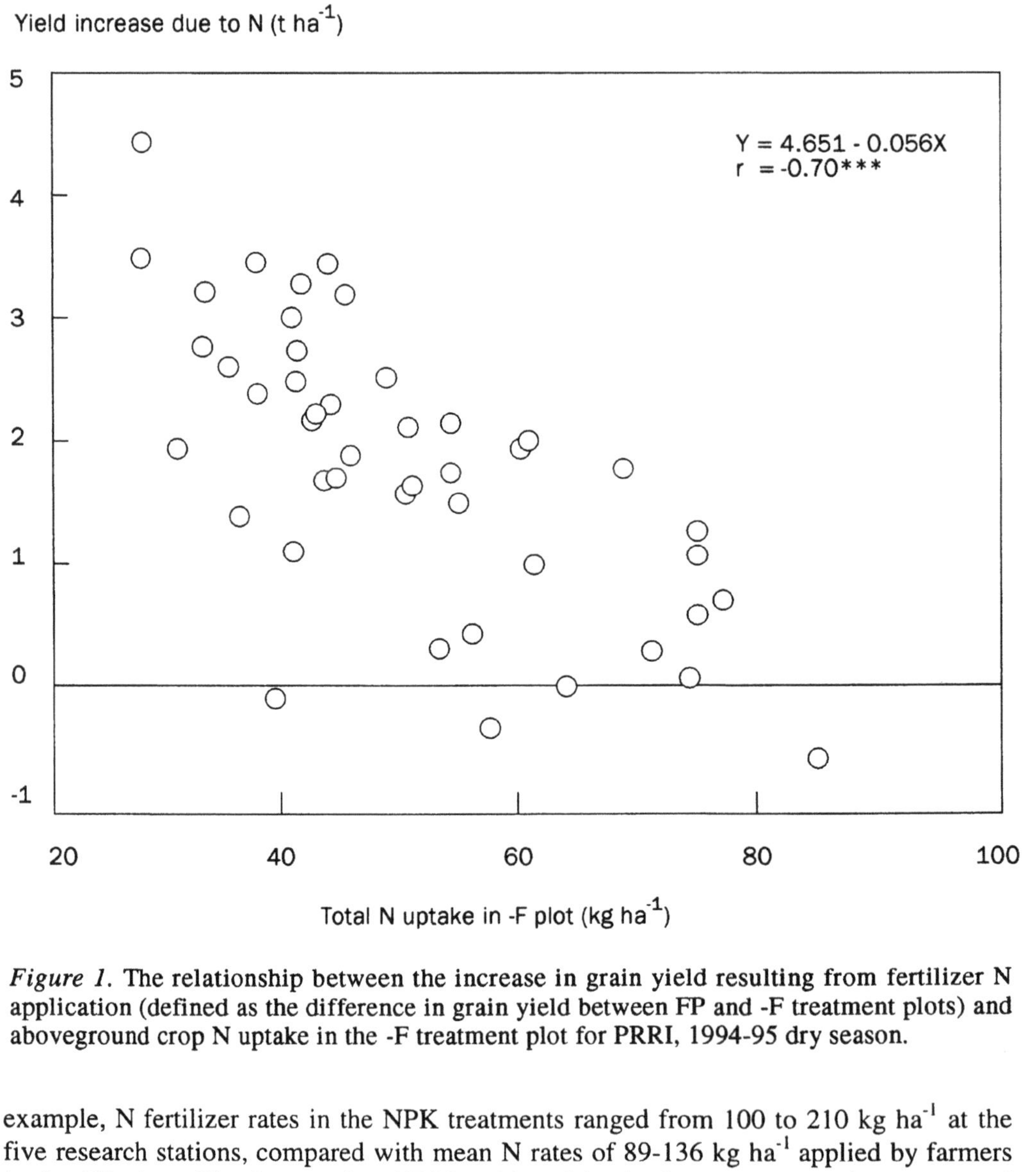

Figure 1. The relationship between the increase in grain yield resulting from fertilizer N application (defined as the difference in grain yield between FP and -F treatment plots) and aboveground crop N uptake in the -F treatment plot for PRRI, 1994-95 dry season.

example, N fertilizer rates in the NPK treatments ranged from 100 to 210 kg ha^{-1} at the five research stations, compared with mean N rates of 89-136 kg ha^{-1} applied by farmers in the FP plots. Simultaneously, AE(N) achieved in the long-term experiments ranged from 17.5 to 24.8 at four of the five sites (data not shown), which was much higher than AE(N) in most farmers' fields (Table 3). The yield gap was not due to differences in INS between the station and the field because GY levels in -F plots on-farm (Table 2) were similar to those in minus NPK treatments on-station (Table 5).

Table. 3. Ranges of N fertilizer rates, partial factor productivity for N [PFP (N)], and agronomic efficiency of applied N fertilizer [AE (N)] for the farmer's practice treatment plots, 1994-95 dry season.

Parameter	Omon (32)[a]	Maligaya (40)	Sukamandi (30)	Suphan Buri (26)	Aduthurai (28)
		N fertilizer rate (kg ha^{-1})			
Mean	90	140	108	114	115
Min	47	70	65	36	63
Max	123	246	176	161	223
Std dev	17	42	32	32	38
		PFP(N) (kg grain kg^{-1} fertilizer N)			
Mean	61	49	60	45	60
Min	42	15	30	24	27
Max	98	92	100	91	104
Std dev	13	17	20	18	22
		AE(N) (Δ kg grain kg^{-1} fertilizer N)			
Mean	8	13	6	11	15
Min	0	0	0	0	0
Max	21	31	32	40	59
Std dev	6	9	5	10	12

[a]Number of farmer's fields sampled. Each farm value is the mean of three replicate plots.

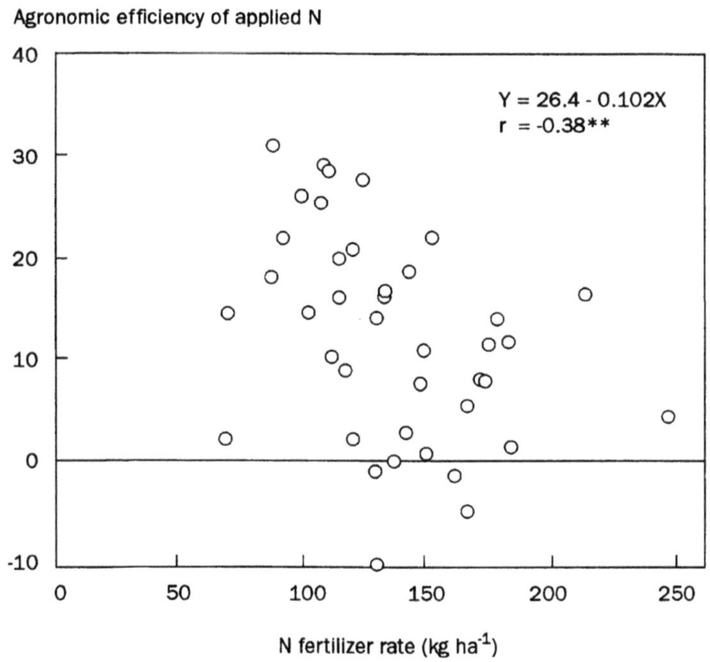

Figure 2. Agronomic efficiency of applied fertilizer N in the FP treatment plot versus rate of fertilizer N application for PRRI, 1995 dry season.

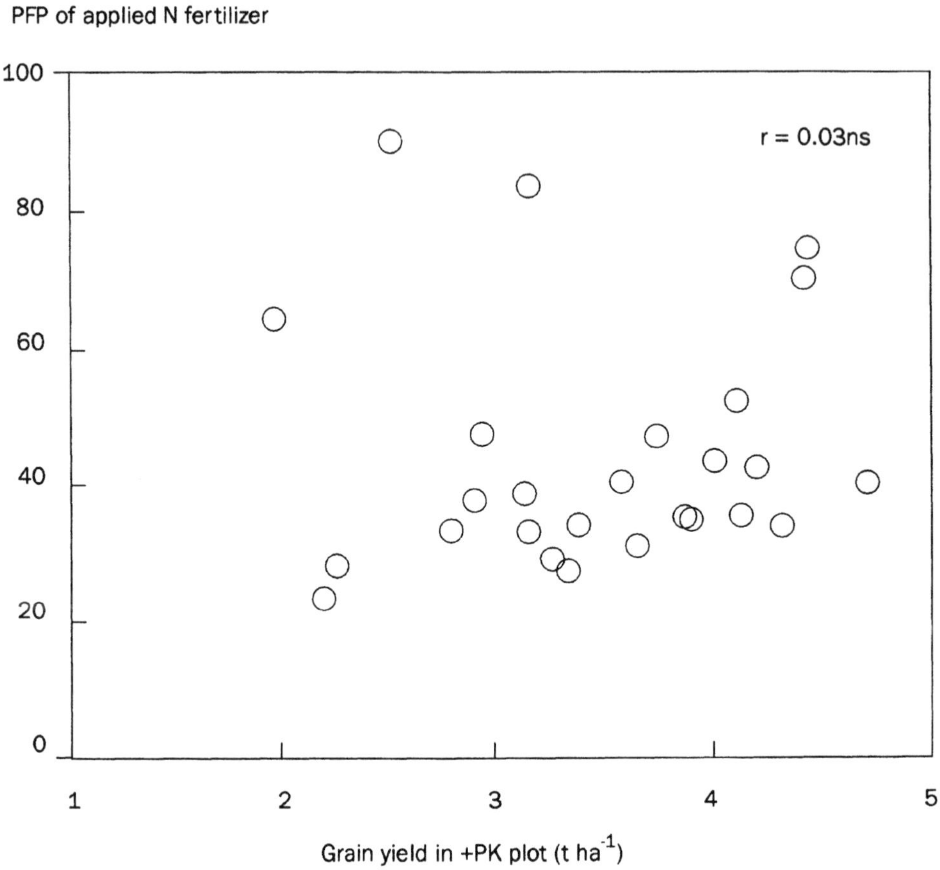

Figure 3. The partial factor productivity (PFP) for applied N fertilizer versus grain yield in the +PK treatment plot for SBRES, 1994-95 dry season.

Table 4. Ranges of partial factor productivity for summed N, P, and K fertilizers [PFP(NPK)], and mean P and K fertilizer inputs in the farmer's practice treatment plots, 1994-95 dry season.

Parameter		Omon (32)[a]	Maligaya (44)	Sukamandi (30)	Suphan Buri (26)	Aduthurai (28)
PFP(NPK)	Mean	51	44	42	37	38
	Min	32	14	20	22	20
	Max	765	124	66	62	67
	Std. dev.	(11)	(22)	(11)	(11)	(13)
Fertilizer P (kg P ha^{-1})	Mean	12	12	19	20	23
	Std. dev.	(6)	(7)	(17)	(9)	(8)
Fertilizer K (kg K ha^{-1})	Mean	6	20	23	0	44
	Std. dev.	(4)	(14)	(24)	(0)	(29)

[a]Number of farmers' fields sampled. Each farm value is the mean of three replicate plots.

Table 5. Grain yield (kg ha^{-1})[a] in the minus NPK control and plus NPK treatments of the on-station long-term fertility experiment, 1994-95 dry season.

Treatment	Omon	Maligaya	Sukamandi	Suphan Buri	Aduthurai
Minus NPK	3966	3320	4456	2656	4956
	(586)	(232)	(260)	(352)	(598)
NPK	6446	8880	7332	3099	7584
	(298)	(1054)	(235)	(406)	(861)

[a]Mean and standard deviation of four field replicates.

Discussion

The results provide initial evidence that farmers are not applying fertilizer at optimal rates. Some apply too much N fertilizer, others too little in relation to INS of their fields, resulting in low AE(N) or GY. An initial interpretation of fertilizer P and K rates also suggests inefficient applications. Thus, the congruence between nutrient supply and demand should be evaluated for each nutrient, especially where yields are limited by multiple nutrient deficiencies.

According to the farmers, one reason for the lack of congruence between INS and their N fertilizer rates is that while they know which fields have greater or less native soil fertility, they have not had the means to quantify that. Unfertilized control plots provide a laborious but direct measure of plant-available N and so appear to be a useful index for researchers. Unfertilized plots may not be useful for measuring INS when plant N uptake or GY is limited by pests or poor weather instead of available soil N. In this study, however, widespread pest damage in the unfertilized plots was not observed and weather during the dry season was favorable for plant growth.

A central index both to understanding long-term productivity trends and to optimizing N fertilizer use efficiency appears to be PFP(N). Its components [Y_o/N_r and AE(N)] contain terms related to the availability of indigenous soil N and fertilizer N, respectively. Quantification of the contributions of these two components to PFP (N) provides a more complete understanding of long-term productivity trends and fertilizer N use efficiency. For example, if a long-term productivity decline coincides with declining Y_o/N_r, the productivity decline may be attributed to a decrease in INS, provided that N fertilizer rates have not increased significantly and other factors such as pests and water do not limit yields. In contrast, if the productivity decline is associated with declining AE(N), changes in crop management may be responsible for declining productivity.

Caution must be taken, however, in interpreting these parameters when crop management is altered. For example, project farmers are currently applying N fertilizer inefficiently and at levels inadequate to reach the yield potential of 10 t ha^{-1} generally attributed to modern tropical varieties (Peng et al 1994). In the second phase of the project, emphasis will be placed on achieving desired yield targets while maximizing AE from applied nutrients. Participating farmers may consequently increase both their fertilizer N application rates and AE(N). Thus, the magnitude of Y_o/N_r will decrease and PFP(N) will move toward optimal values of AE(N), which are well below PFP(N) values measured in the present study. In effect, better use of N fertilizer will be reflected in decreasing PFP(N) values.

Caution must also be taken in using PFP(N) and AE(N) to compare N use efficiency among different sites. By itself, PFP(N) does not allow differentiation between inadequate N fertilizer input to soils with low INS or high N fertilizer input to soils with either high or low INS. Similarly, AE(N) does not describe GY and, given efficient N fertilizer application, it does not distinguish between soils with either low or high INS.

Close examination of the data in Table 3 indicates that use of PFP(N), AE(N), or Y_0/N_r separately does not allow accurate comparison of fertilizer use efficiency across domains. For example, PFP(N) was similar at Sukamandi and Aduthurai, but Y_0/N_r accounted for most of PFP(N) at Sukamandi because INS was relatively high and AE(N) was the lowest of all domains. In contrast, higher N fertilizer efficiency at Aduthurai resulted in a larger contribution of AE(N) to PFP(N) in this domain and INS was lower than at Sukamandi.

When considered in combination with GY and N fertilizer rates, however, these efficiency parameters help identify the reasons for differences in N use efficiency between fields in the same cropping season, and between different years or seasons in the same field as suggested by Cassman et al (1996b). Moreover, use of this analytical framework allows identification of farming practices associated with improved on-farm AE(N) or INS, particularly when the evaluation includes several cropping cycles. Toward this goal, scientists at the Mega Project sites intend to continue data collection on GY, N rates, PFP(N), and AE(N) at the five project sites for several more crop cycles to document long-term productivity trends and present farmers' practices. Soil samples are also being analyzed and archived so that relationships among soil properties, SNSC, and nutrient use efficiency can be studied. Because long-term on-station experiments of intensive irrigated rice cropping suggest that a yield decline can occur after 5-10 yr when management and fertilizer inputs are held constant (Cassman et al 1995), these on-farm studies will help confirm whether the soil processes responsible for these trends in the long-term experiments are also occurring in farmers' fields where continuous rice cropping is practiced.

In addition to conducting research and establishing a database on fertilizer efficiency achieved by farmers in intensive rice-growing domains, the Mega Project was designed to facilitate technology transfer from scientists to farmers. The project will enable rigorous testing of promising new approaches for improved nutrient management under on-farm conditions. The performance of these new technologies can be evaluated in each farmer's field in relation to a long-term database of productivity and plant, soil, and economic factors. Such comparisons have not been made previously, and they are crucial for a definitive study of the factors that govern soil quality and productivity in a given cropping system.

References

Appel T (1994) Relevance of soil N mineralization, total N demand of crops and efficiency of applied N for fertilizer recommendations for cereals-theory and application. Z. Pflanzenernähr. Bodenkd. 157:407-414

Cassman K G, Pingali P L (1995) Extrapolating trends from long-term experiments to farmers' fields: The case of irrigated rice systems in Asia. Pages 63-84 in Barnett V, Payne R, Steiner R (Eds.) Agricultural sustainability: economic, environmental, and statistical considerations. John Wiley & Sons, London

Cassman K G, De Datta S K, Olk D C, Alcantara J, Samson M, Descalsota J, Dizon M (1995) Yield decline and the nitrogen economy of long-term experiments on continuous, irrigated rice systems in the tropics. Pages 181-222 in

Lal R, Stewart BA (Eds.) Soil management: experimental basis for sustainability and environmental quality. CRC Press, Boca Raton, FL

Cassman K G, Dobermann A, Sta Cruz P C, Gines G C, Samson M I, Descalsota J P, Alcantara J M, Dizon M A, Olk D C (1996a) Soil organic matter and the indigenous nitrogen supply of intensive irrigated rice systems in the tropics. Plant Soil 182: 267-278.

Cassman K G, Gines G C, Dizon M A, Samson M I, Alcantara J M (1996b) Nitrogen use efficiency in tropical lowland rice systems: contributions from indigenous and applied nitrogen. Fields Crops Res. 47:1-12.

Crasswell E T, Vlek P L G (1979) Fate of fertilizer nitrogen applied to wetland rice. Pages 175-182 in Nitrogen and rice. International Rice Research Institute, P.O. Box 933, Manila, Philippines.

De Datta S K, Buresh R J (1989) Integrated nitrogen management in irrigated rice. Adv. Agron. 10:143-169.

Olk D C, Cassman K G, Randall E W, Kinchesh P, Sanger L J Anderson JM (1996a) Changes in chemical properties of soil organic matter with intensified rice cropping in tropical lowland soils. Eur. J. Soil Sci. 47: 293-303

Olk D C, Cassman K G, Simbahan G, Sta Cruz P C, Abdulrachman S, Nagarajan R, Pham Sy Tan, Satawathananont S. (1996b) Congruence of N fertilizer management by farmers and soil N supply in tropical irrigated lowland rice systems. Pages 29-38 in Maximizing sustainable rice yields through improved soil and environmental management. Proceedings of the Symposium of the Paddy Soil Fertility Working Group, ISSS, Khon Kaen, Thailand.

Peng S, Khush G S Cassman K G (1994) Evolution of the new plant ideotype for increased yield potential. Pages 5-20 in Cassman K G (Ed.) Breaking the yield barrier: proceedings of a workshop on rice yield potential in favorable environments. International Rice Research Institute. P.O. Box 933, Manila, Philippines.

Peng S, Garcia F V, Laza R C, Sanico A L, Visperas R M, Cassman K G (1996) Increased nitrogen use efficiency at high yield levels using a chlorophyll meter. Field Crops Res. 47: 243-252.

Nitrogen placement technologies for productivity, profitability, and environmental quality of rainfed lowland rice system

S.K. MOHANTY[1], U. SINGH[2], V. BALASUBRAMANIAN[3] and K.P. JHA[1]

[1]*Central Rice Research Institute, Cuttack, 753006, Orissa, India*

[2]*International Fertilizer Development Center-International Rice Research Institute Collaborative Program, P.O. Box 933, Manila 1099, Philippines*

[3]*International Rice Research Institute, P.O. Box 933, Manila 1099, Philippines*

Key words: variable water regime, urea supergranule, deep placement, hand applicator, seedrow N banding, N use efficiency

Abstract

The recovery of applied N by rice is low due to several loss processes operating in the ricefields. Split application of fertilizer suggested for increasing N use efficiency is often not practical in rainfed lowland rice due to adverse soil-water situations. Hence, the entire amount of N required has to be applied in one single application when the water regime is favorable. A single broadcast application, however, increases N loss. Deep placement of urea supergranules (USG) has been proven to improve N fertilizer efficiency. The placement technology is best suited to conditions where the predominant N loss mechanism is ammonia volatilization rather than leaching or denitrification. Deep placement of USG thus has greater benefit over surface split application on soils with moderate to heavy texture, low permeability and percolation rate, and high cation exchange capacity and pH. Environments and management factors conducive to high ammonia volatilization potential would benefit most from deep placement technology. Improved N recovery and efficiency of USG has been well documented in lowland rice, but its market availability and placement methods pose problems. The technology has limited adoption because USG is not commercially available or manufactured in most countries and labor requirement is high with hand placement. Manual application creates more difficulties in handling the granules, besides taking 36-42 more hours per hectare, than two split broadcast applications of prilled urea. Applicators developed so far have not worked satisfactorily under standing water conditions and in direct seeded rice conditions due to hardness of the soil. Hence, it is necessary to develop a suitable applicator to overcome these difficulties. Alternatively, for direct-seeded rice, N fertilizers can be banded in subsoil near seed rows. The placement technology, if adopted by farmers from potential lowland areas in eastern India, is expected to give an additional production of 5.6 million t of rice.

Introduction

The projections for much higher rice yields to meet future demands imply a tremendous increase over the current 10-million-t-per-annum N fertilizer required for rice production (IFA-IFDC-FAO 1992). Numerous researchers have reported that the recovery of applied N by lowland rice with reliable water supply is invariably low and hardly exceeds 30-50% (Prasad and De Datta 1979, Mohanty et al 1989, De Datta and Buresh 1989). This low recovery is attributed to losses of N from the soil-water-plant system due to ammonia volatilization, nitrification-denitrification, leaching, runoff, seepage, and NH_4^+ fixation by clays (Prasad and De Datta 1979, Mohanty and Dash 1982, Zhu et al 1989, Singh and Buresh 1994).

Many strategies have been developed to increase the efficiency of urea N use from fertilizer through proper timing, rate, placement, modified forms of fertilizer, and use of nitrification and urease inhibitors (De Datta and Patrick 1986, Fillery and Vlek 1986, Mohanty et al 1989, Mahapatra et al 1990, Zhu 1990, Subbiah 1992, Rao and Moorthy

V. Balasubramanian, J. K. Ladha and G. L. Denning (eds.), Resource Management in Rice Systems: Nutrients, 57–77.

1993, Kumar et al 1995, Sharma 1995, Singh et al 1995). The response to added nutrients varies markedly due to differences in weather, genotype, soil, agronomic practices, water regime, pest management, and crop history. Despite these differences, it is necessary to develop strategies to improve fertilizer use efficiency.

Deep-point placement of urea fertilizer is probably the most effective application method in reducing N loss except in soils with high percolation rates (Katyal et al 1985). Existing nutrient management practices developed for transplanted rice could be adapted for direct-seeded rice. However, water management and USG placement need major research and development efforts to suit direct-seeded rice. This paper discusses N fertilizer placement technologies for productivity, profitability, their effect on environmental quality, and their limitations in rainfed lowland rice ecosystem.

Placement technologies

Deep-point placement of USG at 5-10 cm depth (reduced soil layer) is one of the most efficient N management techniques developed for rice. It is, however, labor-intensive and hence, recent research has focused on the development and evaluation of less labor-intensive methods for deep placement (De Datta and Buresh 1989), such as pneumatic injection of urea or USG, mechanical deep placement of urea solution, injection as a mud slurry, and use of a USG dispenser (De Datta and Buresh 1989). Co-situs placement of controlled-release urea N fertilizers (CRUNF) provides an innovative way to improve N use efficiency and reduce labor requirement (Shoji and Kanno 1994, Singh et al 1995).

Co-situs placement of CRUNF

The release of urea from CRUNF is predictable depending on temperature, thickness of polymer coating, and moisture permeability of the coating material (Singh et al 1995). Hence, co-situs placement of CRUNF with seeds is possible without salt injury to seeds and plant roots. Co-situs placement brings the fertilizer closer to the emerging roots, thus inducing more N uptake, fertilizer recovery, and increased efficiency. Better recovery and higher efficiency can also be achieved by matching N release from CRUNF with plant N demand.

Ongoing trials at research farms of the International Rice Research Institute in Los Baños, Philippines, and of the Narendra Deva University of Agriculture and Technology in Kumarganj, India, are evaluating the co-situs placement of CRUNF in irrigated and rainfed lowland rice and under transplanted versus direct-seeded conditions.

Band placement of urea solution

Band placement of urea solution into the anaerobic soil layer has been as effective as deep placement of USG in reducing volatilization losses and improving grain yield (Schnier et al 1988, 1993). Devasenapathy and Palaniappan (1995) evaluated the efficiency of a urea solution applicator and found that band placement of urea solution with the applicator at 50 kg N ha^{-1} resulted in higher yields, agronomic efficiency, and apparent recovery than the conventional three-split prilled urea application. The technology, however, has not been tested at the farm level due to its site specificity and greater labor demand than conventional urea application.

Deep-point placement of USG

Agronomic, economic, and environmental advantages of deep-point placement of USG have been well established (Cao et al 1984; Wetselaar 1985; INSFFER 1985, 1986; Kumar et al 1989; Chalam et al 1989; Sarvanan et al 1988; Stangel 1989; Savant et al 1991,1992; Shukla et al 1993; Tyagi et al 1994; FAAS 1994; Misra et al 1995). Experimental results show that with USG, it is possible to save 20-40% of the urea N for the same grain yield, compared with conventional urea applications (Yamada et al 1979, Kumar et al 1989). Deep-point placement of urea-diammonium phosphate supergranules (UDAP-SG) had increased grain yields and improved N and P fertilizer use efficiency in research plots and farmers' fields (Daftardar and Savant 1995, Singh et al 1995). Deep-point placement (5-10 cm depth) in anaerobic soil layer (i) limits the concentration of N in floodwater and in the surface oxidized layer; (ii) decreases N losses through runoff (also applies to P with UDAP-SG), NH_3 volatilization, and denitrification; (iii) minimizes weed use of the applied fertilizer; (iv) minimizes NH_4^+ and P fixation and immobilization; (v) ensures prolonged N availability up to flowering; (vi) stimulates biological N_2 fixation (BNF) because floodwater concentration remains low (Roger and Watanabe 1986); and (vii) reduces tillering, but gives all productive tillers and big panicles (Savant and Stangel 1997).

In an intermediate lowland condition (Table 1), where different N management practices were compared, placement of USG gave significantly higher grain and straw yields than urea split applications. Similarly, total N uptake and N use efficiency were higher with USG than with other treatments. USG deep placement resulted in additional grain yield of 1080, 510, and 350 kg ha^{-1} over prilled urea split in alternate wetting and drying, shallow lowland, and intermediate lowland, respectively (Table 2). Grain yield

Table 1. Effect of urea and urea-based fertilizers on grain and straw yields and other related parameters (*Source*: Patel et al 1989).

Treatment	Grain yield (t ha^{-1})	Straw yield (t ha^{-1})	Total N uptake (kg ha^{-1})	N use efficiency	Apparent N recovery (%)
Control (no N)	1.8	1.6	20.6		
Urea split	4.4	4.0	59.2	21.8	32.1
FYM + urea (1:1 on N basis)	4.7	3.8	56.8	24.3	30.1
Urea supergranule	5.1	5.3	68.8	27.6	39.7
Neem cake-coated urea	4.8	4.7	57.9	25.3	31.0
Lac-coated urea	4.8	4.7	55.7	18.3	29.2
LSD (5%)	0.7	0.6	11.6	-	-

Table 2. Productivity of rice (kg ha^{-1}) with surface application and placement of N fertilizers in different ecosystems of rainfed lowland in India (*Source*: Mohanty et al 1989).

Treatment	Alternate wetting and drying	Shallow lowland	Intermediate lowland
Control (no N)	1580	2270	3020
PU (surface application)	3720	2920	4770
USG (placement)	4800	3430	5120
Yield advantage due to placement of USG over surface application	1080	510	350
Rate of N application (kg N ha^{-1})	60	60	40

Table 3. Comparison of deep placement of USG with conventional and chlorophyll meter-based broadcast application of prilled urea, Research Institute for Rice, Sukamandi, Indonesia (*Source*: Balasubramanian et al, this volume).

N treatment	Yield (kg ha^{-1})	N uptake (kg ha^{-1})	N use efficiency	Apparent N recovery (%)
1995-96 WS				
Zero N. control	5272 b	–	–	–
Prilled urea, 55 kg N ha^{-1}	6233 a	–	17 b	–
Prilled urea, 110 kg N ha^{-1}	6610 a	–	12 b	–
USG, 55 kg N ha^{-1}	6511 a	–	22 ab	–
USG, 110 kg N ha^{-1}	6440 a	–	11 b	–
Chlorophyll meter, 30 kg N ha^{-1}	6188 a	–	30 a	–
1996 DS: Rice cv Memberamo				
Zero N, control	4773 c	92.6 d	–	–
Prilled urea, 90 kg N ha^{-1}	6598 ab	165.3 b	20 b	81 b
USG, 55 kg N ha^{-1}	6423 ab	146.7 c	30 a	98 a
USG, 110 kg N ha^{-1}	7063 a	198.5 a	21 b	96 a
Chlorophyll meter, 60 kg N ha^{-1}	6544 ab	140.3 c	30 a	80 b
1996 DS: Rice cv IR64				
Zero N, control	4400 d	91.1 d	–	–
Prilled urea, 90 kg N ha^{-1}	6308 b	155.8 b	21 bc	72 b
USG, 55 kg N ha^{-1}	6291 ab	145.7 bc	34 a	99 a
USG, 110 kg N ha^{-1}	6988 a	189.6 a	24 b	90 a
Chlorophyll meter, 60 kg N ha^{-1}	5283 c	129.4 c	30 a	64 b

In each column, means followed by a common letter are not significantly different at the 5% probability level by DMRT.

and agronomic use efficiency of USG for the wet and dry seasons were equivalent to or better than conventional as well as the chlorophyll meter-based split application of prilled urea (Table 3). Overall, USG application in multilocation trials throughout Southeast Asia has given 15-35% higher grain yield at the same N rate than two split applications of urea.

Mohanty et al (1989) used ^{15}N-labeled urea to show USG recovery improved by 10 and 16% in absolute terms over split application in the wet and dry season, respectively (Table 4). During the dry season, the ^{15}N retention in soil was also higher by 5% with USG application. Mohanty et al reported that 43-46% of ^{15}N applied was unaccounted

Table 4. Percentage recovery and retention of ^{15}N in the soil as affected by split application of prilled urea and placement of USG in dry and wet season rice (*Source:* Mohanty et al 1989).

Fertilizer	Plant			Soil			Unaccounted for
	Grain	Straw	Total	0-5 cm	5-15 cm	Total	
Dry season							
^{15}N-labeled PU	21.9	12.0	33.9	17.4	2.9	20.3	45.8
^{15}N-labeled USG	34.3	16.1	50.4	18.4	6.9	25.3	24.2
SE	2.2	3.2	2.2	1.2	0.4	1.4	-
Wet season							
^{15}N-labeled PU	25.9	7.7	33.6	19.9	3.1	23.0	43.4
^{15}N-labeled USG	33.2	10.6	43.8	14.3	8.2	22.5	33.7
SE	1.8	0.5	1.8	1.6	1.0	2.2	-

for from prilled urea, whereas only 24-33% was unaccounted for from USG source. N loss from deep-placed USG is highly soil- and site-specific.

Changes in soil and floodwater N concentration. Availability of NH_4^+ N in the surface soil (0-15 cm) was initially lower from deep-placed USG compared with prilled urea split and neem-coated urea, but increased up to 45 d after transplanting (Patel et al 1989). Thereafter, soil NH_4^+-N concentrations with USG compared with other sources remained higher even after harvest of the crop. Floodwater NH_4^+ N concentration was the lowest with USG, followed by neem-coated urea and prilled urea, indicating that potential for NH_3 volatilization loss is in this order: prilled urea > neem-coated urea > USG (Patel et al 1989). Deep placement of both USG and prilled urea resulted in a dramatic reduction in floodwater N (urea N and NH_4^+ N) concentration and consequently negligible NH_3 volatilization loss (Fig. 1). Deep placement of urea fertilizer is much better than surface application in terms of yield, N uptake, and N availability in lowland soils.

Site specificity of deep-placement technology

The Indian Farmers Fertilizer Cooperative Limited conducted a number of agronomic trials in farmers' fields throughout the country during 1981-85. Three treatments were used: urea applied in three splits (1/3 + 1/3 + 1/3), USG placement at 5-10 cm depth, and USG broadcast and incorporated at puddling at 60 kg N ha^{-1} (Kumar and Kaore 1989). The results showed that deeply placed USG gave higher yields than USG broadcast and incorporated and urea applied on the surface. Therefore, adoption of placement technology is possible in all rice-growing states across India, except where the soil is highly permeable, with percolation rates more than 5 cm d^{-1}, because leaching loss of applied urea N is excessive, as in Punjab soils (Katyal et al 1985).

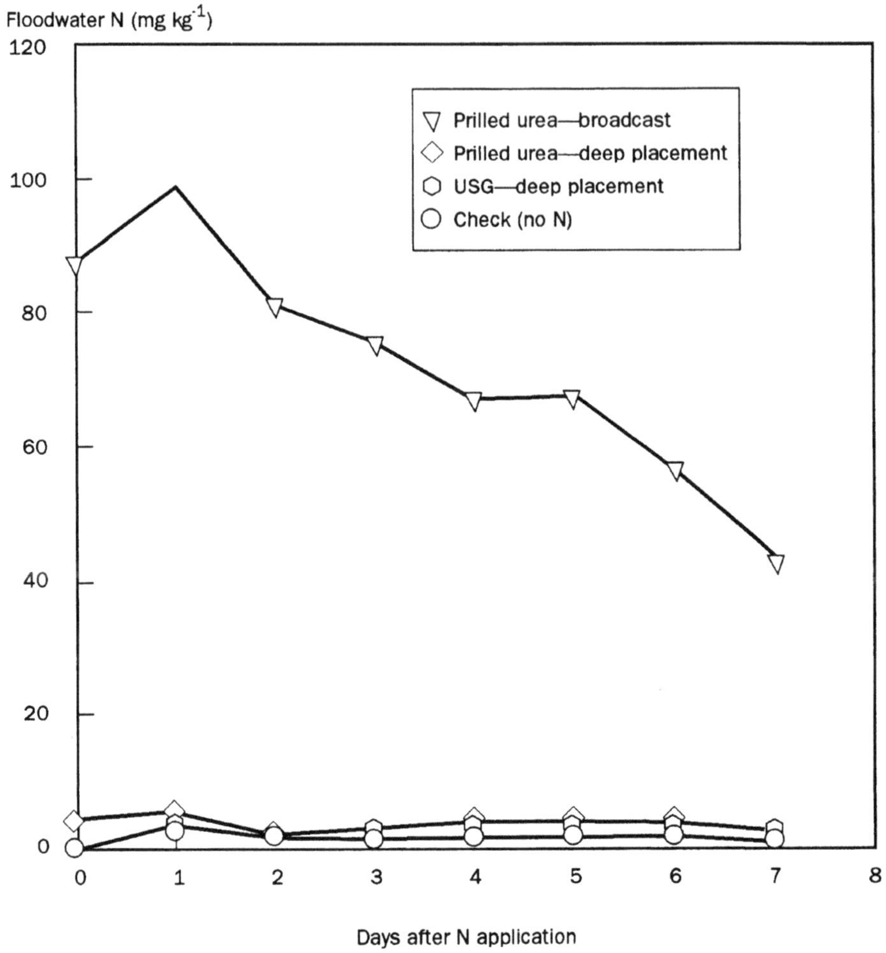

Figure 1. Changes in floodwater N (NH_4-N and urea N) concentration with time for broadcast and deep placed urea. IRRI farm, 1993 wet season.

Recent work (Misra et al 1995) in farmers' fields in the coastal districts of Orissa state showed that it is possible to increase rice yield with deep placement of USG with varying amounts of clay in the soil (Table 5). This technology is adoptable in clay loam, loam, and sandy loam soils with clay content more than 20%. In loamy sand with a low clay content, however, the yield increase due to USG deep placement was not significant. Also, in soils with low CEC, the yield advantage of deep placement over surface application was marginal, because such soils generally have low clay content, high percolation rate, and hence high leaching loss from USG. Misra et al (1995) further suggested that soil organic matter may not be important parameters to influence the placement technology. Alkaline and saline soils with high pH have high potential for NH_3 volatilization loss and hence would benefit greatly from deep-placement technology.

Table 5. Some selected properties of the soils and grain yields of rice as affected by fertilizer management practices in farmers' fields at 58 kg N ha^{-1} level in coastal areas of Orissa, India (*Source:* Misra et al 1995).

Location	Soil texture	Clay (%)	Organic (%)	pH	CEC (cmol kg^{-1} soil)	Percolation rate of soil before puddling (cm h^{-1})	Grain yield (kg ha^{-1})		
							Deep placement of USG	Split application of PU	Yield advantage due to USG placement
Sadasivpur	Clay loam	29	0.70	6.05	22.1	0.59	4980	4330	650
Sadasivpur	Clay loam	28	0.69	5.85	16.5	0.63	4530	4210	320
Sadasivpur	Clay loam	31	0.63	5.60	19.4	0.54	4930	4320	610
Sadasivpur	Loam	16	0.49	5.70	12.9	0.82	4930	4380	550
Kudab	Clay loam	31	-	7.10	20.0	0.71	2640	1940	700
Kudab	Clay loam	34	0.31	7.40	24.3	0.62	3160	2560	600
Kudab	Clay loam	22	0.36	6.50	13.0	0.81	3680	2660	1020
Ranital	Loamy sand	5	0.51	5.50	5.5	1.06	2340	2330	10
Bhubaneswar	Loamy sand	12	0.47	5.80	5.2	0.51	5000	4770	230
Lingpur	Loam	11	0.39	5.95	10.6	0.41	2830	2250	580
Saradeopur	Sandy loam	14	0.70	6.30	8.9	1.12	2110	1720	390
Saradeopur	Sandy loam	20	0.60	5.85	14.0	1.16	5330	5170	160
Saradeopur	Silty loam	9	0.99	6.00	8.3	0.98	4330	3850	480
Arawa	Loam	18	0.71	6.80	11.7	1.42	6100	5650	450
Arawa	Sandy clay loam	24	-[a]	6.06	13.4	0.77	6830	6510	320
Mean	-	-	-	-	-	-	4248	3777	471

[a]- = not done.

Experiments in the Philippines (Table 6) and in India (Table 7) also showed that the fertilizer placement technology was adoptable in different soil types belonging to such textural classes as sandy loam, clay loam, sandy clay loam, and silty clay loam, with a yield advantage varying from 300 to 940 kg ha^{-1}. Grain yield and agronomic efficiency of USG deep placement were superior to conventional farmer's split method, and identical (grain yield) or superior (efficiency) to chlorophyll meter-based prilled urea split application on a clay loam. No statistical difference was observed among the three application methods on a sandy loam (Table 6).

Table 6. Grain yield and agronomic N use efficiency in clay loam vs sandy loam soils. Maligaya, Philippines, 1996 dry season[a] (*Source:* Balasubramanian et al, this volume).

Particular	N treatment	Clay loam (6)	Sandy loam (6)
NPK (kg ha^{-1})	Control	0-26-28	0-21-33
	Farmers' method	121-26-28	131-21-33
	Chlorophyll meter	165-26-28	135-21-33
	USG deep-placed	115-26-28	115-21-33
Yield (kg ha^{-1})	Control	3470 c	4027 b
	Farmers' method	6092 b	6002 a
	Chlorophyll meter	6893 a	6515 a
	USG deep-placed	6880 a	6348 a
AEN	Farmers' method	22 b	15 a
	Chlorophyll meter	21 b	18 a
	USG deep-placed	30 a	20 a

[a]Numbers in parentheses represent the number of farms in each category. In each column, means followed by a common letter are not significantly different at 5% probability level by DMRT.

Table 7. Comparative grain yield of urea and urea supergranule (USG) in different soil types.

Location	Soil type	Texture	N level (kg ha^{-1})	Grain yield (kg ha^{-1})		Yield advantage due to placement of USG (kg ha^{-1})	Reference
				Urea	USG		
Hissar, Himachal Pradesh	Typic Ustochrepts	Sandy loam	120	5950	6426	476	Tyagi et al (1994)
Aduthurai, Tamil Nadu	Entic Chromustert	Clay loam	90	5200	5500	300	Sarvanan et al (1988)
Varanasi, Uttar Pradesh (UP)	Typic Paleastalf	Sandy clay loam	116	4130	5070	940	Rai et al (1991)
Bilaspur, Madhya Pradesh	Typic Paleastalf	Sandy loam	150	4249	4810	561	Shukla et al (1993)
Pantnagar, UP	Aquic Hapluloll	Silty clay loam	120	6280	6770	490	

Extrapolation of deep-placement technology

There is ample evidence that deep placement of N is better than surface application, particularly in soils that have low permeability and percolation rate, high clay content, and moderate to high cation exchange capacity. Thus, to achieve maximum impact from deep placement, it is vital to identify sites and conditions conducive to large N losses, especially through NH_3 volatilization.

Use of CERES-Rice model

The CERES-Rice model has been successfully used to predict yield and N dynamics under lowland conditions (Godwin et al 1990, Singh and Thornton 1992, Singh and Buresh 1994). It can simulate such fertilizer N management strategies as split N application, N sources, methods of N incorporation and placement, and effects of drainage events. The performance of the model with regard to changes in floodwater urea N and NH_4^+ N on surface broadcast application of prilled urea and deep-point placement of USG is presented in Figures 2 and 3, respectively.

Based on sensitivity analyses, deep-placement technology would have the highest benefit on soils with percolation rate of less than 20 mm d^{-1} (Fig. 4). At higher percolation rates, when N losses due to leaching and denitrification become more dominant, deep placement would not have marked advantage over split application of urea. The percolation rate was negatively correlated ($r = -0.57^*$, $n = 16$) to yield advantage with deep placement (Misra et al 1995). The sensitivity analysis with CEC (changing CEC without changes in other soil properties) clearly indicated that deep placement would have clear advantage in reducing volatilization loss and improving grain yield in soils with low CEC. Soils with low CEC have higher NH_3 volatilization potential (Fig. 5). However, when increase in CEC is accompanied by increase in clay content and reduction in percolation rate, then soils with high CEC may have greater yield advantage with deep placement of fertilizer ($r = 0.55^*$, $n = 16$) (Misra et al 1995).

An environment with a high floodwater evaporation rate either due to high temperature and solar radiation regime or poor crop growth and canopy cover would have a high NH_3 volatilization loss potential, as predicted by sensitivity analysis with the CERES-Rice model (Fig. 6). In such an environment, deep placement will have greater impact on improving yield and reducing N loss. Environmental and management conditions that lead to prolonged transplanting shock, poor establishment, and reduced initial growth would gain more from deep-placement technology. The rainfed lowland environment thus could benefit greatly from deep-placement technology.

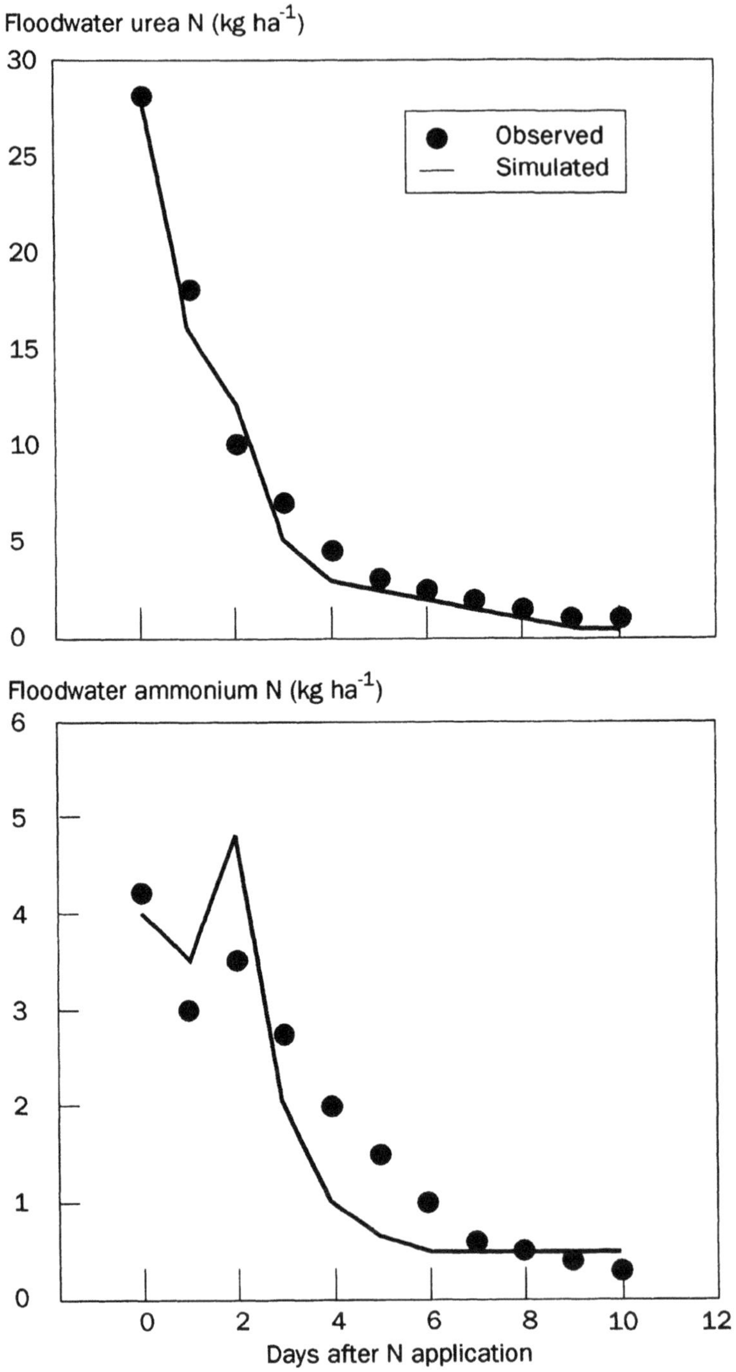

Figure 2. Changes in floodwater N on broadcast application of 30 kg urea N ha^{-1} at 18 DT.

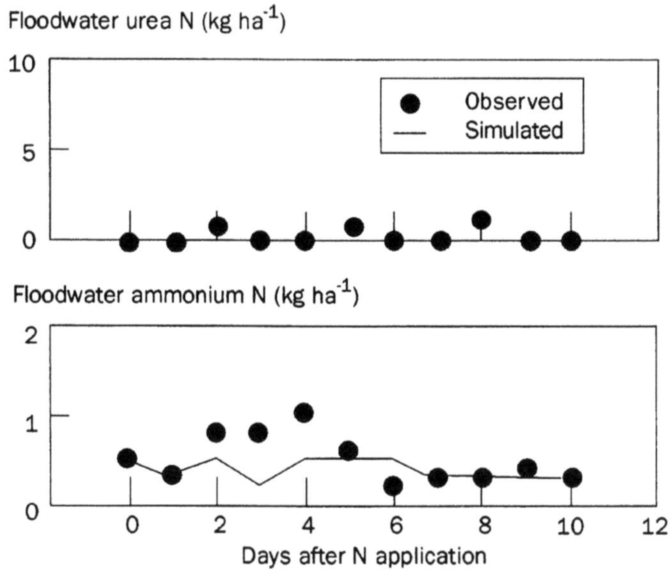

Figure 3. Changes in floodwater N on deep-point placement of 58 kg urea N supergranule at 1 DT.

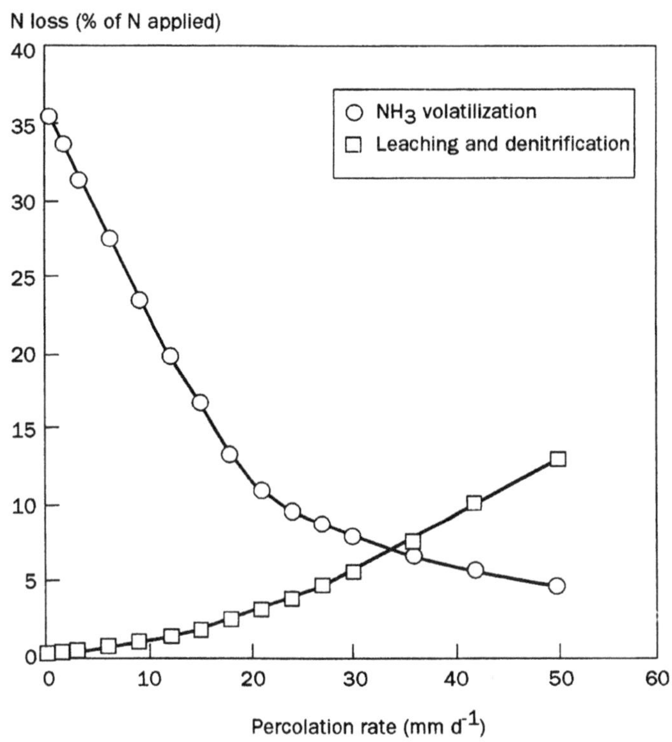

Figure 4. Effect of percolation rate on NH_3 volatilization and leaching + denitrification losses at 140 kg N ha^{-1} applied in two splits (2/3 at 18 DT and 1/3 at 5 DBPI).

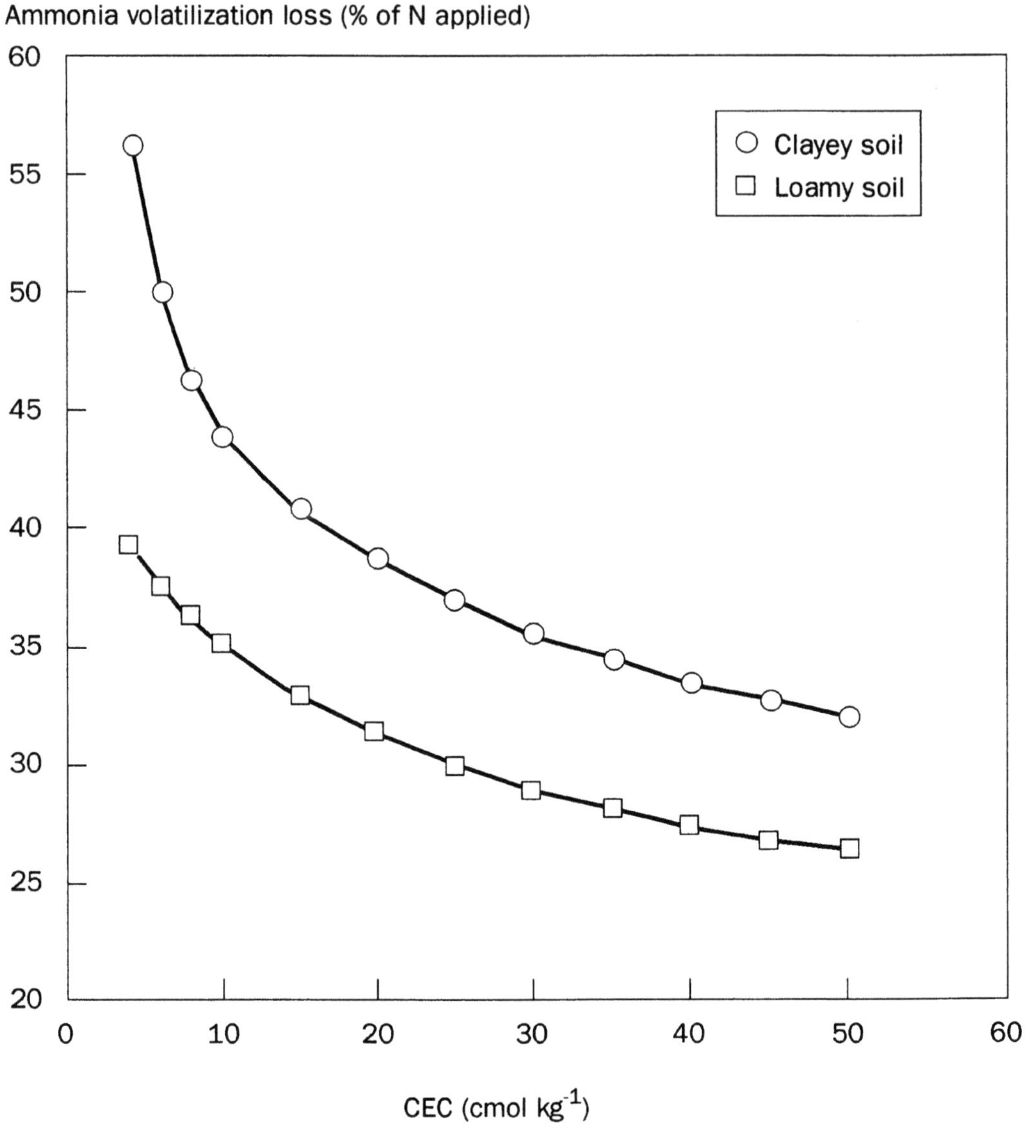

Figure 5. Sensitivity of ammonia volatilization loss to changes in CEC on loamy and clayey soil.

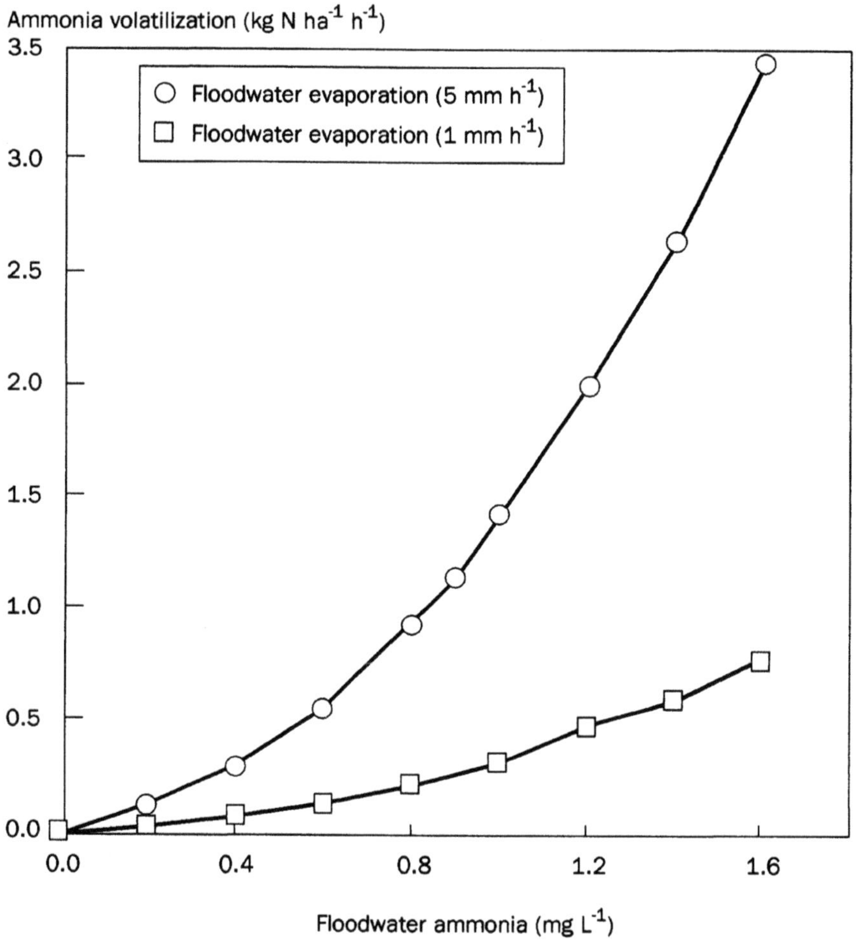

Figure 6. Sensitivity of ammonia volatilization to floodwater ammonia concentration and evaporation regime.

Potential of deep-placement technology for lowland rice in India

Reported and model-generated information has helped identify areas with potential benefits from deep-placement technology. Beneficial results have been reported from the humid Bengal-Assam region, the subhumid to humid eastern and southeastern region, the subhumid Ganga alluvial plains, the humid eastern Himalayan region, the humid to semi-arid western ghat and Karnataka plateau, and the semiarid Lava plateau in India (Table 8). Areas under these zones are the eastern and northeastern states of the country and coastal areas of Andhra Pradesh, Tamil Nadu, Karnataka, Kerala, and Maharashtra states.

However, most of the rainfed lowland areas (82%) are in the eastern Indian states--Assam, West Bengal, Bihar, Orissa, eastern Madhya Pradesh, and eastern Uttar Pradesh. The rainfed lowland ecosystem has been grouped into four subecosystems (Table 9). Among these, the shallow lowlands are favorable lowlands. As the water depth increases with intermediate, semi-deep, and deep water subecosystems, the land becomes less and less favorable for rice growth. However, during the wet season, the farmers have no

Table 8. Agroclimatic regions of India where deep-placement technology has shown a yield advantage.

Agroclimatic region	States	Annual rainfall (mm)	Temperature range (°C)	Major soil groups
Humid Bengal-Assam region	West Bengal and Assam	2000	25 (10 Jan)-41 (25 Jul)	Aqualfs, Orthrepts, Orthents, Ultalfs, Aquepts, Aquents, Ochrepts, Fluvents, Udalfs
Subhumid to humid eastern and south-eastern region	Orissa, Andhra Pradesh and Eastern Madhya Pradesh	1000-1500	28 (16 Jan)-36 (27 Jul)	Aquepts, Aquents, Ustalfs, Ochrepts, Orthents, Fluvents, Tropepts, Aqualfs
Subhumid Ganga alluvial plains	Eastern U.P. and Bihar	1000-1500	23 (6 Jan)-41 (25 Jul)	Ochrepts, Orthents, Aquepts, Udalfs, Ustalfs, Fluvents, Aqualfs
Humid eastern Himalayan region	Arunachal Pradesh, Nagaland, Manipur, Mizoram, Tripura and Meghalaya	2000	24 (11 Jan)-33 (25 Jul)	Aqualfs, Ochrepts, Orthents, Aquepts, Fluvents, Udalfs, Ustalfs
Humid to semiarid Western Ghat and Karnataka plateau	Karnataka, Tamil Nadu, Kerala, Pondicherry	750-2000	29 (20 Jan)-38 (25 Jul)	Fluvents, Udolls, Ustalfs, Tropepts, Ochrepts, Orthents, Usterts, Udalfs
Semiarid Lava plateau	Maharashtra	500-1000	29 (13 Jan)-42 (26 Jul)	Aquepts, Tropepts, Orthents

Table 9. Area (million ha) under different subecosystems of rainfed lowland in India.

State	Shallow lowland (0-30 cm)	Intermediate lowland (30-50 cm)	Semi-deep water lowland (50-100 cm)	Deep water lowland (more than 100 cm)	Total lowland
Assam	0.6	0.3	0.4	0.1	1.4
Bihar	0.8	0.6	0.5	0.7	2.6
Madhya Pradesh	1.5	0.7	0.2	-	2.4
Orissa	0.9	0.5	0.4	0.2	2.0
Uttar Pradesh	0.8	0.6	0.5	0.6	2.5
West Bengal	1.3	0.8	0.5	0.7	3.3
Total (eastern India)	5.9	3.5	2.5	2.3	14.2
Total (India)	8.0	4.0	3.0	2.4	17.4
Percent of total area in eastern India	74	88	83	96	82

choice but to grow rice. The farmers of eastern India are poorer, compared with farmers in other parts of the country. The eastern Indian states receive high rainfall of 1500-2000 mm annually; 80% of this is received during the monsoon (June-September). Therefore, application of fertilizer, particularly surface application, is risk-prone.

Aside from the irrigated ecosystem where deep placement has been reported to be better than surface split application (Sarvanan et al 1988, Mahapatra et al 1990, Rai et al 1991, Shukla et al 1993, Tyagi et al 1994), deep placement also has potential in more than 12 million ha in eastern India and rainfed areas elsewhere, based on soil and site characteristics.

Profitability of deep-placement technology

Both increase in grain yield and N fertilizer savings as well as labor requirement must be considered when evaluating the profitability of deep-placement technology. As evident from the marked difference in grain yield response to N for deep placement of USG and surface-split application of prilled urea (Fig. 7), the profitability should be considered at appropriate N rates.

The price of rice produced as a function of N applied with deep placement of USG ($\$y = 838.5 + 15.7\ x - 0.05\ x^2 - 0.0001\ x^3$, $r^2 = 0.99$) and with split application of prilled urea ($\$y = 879.8 + 8.1\ x - 0.0009\ x^2 - 0.0001\ x^3$, $r^2 = 0.99$) vs price of USG N ($\$y = 0.77\ x$) and urea N ($\$y = 0.62\ x$) as presented in Figure 7, showed the following: (i) price of N fertilizer was very low in comparison with the gross rice price, (ii) price difference in USG and urea was \$ 0.15 kg^{-1} N applied, (iii) maximum USG profit gain over urea (\$ 237) occurred at 80 kg N ha^{-1} rate, (iv) a gross profit of \$1700 with urea was obtained with additional 45 kg N ha^{-1} application over the USG N rate, and (v) under low-input scenario, 33 kg USG N ha^{-1} and 58 kg urea N ha^{-1} produced equivalent yield or gross profit. Assuming an average reduction of 30 kg N ha^{-1} in N fertilizer need with the use of USG on rice, potential annual savings of urea fertilizer could be as high as 6 million t for Asia.

On an average, the grain yield increase due to USG deep placement over surface application of prilled urea in India has been 470 kg ha^{-1}. The expected increase in total rice production from rainfed lowlands in eastern India would be 5.6 million t (12 million ha x 0.47 t ha^{-1}). The farmer is expected to get an additional \$53.7 ha^{-1} (470 kg ha^{-1} x \$ 0.11 kg^{-1} rice).

Practical difficulties in adoption of deep-placement technology

Despite clear-cut profit gains from USG deep-placement technology through yield increases and fertilizer N savings, USG deep placement has not spread to farmers' fields mainly due to (i) nonavailability of USG on commercial scale, (ii) problems associated with breakage of USG during transportation and handling, (iii) high labor requirement for application, (iv) practical difficulty in handling the material during its use in the field, (v) lack of suitable deep-placement applicators and machines, and (vi) lack of cash for one-time purchase of USG. The additional labor cost for hand deep placement over two split

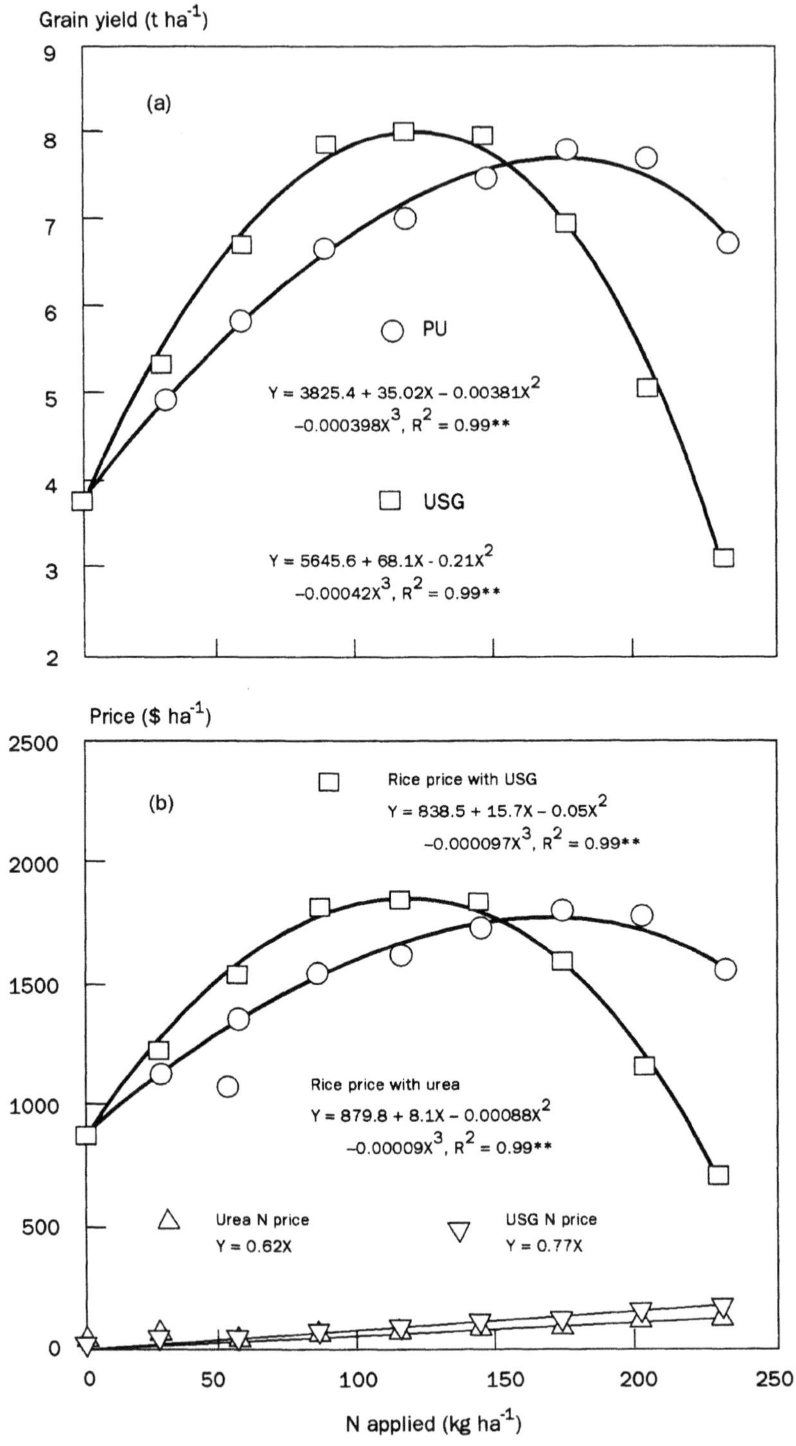

Figure 7. N response function for rice grain yield (a) and fertilizer N price (b) with USG deep-point placement and three equal splits (basal, mid-tillering, panicle initiation) prilled urea. IRRI farm, 1993 dry season.

broadcast applications of N by hand has been estimated to be 36-42 h ha^{-1} (Lapitan et al 1985). Staggered application of USG which allowed farmers to organize family labor for deep placement did not reduce rice grain yield (Rao et al 1989).

Labor requirement for deep placement

The additional labor requirement for deep placement may decrease as workers get more practice. It is also anticipated that some labor saving from weeding would result because deep placement of N (and P) would not promote weed growth.

Nonavailability of USG

The availability of USG is both policy- and demand-driven. To meet the increased demand for USG in Java, Indonesia, the supergranules were produced in decentralized factories for quick distribution to 1.3 million ha of ricefields in 1996. Although deep placement of USG is considered a better N management strategy than surface-split application of prilled urea for rainfed lowland rice in eastern India, its implementation would pose problems because eastern India consumes only 13.3% of the total N fertilizer used in the country. To recover the 10-15% higher manufacturing cost of USG (Tandon 1989), the fertilizer industries would need commercial-scale manufacturing of USG. Policymakers and extensionists could play a critical role in implementing deep-placement technology in eastern India.

Practical difficulties in deep-point placement

To achieve appropriate efficiency, USG has to be point-placed at the 5-10 cm soil depth in between four hills in alternate rows of rice plant. This requires skill and time. It has been estimated that point placement of USG in 1 ha of land will require 40-48 h (Lapitan et al 1985, Tandon 1989). Labor shortage is common during this peak time for land preparation, planting, and deep placement. Deep placement of USG with wet hands causes discomfort and dissolution of the granules. Deep-placement machines tested so far proved less efficient and less reliable than properly hand-placed USG (De Datta 1986). A plunger-type, completely hand-operated USG applicator made of polyvinyl chloride (PVC) has been developed and successfully field-tested by Savant et al (1991, 1992). It is highly promising for deep placement of USG and has been widely used by Indonesian farmers. It is easier to use an applicator in transplanted fields than in direct-seeded fields because the hole left in the soil is not covered automatically due to soil hardness, resulting in higher concentration of NH_4^+ in floodwater. In the future, a major portion of lowland rice will be direct seeded because labor for transplanting is scarce.

Overcoming the difficulties in deep placement

Experiments at the Central Rice Research Institute, Cuttack, for 3 yr under direct-seeded conditions, using 40 kg N ha^{-1} (Sharma, 1994), indicated that drilling of prilled urea and placement of USG in the plow furrows were more effective than broadcast incorporation before sowing (Table 10). All four USG placement methods had relative grain yield difference of less than $\pm 3\%$. Thus, modified placement methods could offer labor savings without yield reduction in row-seeded rice. This technology, however, needs testing in farmers' fields.

Environmental quality

A large proportion of applied fertilizer N is unaccounted for (Table 4) due to different loss pathways such as leaching, NH_3 volatilization, denitrification, and runoff. These losses pollute the groundwater, surface waters, and the atmosphere.

Field experiments conducted by Patel et al (1990) under different hydrological situations such as alternate wetting and drying, shallow water submergence, and intermediate deep water situations, using different sources of N including USG, showed that leaching losses were less than 1% of applied N. The concentration of $nitrate^-$ in the leachate in general varied from 0.01 to 0.09 mg L^{-1} and from 0.01 to 0.08 mg L^{-1} in USG treatment. They further suggested that leaching loss of N in silty clay and sandy clay loam soil with a high water table is not a major environmental concern. Multilocation trials in high-rainfall areas in India showed that leaching of N was not a major pathway of loss of N and the concentration ofNO_3^- N in the leachate did not exceed 1-2 mg L^{-1} during the growth period of the rice crop (Pande and Mohanty 1986). Because most of the lowlands have a high, seasonal water table and comparatively heavy textured soils, NO_3^- pollution of the groundwater is not likely. On the other hand, runoff losses of N (NO_3^- N, NH_4^+ N, urea N) during the wet season has been as high as 10-20 mg L^{-1} (unpubl. data, IFDC-IRRI). Deep-point-placement of USG would effectively reduce N runoff losses since the floodwater N concentration remains low (Figures 3, 4, and 6; Daftardar and Savant 1995).

Gaseous loss in the form of NH_3 is the main pathway of loss of N from lowland ricefields. Deep-placement technology has been very effective in reducing NH_3 volatilization losses (Crasswell et al 1981, Savant and De Datta 1982, Patel et al 1989, Singh and Thornton 1992). As previously discussed, the effectiveness of deep placement in reducing N losses is highly site-specific. Much of fertilizer N saving from USG deep placement is attributed to reduced NH_3 volatilization loss.

Recent studies have shown that nitrification-denitrification is negligible under submerged condition with USG deep placement because much of NH_4^+ N remains in the anaerobic soil layer. Little information on evolution of CH_4 and N_2O with point placement of N fertilizer is available. It is expected that CH_4 evolution may increase due to deep placement of N because increased biomass production of rice would result in more root exudates for methanogenic bacteria.

Table 10. Effect of methods of application of prilled urea and urea granules (40 kg N ha^{-1}) on grain yield of direct-seeded rice under intermediate deep water conditions (*Source:* Sharma 1994).

Treatment	Relative grain yield (no N = 100)
Prilled urea	
(i) Broadcast and incorporated before sowing	133
(ii) Drilled behind 'V' shaped plow furrow and seeding in the same furrow	145
(iii) Drilled between seed rows	145
(iv) 50% drilled behind plow and 50% through foliar spray	120
Urea supergranules	
(i) Placement behind 'V' shaped plow furrow and seeding in the same furrow	142
(ii) Placement in the plow furrow between seed rows	148
(iii) Placement in moist soil after sowing between seed rows	147
(iv) Placement in shallow water between seed rows	143

Yield of control (no N) = 2500-2700 kg ha^{-1}.

Conclusions

Deep placement of fertilizer results in high grain yield and less environmental pollution than surface application under lowland rice conditions, except in soils conducive to high leaching losses. Placement technology is highly desirable for conditions that promote high NH_3 volatilization and runoff loss: soils with more than 20% clay content, soil pH>7, low permeability and percolation rate and low NH_4^+ sorption, and environment with poor crop establishment and heavy rainfall. It also limits denitrification loss under submerged conditions. Thus, deep-placement technology is economically attractive and environmentally sound under most rainfed lowland conditions.

Applicators developed so far have limited success in the field. There is a need to develop suitable applicators especially for the rainfed lowlands: applicators that work even in standing water (a very common situation during the wet season) and cover the soil after placement. In direct-seeded rice, the placement of fertilizer in seed furrows or between seed furrows at seeding holds promise because currently available applicators fail to cover the fertilizer after placement.

Under lowland rice conditions, USG deep-placement technology is environment-friendly because it is very effective in reducing NH_3 volatilization loss; nitrification-denitrification process is limited with negligible N_2O loss; runoff loss is checked; weed growth is retarded and hence herbicide application is reduced; and BNF is promoted under reduced floodwater N concentration.

References

Balasubramanian V, Morales A C, Cruz R T, Abdul Rachman S (1997) On-farm adaptation of knowledge-intensive nitrogen management technologies for rice systems. Pages 79-94 in Balasubramanian V, Ladha J K, Denning G L (Eds.) Resource management in rice systems: nutrients. International Rice Research Institute, P.O. Box 933, Manila, Philippines and Kluwer Academic Publishers, Dordrecht, The Netherllands.

Cao Z H, De Datta S K, Fillery I R P (1984) Nitrogen-15 balance and residual effects of urea-N in wetland rice fields as affected by deep placement techniques. Soil Sci. Soc. Am. J. 48: 203-208.

Chalam A B, Chakravorti S P, Mohanty S K (1989) Comparative efficiency of urea supergranule and prilled urea for lowland rice (*Oryza sativa*) using ^{15}N. Indian J. Agric. Sci. 59:400-401.

Crasswell E T, De Datta S K, Obscemea W N, Hartantyo M (1981) Time and mode of nitrogen fertilizer application to tropical rice. Fert. Res. 2:247-259.

Daftardar S Y, Savant S K (1995) Evaluation of environmentally friendly fertilizer management for rainfed lowland rice on tribal farmers' fields in India. Pages 173-186 in Fragile lives in fragile ecosystems. Proceedings of the International Rice Research Conference; 13-17 Feb 1995. International Rice Research Institute, P.O. Box 933, Manila, Philippines.

De Datta S K, Patrick W H Jr (1986) Nitrogen economy of flooded rice soils. Martinus Nijhoff Publishers, Dordrecht, The Netherlands.

De Datta S K, Buresh R J (1989) Integrated nitrogen management in irrigated rice. Adv. Soil Sci. 10:143-169.

Devasenapathy P, Palaniappan S P (1995) Band placement of urea solution increases Nuse efficiency in transplanted lowland rice. Int. Rice Res. Notes 20:19.

FAAS--Fujian Academy of Agricultural Sciences (1994). A study report on urea deep-point applicator in paddy field. FAAS, Fuzhou, Fujian, China. 10 p. (mimeo)

Fillery I R P, Vlek P L G (1986). Reappraisal of the significance of ammonia volatilization as an N loss mechanism in flooded rice fields. Fert. Res. 9:79-98.

Godwin D C, Singh U, Buresh R J, De Datta S K (1990). Modeling of nitrogen dynamics in relation to rice growth and yield. Trans. 14th Int. Congr. Soil Sci., Kyoto, Jpn. 4: 320-325.

IFA-IFDC-FAO (1992). Fertilizer use by crop. Food and Agriculture Organization, Rome.

INSFFER--International Network on Soil Fertility and Fertilizer Evaluation for Rice (1985). Summary reports on the fifth international trial on nitrogen fertilizer efficiency in irrigated wetland rice (1981-84). International Rice Research Institute, P.O. Box 933, Manila, Philippines.

INSFFER--International Network on Soil Fertility and Fertilizer Evaluation for Rice (1986) Progress report on the first international trial in comparison of hand and machine applied prilled urea and urea supergranules in lowland rice (1984-85). International Rice Research Institute, P.O. Box 933, Manila, Philippines.

Katyal J C, Singh B, Sharma V K, Crasswell E T (1985) Efficiency of some modified urea fertilizer for lowland rice grown on a permeable soil. Fert. Res. 6:279-290.

Kumar V, Kaore S V (1989) Use of urea supergranule in rice--IFFCO's experience. Pages 138-143 in Soil fertility and fertilizer use. Indian Farmers Fertilizer Cooperative, New Delhi, India.

Kumar V, Shrotriya G C, Kaore S V (Eds.) (1989) Soil fertility and fertilizer use. Vol. 3. Urea supergranules for increasing nitrogen-use efficiency. Indian Farmers Fertilizer Cooperative, New Delhi, India. 143 p.

Kumar K, Thangaraju M, Kannaiyan S (1995).Use of nitrification inhibitors and nitrogen-use efficiency in lowland rice soil. Pages 101-116 in Kannaiyan S (Ed.) Rice management biotechnology. Associated Publishing Co., New Delhi, India.

Lapitan J A, Khan A U, Calabio J, Bautista E (1985) Evaluation of fertilizer placement machines and their current developments. Paper presented at the International Rice Research Conference, 1-5 Jun 1985, International Rice Research Institute, Los Baños, Philippines. 25 p.

Mahapatra P, Panda D, Mohanty S K (1990) Efficiency of urea-based fertilizers for wetland rice (*Oryza sativa*). J. Agric. Sci. Cambridge 113:187-191.

Misra C, Mohanty B C, Das B S, Savant N K (1995) Relationship between some selected soil properties and yield of transplanted rice fertilized with urea briquettes. Oryza 32:178-183.

Mohanty S K, Dash R N (1982) Chemistry of waterlogged soils. Pages 389-396 in Wetlands: ecology and management. National Institute of Ecology and International Scientific Publications, Jaipur.

Mohanty S K, Chakravorti S P, Bhadrachalam A (1989) Nitrogen balance studies in rice using ^{15}N labeled urea and urea supergranules. J. Agric. Sci. Cambridge 113:119-121.

Pande HK, Mohanty S K (1986). Consolidated report on ICAR coordinated project on chemistry of submerged soils under rice cultivation in high rainfall areas. Central Rice Research Institute, Cuttack. 150 p.

Patel S K, Panda D, Mohanty S K (1989) Relative ammonia loss from urea-based fertilizers applied to rice under different hydrological situations. Fert. Res. 19:113-119.

Patel S K, Panda D, Mohanty S K (1990) Leaching loss of nitrogen from urea based fertilizers applied to rice under different hydrological conditions. Oryza 27:342-345.

Prasad R, De Datta S K (1979) Increasing fertilizer nitrogen efficiency in wetland rice. Pages 465-485 in Nitrogen and rice. International Rice Research Institute, P.O. Box 933, Manila, Philippines.

Rai O P, Singh N B, Singh R A (1991) Effect of varying doses of nitrogen applied in different forms on the yield and nitrogen efficiency of rice. Oryza 28:167-170.

Rao M V, Padalia C R, Rao K S, Prasad M, Reddy M D, Manna G B, Chandra D, Singh J (1989) Comprehensive studies on use of urea supergranules. Pages 97-113 in Kumar V, Shrotriya G C, Kaore S V (Eds.) Soil fertility and

fertilizer use Vol. 3. Urea supergranules for increasing nitrogen-use efficiency,. Indian Farmers Fertilizer Cooperative, New Delhi, India.
Rao K S, Moorthy B T S (1993) Effect of different urea formulations on the yield of medium and long duration varieties of rice. Oryza 30:129-132.
Roger P A, Watanabe I (1986) Technologies for utilizing biological nitrogen fixation in wetland rice: Potentialities, current usage and limiting factors. Fert. Res. 9:39-77.
Sarvanan A, Velu V, Ramanathan K M (1988) Effect of source of methods of nitrogen application on volatilization loss of ammonia and yield of rice under submerged of Cauvery Delta, India. Oryza 23: 145-148.
Savant N K, De Datta S K (1982) Nitrogen transformation in wetland rice soils. Adv. Agron. 35:241-294.
Savant N K, Ongkingco P S, Zarate I V, Torrizo F M, Stangel P J (1991) Urea briquette applicator for transplanted rice. Fert. Res. 28:323-331.
Savant N K, Ongkingco P S, Garcia F D, Dhane S S, Khadse R R, Chavan S A, Rao K S (1992) Agronomic performance of urea briquette applicator in transplanted rice. Fert. Res. 32:139-142.
Savant N K, Stangel P J (1997) Urea briquettes containing diammonium phosphate: a potential new NP fertilizer for transplanted rice. Fert. Res. (in press)
Schnier H F, De Datta S K, Mengel K, Marqueses E P, Faronilo J E (1988) Nitrogen-use efficiency, floodwater properties and nitrogen-15 balance in transplanted lowland rice as affected by liquid band placement. Fert. Res. 16:241-255.
Schnier H F, De Datta S K, Fagi A M, Eaqub M, Ahmed F, Tejaswarana R, Mazid A (1993) Yield response of wetland rice to band placement of urea solution in various soils in the tropics. Fert. Res. 36:221-227.
Sharma A R (1994) Fertilizer management in lowland rice under excess water condition. Fert. News 39(5): 35-44.
Sharma S K (1995) Urea supergranules--its performance and prospects for rice production. Pages 117-134 in Kannaiyan S (Ed.) Rice management biotechnology. Associated Publishing Co., New Delhi.
Shoji S, Kanno H (1994) Use of olefin-coated fertilizers for increasing fertilizer efficiency and reducing nitrate leaching and nitrous oxide emissions. Fert. Res. 39(2):147-152.
Shukla R K, Sharma R S, Chipde S J (1993) Improvement of N-use efficiency in transplanted rice. Oryza 30: 259-261.
Singh U, Thornton P K (1992) Using crop models for sustainability and environmental quality assessment. Outlook Agric. 21 (3):209-218.
Singh U, Buresh R J (1994) Fertilizer technology for increased fertilizer efficiency in paddy rice fields. Pages 643-653 in Proceedings of the 15th World Congress of Soil Science. Acapulco, Mexico.
Singh U, Cassman K G, Ladha J K, Bronson K F (1995) Innovative nitrogen management strategies for lowland rice systems. Pages 230-254 in Fragile lives in fragile ecosystems. Proceedings of the International Rice Research Conference, 13-17 Feb 1995. International Rice Research Institute, P.O. Box 933, Manila, Philippines.
Stangel P J (1989) Urea supergranule for transplanted rice. IFDC Research Review. Pages 1-9 in Soil fertility and fertilizer use. Vol. 3. Virendra Kumar et al (Eds.) Indian Farmers Fertilizer Cooperative Limited, New Delhi.
Subbiah S V (1992) Agronomic efficiency of modified urea materials. Oryza 29:374-375.
Tandon H L S (1989) Urea supergranule for increasing nitrogen efficiency in rice. An overview. Pages 10-22 in Soil fertility and fertilizer use Vol. 3. Virendra Kumar et al (Eds.) Indian Farmers Fertilizer Cooperative Limited, New Delhi.
Tyagi R C, Singh H, Rammohan Rao D S, Sing T, Singh B R (1994) Effect of levels, sources and time of nitrogen application in transplanted rice. Oryza 31:297-300.
Wetselaar R (1985) Deep-point placed urea in a flooded soil: a mechanistic view. Pages 7-14 in Special Publication SP-6. International Fertilizer Development Center, Muscle Shoals, USA
Yamada Y, Ahmad A, Alcantara A, Khan H (1979) Nitrogen efficiency study under flooded paddy conditions. Pages 39-74 in Proceedings of the Final INPUTS Review Meeting, East West Center, Honolulu, Hawaii.
Zhu Z L (1990) Management of nitrogen fertilizers for flooded rice in relation to nitrogen transformations. Trans. 14th Int. Congr. Soil Sci. Kyoto, Jpn. 4: 337-342.
Zhu Z L, Cai G X, Simpson J R, Zhang S L, Chen D L, Jackson A V, Freney J R (1989) Processes of nitrogen loss from fertilizers applied to flooded rice fields on calcareous soil in north-central China. Fert. Res. 18:101-115.

On-farm adaptation of knowledge-intensive nitrogen management technologies for rice systems

V. BALASUBRAMANIAN[1], A.C. MORALES[1], R.T. CRUZ[2], and S. ABDULRACHMAN[3]

[1]*International Rice Research Institute (IRRI), P.O. Box 933, 1099 Manila, Philippines*

[2]*Philippine Rice Research Institute (PhilRice), Maligaya, Muñoz, Philippines*

[3]*Research Institute for Rice, Sukamandi 41256, Subang, West Java, Indonesia*

Key words: intensive rice systems, efficient N use, variable application rate, chlorophyll meter, leaf color chart

Abstract

Efficient use of all inputs is vital to achieve and sustain high crop yields, maintain resource quality, and minimize environmental pollution. Fertilizer N is one of the major inputs in rice production. Blanket fertilizer recommendations do not take into account the high field-to-field variability and within-season dynamic changes in indigenous N supply. Since the plant growth reflects the total N supply from all sources, plant N status will be a good indicator of N availability to crops at any given time. The chlorophyll meter (SPAD) and leaf color chart (LCC) are simple, portable diagnostic tools that can measure the crop N status in situ in ricefields to determine the timing of N topdressing. Such decision aids are useful to vary N application rates to rice crops, based on crop demand and indigenous N supply. Although the chlorophyll meter cannot be owned by individual farmers due to its high cost, it is a practically useful tool for field researchers, extension specialists, and crop consultants who do not have access to well-equipped laboratories. On-farm, adaptive research is in progress in three countries to adapt the chlorophyll meter technique for transplanted and wet-seeded rice, local cultivar groups, and soil, crop, and environmental conditions. Initial results indicate that the SPAD threshold value of 35 is good for transplanted rice in dry season. The threshold has to be reduced to 32 for wet-seeded rice in dry season and for all rice during wet season with cloudy weather and low radiation. Thus, when calibrated with local cultivar groups and crop conditions, it can be used to accurately monitor crop N status and to advise farmers on N topdressing for rice. It can also be used effectively to verify the adequacy of existing N fertilizer recommendations to rice by the in situ monitoring of foliar N status of crops fertilized with current recommendations and to refine them to further improve N fertilization of rice. The LCC is not as accurate as the chlorophyll meter in determining the leaf N status in rice crops. However, LCC can be calibrated with the chlorophyll meter to fix the critical color shade for local rice cultivar groups and crop conditions. Farmers can then use the LCC to qualitatively assess foliar N status and adjust N topdressing to their rice crops. Initial feedback on the use of LCC from farmer cooperators in the Philippines is highly encouraging. Both methods are affected by factors such as varietal group, plant density, crop stress that causes leaf chlorosis, soil nutrient status, and climate; therefore, they have to be adapted to specific soil, climatic, and crop conditions. Adequate training is necessary for both extension agents and farmers to properly use the new tools for increasing the efficiency of N fertilizer use on rice. Wider farmer adoption of the two diagnostic tools discussed in this paper will minimize overfertilization of rice, increase profitability, and decrease fertilizer-related pollution of the environment.

Introduction

We need to produce 70% more rice by the year 2025 to feed the growing population (Hossain and Fischer 1995). To achieve this target, farmers' average rice yields have to be increased from 4.9 to 8.0 t ha^{-1} in irrigated land and from 2.3 to 3.5 t ha^{-1} in rainfed lowland (Cassman and Pingali 1995). The efficiency of all rice production inputs (water, nutrients, pesticides, labor, energy) must be improved to maintain sustainability at high yield levels.

Fertilizer N is one of the major inputs in rice production. Presently, irrigated rice farmers use about 100 kg N ha^{-1} to achieve a mean grain yield of 4.9 t ha^{-1}. For a

V. Balasubramanian, J. K. Ladha and G. L. Denning (eds.), Resource Management in Rice Systems: Nutrients, 79–93.

projected increase in yield from 4.9 to 8.0 t ha^{-1} in irrigated land, it is necessary to apply 280 kg N ha^{-1} at 33% fertilizer N recovery efficiency and 187 kg N ha^{-1} at 50% recovery efficiency. The present annual consumption of 15.5 million t of urea applied to irrigated rice has to be increased to 46.5 million t at 280 kg N ha^{-1} and 31.0 million t at 187 kg N ha^{-1} rate for a 63% increase in grain yield (Cassman and Pingali 1995). It is well-known that grain yield response diminishes with increasing rates of N application, and yields could further decrease due to lodging and increased incidence of insect pests and diseases at high N levels. It is important to select rice varieties with higher N use efficiency that are tolerant of lodging and pests and diseases at high N supply (Ladha et al 1996).

At high yield levels, rice farmers also require new techniques that can optimize crop yields, improve N use efficiency, and minimize fertilizer-related pollution of the environment. These techniques must be adapted to farm-level conditions to promote their use by farmers. In this paper, we review two N management techniques that are suitable for intensive rice cultivation and highlight the factors that need to be addressed to improve farmer adoption of such techniques.

Principal approaches for efficient N management in rice systems

Two principal approaches are used to satisfy crop demand for N and to minimize N losses in flooded rice systems:

- *Plant-based strategies through monitoring crop N status.* Samples of whole plant tissue or leaves are used to assess the nutrient status of growing plants. Nitrogen concentration of fully expanded topmost leaf has been used as an index to determine the need for N topdressing in rice (Wallihan and Moomah 1967, Wada et al 1986). The method of leaf sampling and analysis is destructive, time-consuming, and expensive; it is also limited by lack of quality analytical laboratories in developing countries (Turner and Jund 1994, Peng et al 1996). It would be highly useful if simple, quick, and nondestructive diagnostic techniques were available to decide when to apply N fertilizers to rice. Two decision aids--chlorophyll meter and leaf color chart (LCC)--are available to monitor leaf N status in situ and to improve the timing of N topdressing in rice.
- *Soil- and fertilizer-based strategies.* It is possible to adjust fertilizer rates or amounts, as well as timing and methods of application based on soil test results, soil properties, and crop response data from fertilizer trials. Soil sampling and analysis are time-consuming, and often farmers do not receive the soil test recommendations on time to make the necessary changes on fertilizer use for current crops. The soil test results and soil test-crop correlation data are variable depending on the quality of sampling, analysis, and interpretation. Preplant soil tests do not take into account the dynamic changes in soil N supply during the crop growth period.

Decision aids for monitoring crop N status and timing of N application to rice

Two decision aids useful for monitoring leaf N status and improving the timing of N topdressing in rice are the chlorophyll meter and the LCC.

Chlorophyll meter

Chlorophyll meter (or SPAD) is a simple, portable diagnostic tool that measures the leaf chlorophyll content. Meter readings are given in Minolta company-defined SPAD values which indicate the relative amounts of chlorophyll present in plant leaves. Thus, the chlorophyll meter measures the relative greenness or chlorophyll content of leaves (Inada 1963, 1985; Kariya et al 1982; Yadava 1986; Turner and Jund 1994). Because the leaf chlorophyll content is closely related to leaf N concentration, SPAD has been adapted to assess crop N status in ricefields and to determine the need for N topdressing (Inada 1985; Chubachi et al 1986; Takebe and Yoneyama 1989; Takebe et al 1990; Turner and Jund 1991,1994; Peng et al 1993, 1995, 1996).

Relationship between SPAD and leaf N concentration. The linear relationship between weight-based leaf N concentration (N_w) and SPAD readings varies with crop growth stage and variety (Takebe and Yoneyama 1989, Turner and Jund 1994), mostly due to differences in leaf thickness or specific leaf weight (Peng et al 1993). Using rice as the test crop, Peng et al (1995) demonstrated that the confounding effect of leaf thickness or specific leaf weight on SPAD readings can be removed if foliar N concentration is expressed on leaf area basis. The use of area-based leaf N concentration (N_a) for SPAD calibration produces a unique linear relationship at all growth stages.

SPAD threshold value. A SPAD threshold value is the SPAD reading at critical N level below which the crop will suffer from N deficiency resulting in yield loss. SPAD threshold values can be worked out, using the relationship between SPAD readings and leaf area-based N concentration (N_a) The SPAD threshold value is not affected by luxury consumption, because a plant will produce only as much chlorophyll as it needs regardless of how much N is in the plant (Peterson et al 1993). Peng et al (1996) successfully used a single SPAD threshold of 35, which represents an N_a of 1.4 g N m^{-2} of leaf area, to determine the need for N topdressing of rice variety IR72. This threshold appears to be sufficient to achieve the potential yield level in IR72 (Kropff et al 1994) under controlled experimental conditions. Whether the same threshold value can be used for a wide range of crop conditions is to be determined.

Use of SPAD to adjust N fertilizer rates based on soil N supply and crop demand. For rice, the relative contribution of soil N is greater than that of fertilizer N (Broadbent 1984, Wada et al 1986, Kundu and Ladha 1995). Thus, if N fertilizer application is adjusted to soil N-supplying capacity, the efficiency of applied N can be increased. However, it is difficult to determine the indigenous N supply of lowland rice soils due to its high spatial and temporal variability (Cassman et al 1996, Dobermann et al 1996). Plant growth reflects the entire N supply from different sources (soil, water, fertilizer, biological N_2

fixation, etc.). Thus, the monitoring of crop N status by SPAD will be useful to devise variable rate (Blackmer et al 1996) and field-specific N application by taking into account the variable soil N supply and crop demand.

Peng et al (1996) state that SPAD-guided N management improves the congruence of N supply and crop demand, resulting in high grain yields, greater fertilizer-use efficiency, and higher harvest index. They also speculate that the more efficient rice plants resulting from SPAD-based N management would be less susceptible to lodging and more resistant to diseases such as sheath blight and blast. Further research is in progress to clarify the relationship between efficient N management and disease incidence in rice.

Factors affecting SPAD readings. Similar to any other diagnostic tool, SPAD reading is affected by some conditions. These are varietal group, plant density, environmental conditions, status of nutrients other than N in soil and plant, and biotic and abiotic plant stresses that cause leaf chlorosis (Peterson et al 1993, Turner and Jund 1994). For example, phosphorus (P)-deficient rice plants tend to produce dark green leaves, which may give a high SPAD reading (e.g., 39 for P-deficient plants as against 35 for normal plants). Thus, P deficiency confounds SPAD measurements. The SPAD technique works well when other nutrients are adequate in the soil and plant. If SPAD readings are properly calibrated with a varietal group under a given condition (environmental, nutritional, and cultural practices), it serves as a good tool to fine-tune N fertilization.

Need for basal N application in SPAD-guided N management. The level of inherent soil fertility (especially soil N-supplying capacity) determines whether there is need for basal N application in relation to SPAD-guided N management. In poor soils with low organic matter and available N content, it is important to apply a part of the total N just before transplanting or seeding to produce good early growth, seedling vigor, and tillering. It appears that for soils producing grain yields of 3 t ha^{-1} or more without any N fertilizer application, there is no need for basal N application. Ongoing research will develop well-defined soil criteria or indicators to divide the soils into 2-3 groups based on the need for basal N application along with SPAD guidance for N topdressing.

On-farm evaluation of SPAD method. We evaluated the chlorophyll meter technique in a set of on-farm trials conducted with selected national agricultural research systems. Transplanted and wet-seeded rice were used as test crops in the 1995 and 1996 dry and wet seasons in Maligaya, Nueva Ecija, Philippines. A zero N control and three to four N treatments were compared in each farm which served as a replication. Plot size was 10 m x 5 m (50 m^2) separated by good bunds all around. Each set of trials had 5-12 farms (replications). Farmers' practices were used for crop establishment and care. A SPAD threshold of 35 in the dry season and 32 in the wet season were used for both transplanted and wet-seeded rice.

Two on-station trials with 3-6 replications were conducted on transplanted rice at the Research Institute for Rice, Sukamandi, Indonesia, during 1995-96, to evaluate the chlorophyll meter method. We used a SPAD threshold of 35 for these trials.

SPAD readings were taken weekly starting at 14 d after transplanting (DT) or 21 days after seeding (DAS) until flowering. For SPAD-guided N treatment, N application rates were 30 kg ha^{-1} in dry season and 20 kg ha^{-1} in wet season during mid-tillering and between panicle initiation (PI) and flowering stages, and 45 kg ha^{-1} in dry season and 30 kg ha^{-1} in wet season during PI. Farmers' rates and timing of N application were recorded. Grain yields at 14% moisture were recorded from three randomly selected 5-m^2 areas for each plot. The results are discussed in the following case studies.

Case A: Irrigated, transplanted rice (TPR)

1995 dry season on-farm trials, Nueva Ecija. The trials were planned and conducted by S. Peng and F.V. Garcia of Agronomy, Plant Physiology, and Agroecology Division of IRRI. We collaborated with them for the evaluation of the chlorophyll meter technique for the first time in six farmers' fields in Nueva Ecija, Philippines. A SPAD threshold of 35 was used to determine the need for N topdressing. The mean grain yield of 7426 kg ha^{-1} obtained in the SPAD-guided N treatment was higher than that in the farmers' plot (6708-6994 kg ha^{-1}), but the differences in yield among fertilized plots were not significant (Table 1). SPAD-based N management reduced fertilizer use by 5.7%, and increased grain yield by 8.4% and agronomic efficiency of applied N (AEN) by 27.3%, compared with farmers' methods.

1996 DS on-farm trials, Nueva Ecija. Only one of the 12 farmers applied all N in a single dose at 14 DT, eight used two splits at 0-18 and 21-36 DT, and three gave three splits at 0-10, 12-29, and 32-43 DT. Only two farmers used basal N application. The farmers' last N application was 36 DT for two splits and 43 DT for three splits. The farmers' practice of applying large amounts of N at early growth stages resulted in N deficit at later stages (Garcia et al 1996).

Table 1. Comparison of chlorophyll meter technique with farmers' practice for increasing grain yield and efficiency of N use on irrigated transplanted rice. Maligaya, Nueva Ecija, Philippines, 1995 and 1996.[a]

N treatment	N rate (kg ha^{-1})	No. of splits	Yield (kg ha^{-1})	AEN[b]
1995 DS; cv IR64; SPAD threshold = 35; kg P ha^{-1} = 24; kg K ha^{-1} = 28 (6 farms)				
Control	0	–	4476 b	–
SPAD	140	4	7426 a	21 a
Farmers' plot	148	2	6994 a	17 b
Farmers' field	148	2	6708 a	16 b
1996 DS; cv IR64; SPAD threshold = 35; kg P ha^{-1} = 24; kg K ha^{-1} = 30 (12 farms)				
Control	0	–	3748 c	–
SPAD	150	5	6704 a	20 b
Farmers' plot	120	2	6047 b	19 b
UT/DP	115	1	6614 a	25 a
1995 WS; cv. IR64; SPAD threshold = 32; kg P ha^{-1} = 18; kg K ha^{-1} = 27 (12 farms)				
Control	0	–	2738 c	–
SPAD	69	3	3939 a	19 a
Farmers' plot	89	2	3593 b	10 b
Farmers' field	89	2	3640 b	11 b

[a]In each column, means followed by a common letter are not significantly different at the 5% probability level by DMRT.

[b]AEN = agronomic efficiency of applied N.

The SPAD method required three splits in one field at 14-33 DT; four splits in two fields at 14 to 32-48 DT; five splits in six fields at 14 to 39-49 DT; six splits in two fields at 14 to 48-56 DT; and seven splits in one field at 14 to 55 DT. The SPAD method suggests the last application to be around 49 DT for IR64 which matures in 110 d.

The SPAD treatment produced the highest yield of 6704 kg ha^{-1} compared with 6047 kg ha^{-1} in the farmers' plot (Table 1), but it also consumed more N (150 kg ha^{-1}) than did the farmers' treatment (120 kg N ha^{-1}). Thus, there was no difference in AEN (20 vs 19) between the two methods of N topdressing. A single deep placement of urea tablets at 14 DT (N rate of 115 kg ha^{-1}) gave a grain yield of 6614 kg ha^{-1}, comparable with that of the SPAD treatment (AEN of 25) (Table 1).

Soil type vs SPAD method. The 12 farms were distributed in two soil types--six farms each in clay loam and sandy loam. Farmers applied slightly less N in clay loam than in sandy loam soils, contrary to the SPAD method which used more N in clay loam than in sandy loam soils (Table 2A). In both soil types, the SPAD treatment gave significantly higher grain yield than did farmers' practice, but the AEN values were not significantly different between the two methods due to higher N fertilizer use in the SPAD method. Urea tablet deep placement gave similar yields as the SPAD, but at much lower N rate.

Soil N-supplying capacity (zero N control yield) vs SPAD method. The sites were divided into two groups based on soil N-supplying capacity as reflected by zero N control yields for comparison of treatments. In seven sites with control yields of more than 3.3 t ha^{-1}, farmers applied relatively more NPK nutrients, which could have built up the native soil fertility for high yields (Table 2B). Despite significantly higher grain yields in SPAD-guided N management, the AEN values were similar for both SPAD and farmers' methods. Deep placement of urea tablets seemed to be more efficient in soils with relatively smaller N-supplying capacity.

Number of farmers' N splits vs SPAD method. Eight farmers applied N in two splits while three farmers used three splits (Table 2C). Farmers using three N splits also applied slightly more P and K, which reflects a better level of crop management resulting in higher grain yields in these farms. Mean grain yield and AEN of the SPAD treatment were higher than those of the farmers' practice in both groups. Because farmers applied the last dose of N at 36 DT (two splits) and at 43 DT (three splits), the farmers' crop could have suffered from N deficit in later growth stages (as pointed out by Garcia et al 1996), resulting in less grain yields. The efficiencies of SPAD and urea tablet deep placement were higher in better managed farms using three N splits (Table 2C).

1995 wet season on-farm trials, Nueva Ecija. In the 1995 wet season, only one of 12 farmer cooperators applied all N in a single dose at 11 DT; all others used two splits at 8-23 and 28-46 DT. The SPAD method required two splits in four fields, three splits in three fields, and four splits in five fields, when a threshold value of 32 was used.

Table 2. Comparison of chlorophyll meter method with farmers' practice and urea tablet deep placement (UT/DP) for increasing grain yield and N use efficiency on irrigated transplanted rice, as affected by soil type, level of zero N control yield, and number of farmers' N splits. Maligaya, Nueva Ecija, Philippines, 1996 DS (cv IR64; SPAD threshold = 35).[a]

2A. Soil type

Particular	N treatment	Clay loam (6)	Sandy loam (6)
NPK (kg ha^{-1})	Control	0-26-28	0-21-33
	Farmers' method	121-26-28	131-21-33
	SPAD	165-26-28	135-21-33
	UT/DP	115-26-28	115-21-33
Yield (kg ha^{-1})	Control	3470 c	4027 b
	Farmers' method	6092 b	6002 a
	SPAD	6893 a	6515 a
	UT/DP	6880 a	6348 a
AEN	Farmers' method	22 b	15 a
	SPAD	21 b	18 a
	UT/DP	30 a	20 a

2B. Level of zero N control yield

Particular	N treatment	Control yield <3.3 t ha^{-1} (5)	Control yield >3.3 t ha^{-1} (7)
NPK (kg ha^{-1})	Control	0-20-25	0-25-33
	Farmers' method	123-20-25	128-25-33
	SPAD	158-20-25	146-25-33
	UT/DP	115-20-25	115-25-33
Yield (kg ha^{-1})	Control	3072 c	4231 c
	Farmers' method	5666 b	6319 b
	SPAD	6458 a	6880 a
	UT/DP	6598 a	6626 ab
AEN	Farmers' method	21 b	16 a
	SPAD	21 b	18 a
	UT/DP	31 a	21 a

2C. Number of farmers' N splits

Particular	N treatment	Two splits (8)	Three splits (3)
NPK (kg ha^{-1})	Control	0-23-27	0-28-41
	Farmers' method	128-23-27	135-28-41
	SPAD	146-23-27	150-28-41
	UT/DP	115-23-27	115-28-41
Yield (kg ha^{-1})	Control	3884 c	3267 c
	Farmers' method	6064 b	6437 b
	SPAD	6545 a	7490 a
	UT/DP	6524 a	7150 a
AEN	Farmers' method	17 ab	23 c
	SPAD	18 b	28 b
	UT/DP	23 a	31 a

[a]Numbers in parentheses represent the number of farms in each category. In each column, means followed by a common letter are not significantly different at the 5% probability level by DMRT.

The SPAD-guided N treatment produced significantly higher grain yield (3939 kg ha^{-1}) than did farmers' practice--plot (3593 kg ha^{-1}) or field (3640 kg ha^{-1}) (Table 1). The SPAD method gave 7.6-8.8% more grain yield with 22.4% less N fertilizer input compared with farmers' practice. The AEN of the SPAD method (19) was significantly higher than that of farmers' practice (10-11).

1995-96 wet- and dry-season on-station trials, Sukamandi. In replicated trials conducted at the Research Institute for Rice, Sukamandi, Indonesia, in situ monitoring of crop N status by chlorophyll meter indicated overapplication of N to rice. The SPAD-guided N treatment used only 30 kg N ha^{-1} at one site during the 1995-96 wet season and 60 kg N ha^{-1} at another site during the 1996 dry season, compared with the recommended rate of 90 kg N ha^{-1} (Table 3). During the wet season, the SPAD treatment yielded 6188 kg ha^{-1}; it had the highest AEN (30). Deep-placed urea tablets at 55 kg N ha^{-1} gave 6511 kg ha^{-1} grain yield and had an AEN of 22. All other N treatments were less efficient.

During the 1996 dry season, two varieties, Memberamo and IR64, were used for evaluating the N management techniques. The zero-N control yield was higher for Memberamo than for IR64 (Table 3), suggesting the higher efficiency of Memberamo in exploiting native soil N supply. Grain yields of both varieties were similar for all N treatments, except for the SPAD-guided N management. A SPAD threshold value of 35 appears to work well with Memberamo but not with IR64 in this trial. For the SPAD treatment, the AEN value was 30 for Memberamo and 15 for IR64. This observation needs to be confirmed.

Table 3. Comparison of chlorophyll meter-based N management with fixed-schedule N application at two rates (55 and 110 kg N ha^{-1}) either by broadcasting of prilled urea and deep placement of urea tablets in transplanted rice. Research Institute for Rice, Sukamandi, Indonesia, 1995-96 wet and dry seasons (SPAD threshold = 35).[a]

Treatment (kg N ha^{-1})	Yield (kg ha^{-1})	AEN	Grain N (%)	Straw N (%)	N uptake (kg ha^{-1})
1995-96 wet seasons					
Control, 0	5272 b	–			
PU, 55	6233 a	17 b			
PU, 110	6610 a	12 b			
UT, 55	6511 a	22 ab			
UT, 110	6440 a	11 b			
SPAD, 30	6188 a	30 a			
1996 dry season; rice cv Memberamo					
Control, 0	4773 c	–	1.14 c	0.58 d	92.6 d
PU, 90	6598 ab	20 b	1.42 a	0.68 b	165.3 b
UT, 55	6423 b	30 a	1.26 b	0.61 cd	146.7 c
UT, 110	7063 a	21 b	1.39 a	0.77 a	198.5 a
SPAD, 60	6544 ab	30 a	1.27 b	0.65 bc	140.3 c
1996 dry season; rice cv IR64					
Control, 0	4400 d	–	1.16 d	0,58 c	91.1 d
PU, 90	6308 b	21 bc	1.33 ab	0.68 b	155.8 b
UT, 55	6291 b	34 a	1.25 c	0.67 b	145.7 bc
UT, 110	6988 a	24 b	1.38 a	0.78 a	189.6 a
SPAD, 60	5283 c	15 c	1.28 bc	0.67 b	129.4 c

[a]PU=prilled urea, UT=urea tablets. In each column, means followed by a common letter are not significantly different at the 5% probability level by DMRT.

The deep placement of urea tablets at 55 kg N ha^{-1} was highly efficient, as reflected by the high grain yield and agronomic efficiency of both varieties (Table 3). Placement of 110 kg N ha^{-1} as urea tablets produced the highest mean grain yields, but the efficiency of N fertilizer was significantly less than that of urea deep placement at 55 kg N ha^{-1}.

Case B: Irrigated, broadcast or row wet-seeded rice (B- or R-WSR) in dry season

1995 dry-season on-farm trials with B-WSR, Nueva Ecija. Five on-farm trials were conducted on B-WSR in collaboration with IRRI scientists. Using a threshold value of 35, the SPAD method consumed more fertilizer (189 kg N ha^{-1}) than did farmers' method (153 kg N ha^{-1}) for similar grain yields (Table 4). Thus, the AEN was significantly lower for SPAD method than for farmers' practice in B-WSR.

1996 dry-season on-farm trials with B-WSR, Nueva Ecija. Six farmers participated in this trial. A SPAD threshold of 35 was used again. The SPAD treatment produced an average grain yield of 6297 kg ha^{-1} for a mean fertilizer application of 190 kg N ha^{-1}. The farmers' practice produced 5947-6348 kg ha^{-1} at a mean N rate of 134 kg ha^{-1}. Again, the AEN values of the two methods did not differ significantly (Table 4).

Table 4. Performance of chlorophyll meter method compared with farmers' practice in irrigated broadcast-wet-seeded rice. Maligaya, Nueva Ecija, Philippines, 1995 and 1996 dry seasons.[a]

N treatment	N rate (kg N ha^{-1})	No. of N splits	Yield (kg N ha^{-1})	AEN
1995 dry season; cv IR64; SPAD threshold = 35; kg P ha^{-1} = 19; kg K ha^{-1} = 33 (5 farms)				
Control	0	–	4493 b	–
SPAD	189	6	6963 a	13 b
Farmers' plot	153	3	7142 a	18 a
Farmers' field	153	3	6996 a	17 a
1996 dry season; cv IR64; SPAD threshold = 35; kg P ha^{-1} = 19; kg K ha^{-1} = 30 (6 farms)				
Control	0	–	4273 b	–
SPAD	190	6	6297 a	11 b
Farmers' plot	134	2	5947 a	12 ab
Farmers' field	134	2	6348 a	15 a

[a]In each column, means followed by a common letter are not significantly different at the 5% probability level by DMRT.

Table 5. SPAD readings as influenced by growth stage and rate and time of N application in irrigated broadcast-wet-seeded rice. Maligaya, Nueva Ecija, Philippines, 1996 dry season.[a]

N treatment	No. of splits	SPAD reading			
		Max. tillering	PI	Flowering	Milking
Control	–	22 a	24 a	24 a	25 a
Farmers' method	2	31 b	29 b	29 b	30 b
SPAD	5-7	30 b	32 c	32 c	32 c

[a]Mean of 6 farms. PI=panicle initiation. In each column, means followed by a common letter are not significantly different at the 5% probability level by DMRT.

In the B-WSR trials on six farms, the SPAD-guided N treatment maintained a SPAD reading of 32, which is significantly higher than that of farmers' practice at all growth stages, except at maximum tillering (Table 5). This observation suggests that the critical SPAD value for B-WSR is around 32.

1996 dry-season on-farm trials with B- or R-WSR, Nueva Ecija. Another trial conducted in 10 farmers' fields during 1996 dry season compared with the two- or three-split fixed-schedule N treatments with the SPAD-guided N management. Using a threshold value of 35, the SPAD method required 212 kg N ha^{-1} in seven splits to produce an average yield of 6042 kg ha^{-1} in B-WSR. Nitrogen fertilizer applied at 150 kg N ha^{-1} in three splits at 21, 35, and 49 DAS gave a grain yield of 5789 kg ha^{-1} (Table 6). The grain yield was 5978 kg ha^{-1} for R-WSR with 150 kg N ha^{-1} applied in two splits. The AEN values were significantly higher (19) for B-WSR with three N splits and R-WSR with two N splits than for the SPAD method or B-WSR with two N splits (AEN = 15).

In the above trial, the SPAD meter readings oscillated around 31-32 for B-WSR with N applied in three splits and R-WSR with N given in two splits, compared with 32-33 for the SPAD treatment at maximum tillering, PI, and flowering (Fig. 1). Even with almost weekly N application in the SPAD treatment, the mean SPAD reading stayed around 32-33 only. This observation confirms that the SPAD threshold value for B-WSR appears to be around 32. The high plant density in B-WSR could have decreased the leaf N concentration and hence the leaf greenness. Further testing is in progress to confirm this point and to develop specific SPAD threshold values for WSR.

The results from cases A and B indicate that

- the SPAD threshold value of 35 is useful for semidwarf indica varieties and Memberamo (Indonesia) in irrigated TPR during the dry season with high radiation;
- the SPAD threshold has to be adjusted downward to 32 for TPR in the wet season with low yield potential due to overcast sky and low radiation;

Table 6. Performance of chlorophyll meter method compared with fixed-schedule N application in two (21 and 42 DAS) or three splits (21, 35, and 49 DAS) in irrigated broadcast- or row-wet-seeded rice (B- or R-WSR). Maligaya, Nueva Ecija, Philippines, 1996 dry season.[a]

N treatment	N rate (kg N ha^{-1})	No. of N splits	Yield (kg ha^{-1})	AEN
cv IR64; SPAD threshold = 35; kg P ha^{-1} = 13^{b}; kg K ha^{-1} = 25 (10 farms)				
Control, B-WSR	0	–	2909c	–
SPAD, B-WSR	213	7	6042a	15b
2 splits, B-WSR	150	2	5157b	15b
3 splits, B-WSR	150	3	5789a	19a
2 splits, R-WSR	150	2	5978a	19a

[a]DAS = days after seeding. In each column, means followed by a common letter are not significantly different at the 5% probability level by DMRT. [b]Yields could have been reduced by low rate of P application (13 kg ha^{-1}).

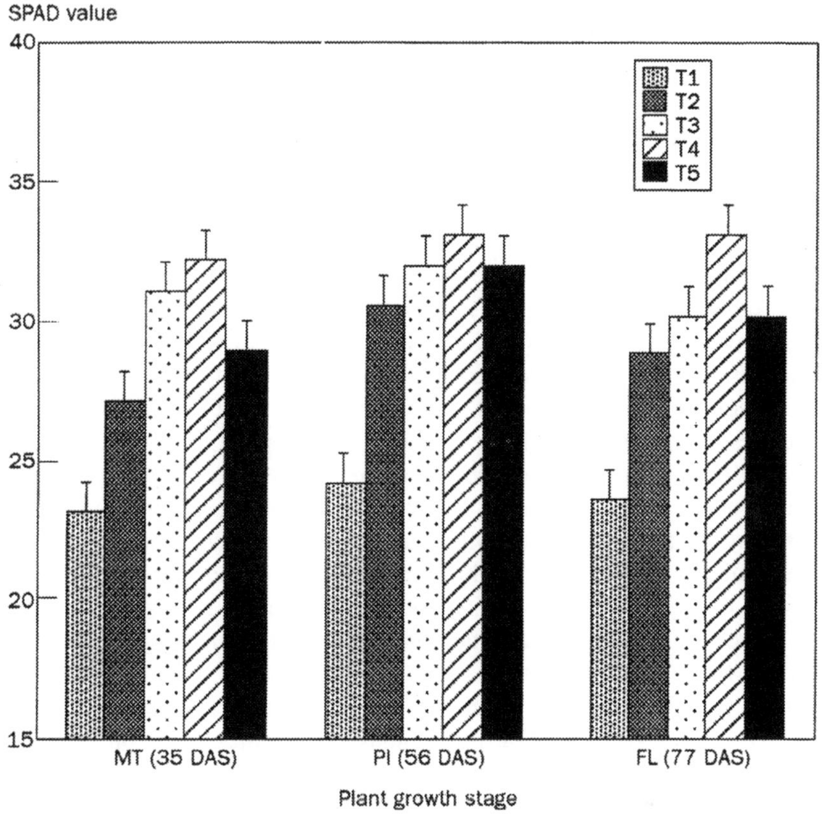

Figure 1. SPAD value as influenced by time and rate of N application and method of establishment in B- or R-WSR cv IR64, 1996 DS (mean of 10 farms). T1 = Minus-N (B-WSR), T2 = fixed N (150 kg N ha^{-1} - 2 splits) - B-WSR, T3 = fixed N (150 kg N ha^{-1} - 3 splits) - B-WSR, T4 = SPAD-N (212 kg N ha^{-1} - 7 splits) - B-WSR, T5 = fixed N (150 kg N ha^{-1} - 2 splits) - R-WSR.

- the SPAD threshold value of 35 overestimates the N requirement for WSR; the optimum threshold for WSR appears to be around 32 and further testing is in progress to confirm it;
- the SPAD meter can be used to verify the crop N status under strategically timed fixed-schedule N applications for B-WSR (e.g., three splits at 21, 35, and 49 DAS); and
- the number of N splits is high (5-6) for the SPAD method, which farmers may not adopt. Further research will be undertaken to reduce the number of splits by optimizing the amount of N applied at critical times, maybe with the aid of Manage-N simulation model (Thiyagarajan et al 1994, Riethoven et al 1995).

Leaf color chart (LCC)

The relatively high cost of the chlorophyll meter (about US$1300-1400 per unit) seems to have limited its acceptance by farmers, even in the USA (Turner and Jund 1994). A simple, easy-to-use, and inexpensive indicator of rice leaf N status is, therefore, needed to determine the field-specific N requirement of rice crops. Farmers use leaf color as a visual and subjective indicator of the need for N fertilizer application (Wells and Turner 1984, Furuya 1987). Visual assessment of leaf greenness is influenced by sunlight variability (sun's intensity and angle) (Turner and Jund 1994).

A LCC developed from a Japanese prototype (Furuya 1987) is being standardized with the chlorophyll meter to assess the relative accuracy of LCC in measuring the greenness of rice leaves. We observed that the difference between adjacent shades of green color in the LCC is equal to 3-4 SPAD units. Thus, the LCC cannot indicate smaller differences in leaf greenness as the chlorophyll meter does. Field observations indicate that the LCC can still be used to determine the time of N topdressing to rice crops, once the critical color shades are established for different cultivar groups and crop conditions with the help of the chlorophyll meter. Results of 1995 trials with two rice varieties--PSBRc 2 and PSBRc 6--at the Maligaya research farm, PhilRice, Philippines, indicate that the LCC and SPAD readings are highly correlated, with regression coefficients ranging from 0.70 to 0.91 at 2, 4, 6, 8, and 10 wk after transplanting (Asis and Cruz, R.T., PhilRice, Maligaya, Philippines, 1995, pers. commun.).

We have distributed more than 50 LCCs to farmer cooperators located around PhilRice, Maligaya, Nueva Ecija, Philippines, and trained them on how to use the new tool for N fertilization of rice. Initial observations indicate that most of them are highly interested in using the LCC to determine the time of N topdressing for rice. More than half of the cooperators feel that they could save N fertilizer either with the same or higher grain yield than what they were getting before with their own practice. Detailed data are being collected to analyze farmers' experience and reactions in using the LCC.

Research needs on chlorophyll meter and LCC techniques

Adaptation of the chlorophyll meter to different crop conditions is important: more adaptive research is in progress to identify appropriate SPAD threshold values for direct-seeded rice and for different varietal groups, to clarify the effect of nutrient imbalance (deficiency or toxicity) on SPAD readings, and to develop a reduced number of strategic split N applications for different rice systems with the combined use of the chlorophyll meter and the Manage-N simulation model. We will develop soil criteria or indicators to determine the kind of soils we need to combine basal N application with SPAD technique for N topdressing.

Peng et al (1996) speculate that the more efficient rice plants resulting from SPAD-based N management would be less susceptible to lodging and more resistant to diseases such as sheath blight and blast. Research must authenticate this observation and produce experimental evidence to clarify the effect of regulated N supply on plant health and disease incidence in rice.

The LCC will be highly complementary to the chlorophyll meter technique. The standardization of LCC with chlorophyll meter, which is already progressing well in 2-3 countries, will enable farmers to use it as a reliable guide to develop need-based, field-specific N application for their rice crops.

Educating and training extension agents and farmers on new techniques

The techniques described in the preceding section are fairly complex and knowledge-intensive. The two decision aids, chlorophyll meter and LCC, are valuable tools for field researchers, extension agents, crop consultants, and farmers. A good deal of training is necessary for all to successfully use these tools for fine-tuning N fertilization. They must recognize the influence of interfering factors on both techniques, learn about the appropriate critical values for different cultivar groups and environmental/crop conditions, and appreciate the need for using the decision aids in combination with other methods to further refine N management. With sufficient training, farmers can improve their N fertilization decisions, for example, when to apply N and how much N to apply at different growth stages of different rice varieties. The Crop and Resource Management Network of IRRI works with national agricultural research and development organizations to promote farmer-participatory evaluation of new N management technologies under actual farming conditions (with variable land preparation, water regimes, cultural practices, and management levels). The process of learning-by-doing in participatory on-farm evaluation trials will strengthen their understanding of the intricacies of new techniques and improve their field application.

Farmers must learn to appreciate not only the economic advantages of new technologies, but also their impact on resource conservation, environmental quality, and human health. Ignorance will spell disaster to farmers and their environment, as well as to the community at large (Gardner 1996).

Incorporating new information into farmers' knowledge base

Most farmers are cautious about new recommendations and are slow to accept changes or take risks, because of several uncertainties inherent in farming (Hapgood 1995). However, each agrarian community has a few market-oriented, progressive farmers who show the way to others by adopting new farming practices at certain risks to increase their productivity and income. Both types of farmers have built a strong knowledge base from practical experiences gained over generations.

It is possible to incorporate new information or knowledge into the base, if farmers are convinced about it by their own experimentation and testing. The new practice or technology should be compatible with existing cropping or farming systems. The process of incorporating a new information into the base is iterative (Pinners and Balasubramanian 1991). It involves a cycle of testing, modification, evaluation, redesign, and so on, until the farmer is convinced about the utility of the new information or technology.

Conclusions

The two promising decision aids that can increase the efficiency of N fertilizer use on rice are chlorophyll meter and LCC. These techniques are being adapted to rice farming conditions with variable cultural practices to improve their adoption by farmers.

On-farm, adaptive research is in progress in three countries to adapt the chlorophyll meter technique for TPR and WSR, local cultivar groups, and soil, crop, and environmental conditions. Initial results indicate that the SPAD threshold value of 35 is good for TPR in dry season. The threshold has to be reduced to 32 for WSR in dry season and for all rice during wet season with cloudy weather and low radiation. Further research is needed to clarify the effect of SPAD-regulated N management on plant health, crop resistance to diseases, and evolution of long-term soil quality.

The LCC is a simple, easy-to-use, and inexpensive tool to monitor crop N status in the field and to determine the timing of N topdressing to rice. It is being calibrated with the chlorophyll meter to make it a complementary tool for fine-tuning N management in rice. Initial feedback from farmer cooperators in the Philippines is highly encouraging.

Farmers and extension agents must be appropriately and adequately trained to improve their understanding and proper utilization of new techniques. Farmers must also learn to appreciate not only the economic advantages of new technologies but also their impact on resource conservation, environmental quality, and human health.

References

Blackmer T M, Schepers J S, Varvel G E, Walter-Shea EA (1996) Nitrogen deficiency detection using reflected short wave radiation from irrigated corn canopies. Agron. J. 88: 1-5.

Broadbent F E (1984) Plant use of soil nitrogen. Pages 171-182 in Hauck D (Ed.) Nitrogen in crop production. American Society of Agronomy, Madison, Wisconsin, USA.

Cassman K G, Pingali P L (1995) Intensification of irrigated rice systems: learning from the past to meet future challenges. GeoJournal 35 (3): 299-305.

Cassman K G; Gines H C, Dizon M A, Samson M I, Alcantara J M (1996) Nitrogen-use efficiency in tropical lowland rice systems: relative contributions of indigenous soil resources and applied nitrogen inputs. Field Crops Res. 47: 1-12.

Chubachi T, Asano I, Oikawa T (1986) The diagnosis of nitrogen nutrition of rice plants (Sasanishiki) using chlorophyll meter. Soil Sci. Plant Nutr. 57: 190-193.

Dobermann A, Cassman K G, Peng S, Tan P S, Phung C V, Sta. Cruz P C, Bajita J B, Adviento M A A, Olk D C (1996) Precision nutrient management in intensive irrigated rice systems. In Maximizing sustainable rice yields through improved soil and environmental management. Symposium of the Paddy Soil Fertility Working Group of ISSS, Khon KAEN, Thailand, 11-17 Nov 1996. 24 p. (mimeo)

Furuya S (1987) Growth diagnosis of rice plants by means of leaf color. Jpn. Agric. Res. Q. 20: 147-153.

Garcia F V, Peng S, Gines H C, Laza R C, Sanico A L, Visperas R M, Cassman K G (1996) Chlorophyll meter-based nitrogen management improves nitrogen use efficiency of irrigated rice in farmers' fields. Pages 187-190 in Ishii R, Horie T (Eds.) Crop research in Asia. Proceedings of the 2nd Asian Crop Science Conference, 21-23 Aug 1995, Fukui, Japan.

Gardner G (1996) Preserving agricultural resources. Pages 78-94, 208-213. in Brown L.R. et al (Eds.) State of the world 1996. World Watch Institute and Norton and Company, New York, USA.

Hapgood F (1995) High-tech harvest Inc. Technology No. 3. p 52-56.

Hossain M, Fischer K S (1995) Rice research for food security and sustainable agricultural development in Asia. GeoJournal 35 (3): 286-298.

Inada K (1963) Studies on a method for determining deepness of green color and chlorophyll content of intact crop leaves and its practical applications. 1. Principle for estimating the deepness of green color and chlorophyll content of whole leaves. Proc. Crop Sci. Soc. Jpn. 32: 157-162.

Inada K (1985) Spectral ratio of reflectance for estimating chlorophyll content of leaf. Jpn. J. Crop Sci. 54: 261-265.

Kariya K, Matsuzaki A, Machida H (1982) Distribution of chlorophyll content in leaf blade of rice plant. Jpn. J. Crop Sci. 51: 134-135.
Kropff M J, Cassman K G, Peng S, Mathews R B, Setter T L (1994) Quantitative understanding of yield potential. Pages 21-38 in Cassman K G (Ed.) Breaking the yield barrier. Proceedings of a workshop on rice yield potential in favorable environments. International Rice Research Institute, P.O. Box 933, Manila, Philippines.
Kundu D K, Ladha J K (1995) Efficient management of soil and biologically fixed N_2 in intensively cultivated rice fields. Soil Biol. Biochem. 27: 431-439.
Ladha J K, Kundu D K, van Coppennole M G A, Peoples M B, Carangal V R, Dart PJ (1996). Legume productivity and soil nitrogen dynamics in lowland rice-based cropping systems. Soil Sci. Soc. Am. J 60: 183-192.
Peng S, Garcia F V, Laza R C, Cassman K G (1993) Adjustment for specific leaf weight improves chlorophyll meter's estimation of rice leaf nitrogen concentration. Agron. J. 85: 987-990.
Peng S, Laza R C, Garcia F V, Cassman K G (1995) Chlorophyll meter estimates leaf area-based nitrogen concentration of rice. Commun. Soil Sci. Plant Anal. 26: 927-935.
Peng S, Garcia F V, Laza R C, Sanico A L, Visperas R M, Cassman K G (1996) Increased N-use efficiency using a chlorophyll meter on high-yielding irrigated rice. Field Crops Res. 47:243-252.
Peterson T A, Blackmer T M, Francis D D, Schepers J S (1993) Using a chlorophyll meter to improve N management. A Nebguide in Soil Resource Management: D-13, Fertility. Cooperative Extension, Institute of Agriculture and Natural Resources, University of Nebraska, Lincoln, USA.
Pinners E, Balasubramanian V (1991) Use of the iterative diagnosis and design approach in the development of suitable agroforestry systems for a target area. Agrofor. Syst. 15: 183-201.
Riethoven J J M, ten Berge H F M, Drenth H (Eds.) (1995) Software developments in the SARP Project: a guide to applications and tools. SARP research proceedings. DLO Research Institute for Agrobiology and Soil Fertility Wageningen, The Netherlands and International Rice Research Institute, P.O. Box 933, Manila, Philippines. 301 p.
Takebe M, Yoneyama T (1989) Measurement of leaf color scores and its implication to nitrogen nutrition of rice plants. Jpn. Agric. Res. Q. 23: 86-93.
Takebe M, Yoneyama T, Inada K, Murakam T (1990) Spectral reflectance ratio of rice canopy for estimating crop nitrogen status. Plant Soil 122: 295-297.
Thiyagarajan T M, Sivasamy R, ten Berge H F M (1994) Time courses of leaf nitrogen concentration required to attain target yields in transplanted rice. Pages 267-282 in ten Berge H F M, Wopereis M C S, Shin J C (Eds.) Nitrogen economy of irrigated rice: field and simulation studies: SARP research proceedings. DLO Research Institute for Agrobiology and Soil Fertility, Wageningen, The Netherlands and International Rice Research Institute, P.O. Box 933, Manila, Philippines.
Turner F T, Jund M F (1991) Chlorophyll meter to predict nitrogen topdress requirement for semidwarf rice. Agron. J. 83: 926-928.
Turner F T, Jund M F (1994) Assessing the nitrogen requirements of rice crops with a chlorophyll meter method. Aust. J. Exp. Agric. 34: 1001-1005.
Wada G, Shoji S, Mae T (1986) Relationship between nitrogen absorption and growth and yield of rice plants. Jpn. Agric. Res. Q 20: 135-145.
Wallihan E F, Moomah J C (1967) Selection of index leaf for studying the critical concentration of nitrogen in rice plants. Agron. J. 59: 473-474.
Wells B R, Turner F T (1984) Nitrogen use in flooded rice soils. Pages 354-362 in Hauck, R .D (Ed.) Nitrogen in crop production. American Society of Agronomy, Madison, Wisconsin, USA.
Yadava U L (1986) A rapid and non-destructive method to determine chlorophyll in intact leaves. Hort. Sci. 21: 1449-1450.

Resource management options for upland areas of Indonesia

Z. ZAINI and A. M. FAGI
Central Research Institute for Food Crops, Bogor, Indonesia

Key words: land clearing, reclamation, conservation farming, intercropping, alley cropping, integrated nutrient management, holistic approach, household/mixed farming, estate crops, greenhouse gas emission

Abstract
The upland areas in Indonesia are highly diverse and fragile. Appropriate measures of land clearing, reclamation, and conservation are the essential first steps for developing intensive and sustainable upland farming systems. Because upland environments are fragile, nutrient management should be an integral part of upland conservation farming. Low-cost technology is the most preferable choice because most farmers in upland areas are relatively poor. Hedgerow planting or alley cropping has several advantages and limitations. Incorporation of green manure derived from trees planted along the contour lines improved both physical and chemical properties of acid upland soils, and thus increased productivity. A combination of green manure, lime, and inorganic fertilizer applications sustained soil fertility and crop productivity. On the other hand, trees along the contour lines compete with seasonal crops planted in alleys between rows of tree crops for sunshine, water, and nutrients. Selection of appropriate tree crops is therefore important. A holistic approach must be followed in designing household farming systems. Contributions of livestock and estate crops to farm income were significant in settlement areas. Resource mapping and resource flow within household units have to be considered in establishing sustainable household farming systems.

Introduction

In Indonesia, the principal approaches followed to meet the demand for food are (a) increasing the productivity of existing lands; (b) expanding the net area cultivated, either by bringing new land into cultivation or by reclamation; and (c) diversifying land use through farming systems. Concern about the intensive cultivation of upland area is growing because of continuous shrinkage of irrigated area due to land allocation for nonagricultural purposes. Combating environmental degradation and improving management of upland areas have, therefore, received more serious attention in recent years.

Improved upland rice varieties are also responsive to fertilizer application, especially N, P, and K. Therefore, fertilizers play an important role in replenishing the nutrients removed by crops in upland ecosystems. Fertilizers increase crop residues, which can be returned to the soil to maintain soil organic matter (Zaini et al 1994, Blair et al 1995). Fertilizers should, however, be a part of an integrated resource management system in which all nutrient sources, biological and natural, are used to increase production in an efficient and environment-friendly way.

This paper reviews some research results on management and utilization of uplands in Indonesia. It brings together pertinent research on upland management that includes nutrients, analyzes and articulates constraints, and explores integrated soil and crop management systems with low external inputs which can be developed and recommended for further testing and evaluation.

V. Balasubramanian, J. K. Ladha and G. L. Denning (eds.), Resource Management in Rice Systems: Nutrients, 95–104.

Limitations of upland ecosystem

By definition, upland agriculture refers to rainfed agriculture in nonbunded field, which can be practiced almost anywhere, regardless of elevation. Researchers have collected much data on the chemical and physical characteristics of upland soils. Except the soils in West Nusa Tenggara, East Nusa Tenggara, and East Timor, where the climate is relatively dry, the soils of the other areas developed mostly in the humid tropics can be classified as acid soils (Santoso and Sibuea 1994). Ultisols and Oxisols are dominant.

The major constraints to crop production in the uplands are grouped into physicochemical, biological, and socioeconomic.

Physicochemical constraints

Erratic rainfall. Plants require an adequate and timely supply of water. Thus, in upland areas that receive erratic rainfall, the risk of using fertilizers is much greater than in those with more predictable rainfall.

Uneven terrain. Soil survey reports indicate large areas of upland with undulating topography and steep slopes. Such lands are difficult to cultivate safely. Cultivation of sloping uplands is recommended up to 15% slope. Because much of the nutrients are contained in the topsoil and the depth of topsoil is shallow, even the slightest amount of erosion can cause significant loss of soil nutrients.

Rapid loss of organic matter. Organic matter is an efficient retainer and exchanger of nutrient ions. Organic matter declines rapidly once the natural vegetation is removed and the soils are brought into cultivation. This is aggravated by improper cultivation of the sloping uplands such as shifting cultivation by slash-and-burn method.

Low inherent soil fertility. Oxisols and Ultisols are highly weathered, low-activity, freely draining soils with low content of weatherable minerals, low nutrient retention, and high pressure of leaching. On Oxisols and Ultisols, food crops are unable to effectively retrieve and utilize applied nutrients. Therefore, upland soils are acidic with high aluminum (Al) saturation, low phosphorus, and low base saturation (Adiningsih et al 1988, Arya et al 1992). Acidity and Al toxicity are considered the most important agronomic problems. Many food crops are sensitive to Al toxicity. The root systems of Al-sensitive plants are generally confined to the top 10-20 cm soil depth and are, therefore, unable to absorb water and nutrients accumulated in the deeper soil layers. Acidity increases as a result of cultivation practices (Zaini et al 1994). Adding fertilizers increases soil nutrients, but the added nutrients are rapidly lost because of low nutrient retention capacity of the soils and high rate of leaching.

Because of these limitations, only 0.26 million ha of the existing upland rice areas are considered fertile with long growing season; the majority of uplands with long growing season (0.79 million ha) are infertile. The remainder of the upland areas (0.09 million ha) have short growing season, about half of which are fertile (Table 1).

Table 1. Upland rice areas in Indonesia by environmental complex (*Source:* Baharsjah and Fagi 1995).

Growing season [a] and soil fertility level [b]	Upland rice area	
	Million ha	%
Long growing season		
Fertile soils	0.26	22.8
Infertile soils	0.79	69.3
Short growing season		
Fertile soils	0.05	4.4
Infertile soils	0.04	3.5
Total	1.14	100

[a]Growing season is the number of months in which rainfall exceeds potential evaporation by 20%.
[b]Inherent soil fertility status: 1-5 (high to moderate), 6-9 (low to very low).

Biological constraints

Weeds, diseases, and insects are major biological constraints to food crop production in upland areas. These adverse factors cause many farmers to limit food crop production to subsistence level. In this environment, traditional rice varieties are associated with low grain yield (McIntosh 1986). Blast disease is the most devastating. Varietal improvement for blast resistance has not been successful because different strains exist in different upland rice production centers.

Socioeconomic constraints

Production programs and associated packages are often unacceptable because farmers lack either the money or the desire or ability to assume risk. Under these conditions, the only new technologies that farmers can accept are new varieties and soil fertility management that involve green manuring and improved crop residue management (Baharsjah and Fagi 1995).

Transportation and communication are limited in upland rice production centers, making the areas highly isolated and not easily reached by extension officers.

Cultural practices for sustainable upland farming

Intensive cultivation of uplands must start with the implementation of soil and water conservation measures. The central idea is that intensive cultivation can be applied if land is treated and protected from soil erosion.

The farming systems research of the Upland Agriculture and Conservation Project introduced treatment-oriented land capability classification using land slope, depth of soil profile, and soil erodibility as criteria (Fagi and Mackie 1988). The scheme used in Brantas (East Java) and Jratunseluna (Central Java) watersheds is shown in Table 2. Hedgerow or alley cropping can be used in all land categories if farmers cannot afford to construct conventional conservation techniques. It takes, however, 2-3 yr to establish natural bench terraces by alley cropping (Baharsjah and Fagi 1995).

Table 2. The scheme of treatment-oriented land capability classification recommended in Brantas and Jratunseluna watersheds through Farming Systems Research, Upland Agriculture and Conservation Project (*Source:* Fagi and Mackie 1988).

Slope (%)	Depth of soil profile					
	>90 cm		90-40 cm		<40 cm	
	Soil erodability					
	Low	High	Low	High	Low	High
<15	B[a]	B	B	B	C[b]	C
15-30	B	B	B	C	C	C
30-45	B	C	C	C	C	D[c]
>45	D	D	D	D	D	D

[a]B: bench terrace, food crops (75%), tree crops (25%); tree crops are planted on terrace edges, fodder crops on terrace edges and risers; small ruminants and poultry are raised in homestead.
[b]C: ridge terrace, food crops (25%), tree crops (75%); others are the same as B.
[c]D: individual terrace; tree crops (100%) are planted on steps according to contour lines in combination with fodder crops.

Soil fertility management

Organic fertilizer application. Alley cropping is becoming popular. Some of its most frequently mentioned advantages are

- tree pruning is a source of nutrient,
- organic compounds from green leaves of certain species can detoxify toxic Al in the soil,
- green leaves can be used as fodder for livestock,
- thick twigs and branches can be used as fuel,
- rows of alley-cropped trees planted on contour lines on sloping lands can prevent soil erosion, and
- it is a low-input system of soil management for the resource-poor farmers.

Effandi et al (1988) and Zaini et al (1993) reported several benefits of alley cropping (one that uses several tree species) on green manure production, on soil properties, and on crop yield (Table 3). *Flemingia* produced the highest amount of green manure, followed by *Calliandra* and *Leucaena*. Continuous application of green manure for five seasons increased soil porosity and consequently decreased bulk density. Organic matter contents proportionally increased with the amount of green manure applied. Green manure application improved N content and water retention.

Table 3. Effect of alley cropping on yield of soybean and soil properties after five seasons (*Source:* Effandi et al 1988).

Hedgerow plant	Green manure fresh weight ($t\ ha^{-1}\ yr^{-1}$)	Grain yield ($t\ ha^{-1}$)	Bulk density ($g\ cm^{-3}$)	Drainage pores (%)		Available water (%)	C (%)	N (%)
				Rapid	Slow			
Leucaena	6.09	0.57	1.12	20.4	4.6	12.5	2.63	0.18
Flemingia	25.80	0.66	1.08	25.6	5.3	14.6	2.85	0.20
Calliandra	25.79	0.67	1.10	24.1	4.8	13.6	2.73	0.16
No hedgerow	-	0.41	1.28	13.7	4.0	11.7	1.31	0.07

Swift et al (1994) suggest the application of fertilizer, lime, and organic matter to upland ecosystems in the humid tropics to maintain and improve soil fertility. A study conducted at Sitiung, West Sumatra, for three crop seasons concluded that (a) Al saturation decreased with increased level of limestone applied; (b) organic C and N contents increased with increased amount of limestone; (c) hedgerow crops that produce green manure further decreased Al saturation and increased organic C and N contents, compared with control plots; and (d) grain yield of upland rice in the third season dropped when no green manure was applied (Table 4).

Inorganic fertilizer application. Direct application of phosphate rock (PR) is a low-cost technology because it is much cheaper than triple superphosphate (TSP). PR was as effective as TSP in increasing grain yields of Cisadane and IR36 rice varieties planted in acid soil under irrigated condition (Adiningsih et al 1988).

A six-crop-season experiment was conducted on Ultisols in North Lampung, Sumatra, to compare the effects of PR, TSP, and diammonium phosphate (DAP) on maize yield. PR gave the lowest maize yield in the first season, but its effectiveness increased with time. Its relative agronomic efficiency (RAE) was similar to that of TSP in the fifth and sixth seasons. The RAE of DAP was the highest in the second season, but it gradually dropped to the lowest level in the sixth season (Table 5).

Table 4. Effects of various hedgerow plants and fertilizer application on grain yield of upland rice, organic C and N contents, and Al saturation.[a] Sitiung, West Sumatra, 1992-93 wet season (*Source:* Zaini et al 1993).

Hedgerow plant/fertilizer rate (kg ha^{-1})	1st crop upland rice (t ha^{-1})	3rd crop upland rice (t ha^{-1})	Total N (%)	Organic C (%)	Al saturation (%)
Flemingia					
No input	1.78	1.62	0.22	1.55	54
22.5 N+20 P+25 K+500 limestone	1.96	1.86	0.24	1.52	46
45.0 N+40 P+50 K +1000 limestone	1.68	2.116	0.25	1.81	34
Av	1.81				
Pineapple					
No input	1.64	1.23	0.19	1.32	58
22.5 N+20 P+25 K+500 limestone	1.97	1.486	0.19	1.81	56
45.0 N+40 P+50 K +1000 limestone	2.07	1.54	0.21	1.73	48
Av	1.89				
Banana + Flemingia					
No input	1.80	1.548	0.21	1.89	56
22.5 N+20 P+25 K+500 limestone	1.74	1.829	0.23	1.62	51
45.0 N+40 P+50 K +1000 limestone	1.82	2.117	0.22	1.84	42
Av	1.79				
Coffee + P. purpureum					
No input	1.75	1.10	0.20	1.62	59
22.5 N+20 P+25 K+500 limestone	1.80	1.376	0.20	2.07	54
45.0 N+40 P+50 K +1000 limestone	2.75	1.44	0.22	1.89	49
Av	1.77				
No hedgerow					
No input	1.79	0.837	0.18	1.84	62
22.5 N+20 P+25 K+500 limestone	1.85	1.11	0.17	1.86	56
45.0 N+40 P+50 K +1000 limestone	1.80	1.255	0.19	1.86	49
Av	1.81				

[a]Initial organic C = 2.19%; total N = 0.21%; Al saturation = 61%.

Table 5. Relative agronomic efficiency (%) of P fertilizers for six crop seasons (1973-79) on Ultisols in North Lampung, Sumatra (*Source*: Hakim 1982).

P fertilizer [a]	Season					
	1st	2nd	3rd	4th	5th	6th
TSP	100	100	100	100	100	100
RP	67	91	94	96	101	119
DAP	97	110	90	95	94	86

[a]Fertilizers were applied once to the first crop. TSP=triple superphosphate, RP=rock phosphate, DAP=diammonium phosphate.

Crop management

Despite the problems encountered, Indonesia's uplands have enormous potential. The four largest islands--Sumatra, Kalimantan, Sulawesi, and Irian Jaya--together have 38 million ha suited for upland farming. With appropriate, sustainable farming practices, these areas could make a major contribution to Indonesia's food production (MOA 1996). Previous field studies demonstrated that the use of improved cultivars in combination with liming, PR, and organic matter applications converted infertile acid soil to fertile soil. Therefore, improved varieties are a prerequisite for upland area intensification.

Improved high-yielding varieties. Rice is the only major food crop that grows both in standing water and on moist upland soil. Therefore, upland rice is the first choice of farmers cropping uplands when water is available in the wet season. The use of high-yielding varieties is the most efficient among other component technologies. Significant progress has been achieved in varietal improvement of upland rice. Four improved upland rice varieties released in 1994 have very specific characteristics that make them adaptable to specific physicochemical and biological constraints (Table 6). Gajah Mungkur and Kalimutu are suitable for cropping in areas with a short rainy season. Way Rarem is recommended for acid soils with high active Al and Fe. Jatiluhur can be planted under estate or forest crops.

The adaptability of Way Rarem and Jatiluhur to shading was tested in rubber plantations at different stages of growth in South Sumatra. Local variety Siam was used for comparison. Table 7 shows that Way Rarem was the most adaptable to shading. With proper soil and fertilizer management, Way Rarem yielded 3.25 t ha^{-1} at 15% shading intensity. Jatiluhur and Siam gave 2.4 and 2.0 t ha^{-1}, respectively. Grain yields of all varieties dropped sharply as shading intensity increased to 40-60%. Way Rarem was still superior even at high shading intensity. Because these varieties are shade-tolerant, they are particularly suited for planting between rows of young tree crops, such as rubber, oil palm, and cashew. Economic analysis shows that this cropping system breaks even, economically, if the rice yield is 1.7 t ha^{-1}. Each of those varieties yields 2.5 t ha^{-1} or more.

Rice-based farming systems. Upland rice grown on dry soil and often on slopes does not enjoy the advantage of standing water. Upland rice yields are therefore low, typically less than half of wetland rice yields. Realizing the uncertainty of upland environment, farmers

Table 6. Character traits of improved upland rice varieties released in Indonesia's upland areas, 1994.

Character trait	Gajah Mungkur	Kalimutu	Way Rarem	Jatiluhur
Av yield (t ha^{-1})	2.5	2.5	3.0-4.0	2.5-3.5
Maturity (d)	90-95	90-95	100-110	110-115
Plant height (cm)	90-100	105-110	95-100	95-100
Amylose content (%)	23.2	23.5	27.0	27.6
Other traits	Resistant to blast; moderately tolerant of drought	Moderately resistant to blast; moderately tolerant of drought	Resistant to blast and brown spot; moderately resistant to Al and Fe toxicity	Resistant to blast; tolerant of shade; resistant to seedling fly

Table 7. Effect of shading intensity of rubber plants on grain yields of upland rice varieties in an Ultisol (*Source:* Wibawa and Rosyid 1996).

Variety[a]	Age of rubber plant (yr)	Shading intensity (%)	Grain yield (14% moisture; t ha^{-1})
Way Rarem	2	15	3.25
Jatiluhur			2.40
Siam			2.00
Way Rarem	3-4	40-50	0.70
Jatiluhur			0.60
Siam			No yield
Way Rarem	4-5	55-60	0.60
Jatiluhur			0.25
Siam			No yield

[a]All varieties received 100 kg urea, 125 kg TSP, and 75 kg KCl ha^{-1}. Urea was applied twice, 2 wk after seeding (first weeding) at 50 kg ha^{-1} and 6 wk after seeding (second weeding) at 50 kg ha^{-1}; all the phosphorus and potassium were applied at seeding.

grow more than just rice. Therefore, uplands are more diversified than lowland area. The traditional upland rice-based farming systems are, in most cases, less productive.

Field trials to develop strategies for farming system technology, which include cultural practices, crop management, and livestock raising, were conducted in resettlement areas in Batumarta Dua (South Sumatra) and Air Manganyau (Bengkulu). In these areas, the transmigrants grow food crops and rubber and raise livestock. The data from the trials (Table 8) may be interpreted as follows:

- Farm incomes of cooperators who grew palawija crops and upland rice were more stable. In the 1992-93 crop seasons, noncooperators who grew only upland rice lost their money because of poor growth of upland rice. Cooperators experienced the same with their upland rice, but palawija crops compensated for it.
- Cropping systems technology is location-specific. The better environment at Air Manganyau allows farmers to grow red pepper. Farm incomes of cooperators and noncooperators at Air Manganyau were higher than those at Batumarta Dua because farmers grew red pepper.
- The average production costs (1992-93 to 1994-95) of the introduced technology were comparable with those of the traditional technology.

Table 8. Performance of farming systems under improved and traditional technologies at two sites in the resettlement areas, October 1992-March 1995 (*Source:* AARD-IDRC 1995).

Crop season	Cooperator[a] (improved)		Noncooperator[b] (traditional)	
	Batumarta Dua	Air Manganyau	Batumarta Dua	Air Manganyau
		Production value (× Rp 1000)		
1992-93	1,342.60	2,640.00	739.00	2,526.80
1993-94	1,249.30	939.70	1,303.30	467.00
1994-95	1,054.60	1,321.40	731.90	857.30
		Production cost (× Rp 1000)		
1992-93	1,176.00	1,275.30	949.70	2,078.80
1993-94	899.90	593.00	803.20	411.30
1994-95	766.10	616.00	512.30	536.30
		Net value (× Rp 1000)		
1992-93	166.60	1,364.70	-210.70	448.00
1993-94	349.40	346.70	227.10	55.70
1994-95	288.50	705.40	219.60	321.00
Av food crops	268.20	805.60	78.70	274.90
Av livestock	416.00	474.00	31.10	119.00
Av rubbers	2,553.10	1,431.20	2,249.30	1,163.30
Total	3,237.30	2,710.80	2,359.10	1,557.20

[a]Batumarta Dua: maize + dryland rice - legumes (soybean, cowpea, mungbean).
Air Manganyau: maize + dryland rice - legumes - red pepper - red pepper.
[b]Batumarta Dua: dryland rice.
Air Manganyau: dryland rice - red pepper - red pepper.

- Cooperators maintained their rubber plants and managed their livestock better, hence cooperators gained more money than noncooperators from rubber and livestock.

Potential impact of research findings

Resource quality

Sustainability in agriculture should involve the successful management of resources for agriculture to satisfy changing human needs while maintaining or enhancing the quality of the environment and conserving the natural resources (FAO 1989). The most practical and effective, and least costly, method of conserving upland soils is the use of vegetative cover, both live and dead (Lal 1987). In addition, conservation tillage that involves crop residue management is a technology that provides a cost-effective system of controlling erosion on the highlands (Blevins and Frye 1993, Sanderas 1992).

With the increasing use of high-yielding crop varieties, disposal of crop residues has become a serious problem, especially in the intensive food crop-growing areas. A large quantity of crop residues is piled up every season and normally disposed of by burning, which causes the loss of a potential source of organic matter. Burning straw, commonly practiced in many farms in East Timor (Adnyana et al 1993), destroys most of the nitrogen and sulfur and some of the potassium, and makes silica less available (Blevins et al 1994).

Environmental quality

Acid upland soils are distributed over large areas, mainly in Sumatra, Kalimantan, and Irian Jaya, covering about 39 million ha. These areas receive adequate rainfall for crop production. Developing these soils for crop production can reduce pressure on intensive rice areas, forests, and fragile and marginal upland areas.

Temporary intercropping of food crops, particularly upland rice, with young perennial trees is a common practice in upland areas of Sumatra, Kalimantan, and Sulawesi. A survey in Sumatra showed a significant share of slash-and-burn land clearing of natural forests and secondary vegetation involving smallholder, old-jungle rubber bushes, thereby contributing to greenhouse gas emission. Table 9 gives estimates of the C emissions saved per year if rubber woods are utilized rather than burned at the time of rubber replanting.

Upland annual crops - perennial crops intercropping that use upland rice as a cover crop in the wet season in young rubber plantations in resettlement areas is not only agronomically suitable and economically attractive but also environment-friendly because rubber tree - upland rice intercropping absorbs a considerable amount of methane (AARD-ICRAF 1995).

Production potential

Planted areas of rubber and coconut smallholders in Indonesia are 2.7 and 3.5 million ha, respectively (CBS 1995). If we assume that about 5% of these areas can be replanted each year and that 90% of the farmers will be growing high-yielding upland rice varieties for the first year, about 0.28 million ha of land will be cultivated per year. With moderate inputs, grain production could be about 0.56 million t.

Total harvested area of upland rice in Indonesia is about 1.2 million ha (CBS 1995). Total area planted to high-yielding varieties is only about 5-10%, which indicates that the potential to increase yield productivity is still high. However, efforts should be made to improve and sustain soil fertility. Agronomic research has to be conducted in specific

Table 9. Estimate of CO_2 and CH_4 emission (t ha^{-1}) saved per year when rubber wood is utilized at the time of rubber replanting (*Source*: AARD-ICRAF 1995).

Type of C	C emission when rubber wood is burned in situ[a]	C emission when rubber wood is burned ex situ[b]	C emission when rubber wood is utilized[c]
CO_2	384	269	115
CH_4	7	5	2
CO_2-equivalent in radiative forcing	481	344	137

[a] Old rubber trees are cut and the woods are burned in situ.
[b] Old rubber trees are cut and the woods are burned away from the field (ex situ).
[c] Old rubber trees are cut and the woods are processed outside for several purposes.

locations. Such research requires efforts of local institutions, and every province in Indonesia should have the capability to conduct such research. Adoption of technologies was facilitated by the development of infrastructure and by a decision to strengthen adaptive research in the region through the Assessment Institute for Agricultural Technology established in 1995.

References

AARD-ICRAF--Agency for Agricultural Research and Development-International Center for Research in Agroforestry) (1995) Summary report of Phase I. Alternatives to Slash-and-Burn, Indonesia. 154 p.

AARD-IDRC--Agency for Agricultural Research and Development-International Development Research Centre (1995) Crop-animal systems research. Final report for Phase III (1992/93-1994/95).

Adiningsih J S, Sudjadi M, Setyorini S (1988) Overcoming soil fertility constraints in acid upland soils for food crops based farming systems in Indonesia. Indonesian Agric. Res. Dev. J. 10:49-58

Adnyana M O, Zaini Z, Sukma D, dan A. Drajat (1993) Potensi Pengembangan Terigu di Propinsi Timor Timur: Analisis keuntungan komparatif. Badan Litbang Pertanian, Jakarta.

Arya L M, Dierolf T S, Rusman B, Sofyan A, Widjaja-Adhi I P G (1992) Soil structure effects on hydrologic processes and crop water availability in Ultisols and Oxisols of Sitiung, Indonesia. Soil Management CRSP Bulletin No. 92-03. 51 p.

Baharsjah J S, Fagi A M (1995) Agronomic contribution to poverty alleviation in resettlement areas. Paper presented at the International Seminar on Population Resettlement for Poverty Alleviation, 27 Nov-1 Dec 1995, Jakarta.

Blair G, Levroy R, Konboon Y, Wonprasaid S, Naklang K (1995) Carbon and nutrient pools in rice cropping systems. Pages 161-172 in Fragile lives in fragile ecosystems. Proceedings of the International Rice Research Conference, 13-17 Feb 1995, Manila, Philippines. International Rice Research Institute, P.O. Box 933, Manila, Philippines.

Blevins R L, Frye W W (1993) Conservation tillage: an ecological approach to soil management. Adv. Agron. 51:33-78.

Blevins R L, Frye W W, Wagger M G, Tyler D D (1994) Residue management strategies for the southeast. Pages 63-74 in Hatfield J L, Stewart B A (Eds.) Crop residue management. Lewis Publishing, London.

CBS--Central Bureau of Statistics (1995) Statistical yearbook of Indonesia, 1994. Jakarta.

Effandi D, Suwardjo H, Abdurachman A (1988) Penelitian alley cropping di Kuamang Kuning, Jambi. Hasil penelitian pola usahatani terpadu di daerah transmigrasi Kuamang Kuning. Pusat Penelitian Tanah, Bogor.

Fagi AM, Mackie C (1988) Watershed management in Java's uplands: past experience and future directions. Pages 254-264 in Moldenhauer W C, Hudson N W (Eds.) Conservation farming on steep lands. Soil and Water Conservation Society, Ankeny, Iowa.

FAO--Food and Agriculture Organization (1989) Sustainable agriculture production: for international agriculture research. Report of the Technical Advisory Committee, Consultative Group on International Agricultural Research, Washington, D.C.

Hakim N (1982) Pengaruh pemberian pupuk hijau dan kapur pada podsolik merah kuning terhadap ketersediaan fosfor dan produksi tanaman jagung *(Zea mays* L.). Ph D thesis, Fakultas Pascasarjana, Institut Pertanian Bogor.

Lal R (1987) Managing the soils of Sub-Saharan Africa. Science 236:1069-1079.

McIntosh J L (1986) Ecological issues in preproduction testing and production programs involving upland rice. Pages 461-474 in Progress in upland rice research. International Rice Research Institute, P.O. Box 933, Manila, Philippines.

MOA--Ministry of Agriculture (1996) Upland rice in Indonesia: upland for life. World Food Summit. 26 p.

Sanderas D W (1992) International activities in assessing and monitoring soil degradation. Am. Alternative Agric. 7 91, 2: 195-201.

Santoso D, Sibuea L H (1994) Effect of soil and crop management on chemistry of a Lampung upland soil, Indonesia. Pemberitaan Penelitian Tanah dan Pupuk No. 3:1-7. Pusat Penelitian Tanah, Bogor.

Swift M J, Bohren L, Carter S E, Izac A M, Woomer P L (1994) Biological management of tropical soils: integrating process research and farm practice. Pages 209-228 in Woomer P L, Swift M K (Eds.) The biological management of tropical soil fertility. John Wiley & Sons, United Kingdom.

Wibawa G, Rosyid M J (1996) Budi daya padi gogo di lahan kering perkebunan karet untuk meningkatkan swasembada beras. Makalah poster disajikan pada Simposium dan Kongres VI PERAGI. Jakarta 25-27 Juni 1996.

Zaini Z, Abdullah S, Erythriana (1993) Alley cropping of upland rice and soybean with some hedgerow plants on acid sloping lands of Sitiung. In Adnyana M O,. Djauhari A, Zaini Z, Partohardjono S (Eds.) Farming system technologies for food crops. Central Research Institute for Food Crops, Bogor, Indonesia.

Zaini Z, Abdullah S, Erythriana (1994) The effect of lime sources and phosphorus rates on the amelioration of subsoil acidity. Indonesia J. Crop Sci. 9(2): 40-48.

The role of legume fallows in intensified upland rice-based systems of West Africa

M. BECKER and D. JOHNSON
West Africa Rice Development Association (WARDA), BP 2551 Bouaké 01, Côte d'Ivoire

Key words: fallow, legumes, nitrogen fixation, rice, weeds, West Africa

Abstract
Traditional upland rice-based cropping systems in West Africa rely on periods of fallow to restore soil fertility and prevent the buildup of insect pests and weeds. Demographic growth and increased demand for land are forcing many farmers to intensify their rice production systems. Declining fallow length and increasing number of crops before leaving the land to extended fallow significantly reduce yield. Promising cropping system options include the use of site-specific, weed-suppressing, multipurpose cover legumes as short-duration fallows. Poor adoption of legume technology and the near complete absence of research on fallow management in the extremely diverse upland rice-based systems of West Africa require a multiscale approach for generating and extrapolating fallow technology. Constraints to rice production and the yield gaps related to intensification were determined in 190 farmers' fields in three agroecological zones (farm level). Nitrogen accumulation and weed suppression were evaluated in 54 legume accessions grown for 6 mo during the dry season, under a range of hydrological and soil conditions (plot level). To increase benefits from improved fallow technology, the timing of legume establishment in relation to rice and the effect of removing, burning, mulching, or incorporating fallow residues before the rice crop on crop and weed growth were determined. Farmers' reaction to improved legume fallows was evaluated in group interviews. Best-bet technologies for given rice-based systems are being evaluated at the village level throughout the region in the framework of a regional research network (Rice Cropping Systems Task Force). Legume fallows appear to offer the potential to sustain rice yields under intensified cropping. Absolute effects varied as a function of site, legume species, and management practice. Weed control and multiple-use options are important determinants of potential legume technology adoption. Preferences for various legume phenotypes and management practices depend on farmers' resources and the production system.

Introduction

In contrast to Asia, rice in Africa is largely grown in the uplands. In West Africa, the uplands are the most important rice-growing environment (Terry et al 1994) both in terms of area (60%) and regional production (40%). The majority of food crops, including rice, are produced in extensive production systems wherein farmers traditionally rely on extended periods of fallow to restore soil fertility and to control insect pests and weeds (Johnson and Adesina 1993, Roose 1994). However, current demographic pressure forces farmers to intensify their land use in an unprecedented way (Becker et al 1995a). Surveys of rice-based systems in West Africa indicate a reduction from about 12-15 yr of secondary forest fallow in the early 1980s to 3-7 yr at present (Becker and Assigbe 1995). In fallow rotation systems of the Guinea savanna zone, the number of crop cycles between typically 12-15 yr of fallow has increased from 2-3 to 4-7 during the same period. Finally, in cash-crop rotations (mainly cotton- and soybean-based), sedentary production systems without extended fallow periods are emerging in some areas (Le Roy 1995). Cropping intensification of upland ecosystems, without adaptation of management practices to the new production objectives, occurs at the expense of the resource base

V. Balasubramanian, J. K. Ladha and G. L. Denning (eds.), Resource Management in Rice Systems: Nutrients, 105–120.

quality. Loss of soil nutrients is a major factor contributing to the degradation of the African resource base, and 45.4 million ha were estimated to have come under moderate to severe degradation between 1945 and 1990 due to inappropriate management and nutrient depletion (Oldeman et al 1991). Studies on shifting cultivation, in mainly maize-based systems, indicate that cropping intensification reduces the soil organic matter content (Agboola 1994, Hien et al 1994) with an associated decline in soil N (Gigou 1992) and increases in weed infestation (Nye and Greenland 1960, de Rouw 1994 and 1995, Heinrichs et al 1995). In cash-crop rotations, weed growth and nutrient deficiencies can partially be controlled by herbicides and mineral fertilizers (Adesina et al 1994). In this situation, the physical degradation (wind erosion, compaction) of the soils is rapidly emerging as the major production constraint (Wilson et al 1982, Pieri 1992, Alegre and Cassel 1994).

Given the current intensity of land use and the fragility of upland soils, production gains from subsistence food-crop agriculture are likely to be modest, despite the large area occupied by these systems. Technology generation should realistically aim at sustaining productivity gains and stabilizing intensified subsistence food-based systems. One of the most prominent cropping systems options is the improvement of the quality of fallow vegetation by introducing leguminous cover crops (Balasubramanian and Sekayange 1992, Hoefsloot et al 1993). Despite a large variability, mean N accumulation in cover legume crops is reportedly high (about 100 kg N ha^{-1}) and a major share of this N (about 70%) appears to be derived from biological N_2 fixation (BNF) (Becker et al 1995b, Peoples and Craswell 1992, Peoples et al 1995). The encouraging results of three decades of cover legume research in Asia (Yost and Evans 1986, Carangal et al 1994), Latin America (Lathwell 1980, Lobo Burle et al 1992) and Africa (Hartmans 1981, Tarawali 1991) have resulted in promotion and extension of such technologies to improve soil fertility throughout the tropics (Bunch 1990, IITA 1993, Ladha and Garrity 1994). These efforts, however, are in stark contrast to farmers' poor adoption of improved fallow technologies in tropical food-crop production systems (Becker et al 1995b). Many of the studies tended to focus on benefits of such systems in terms of soil fertility and erosion control, ignoring systems diversity, socioeconomic specifics, weeds, and farmers' perceptions and production objectives.

Worldwide, rice is grown within a wider range of conditions than any other cereal crop, and rice in West Africa is no exception (Becker and Diallo 1992, Adesina 1993). Further, very few studies on improved fallows in West Africa have been conducted on the rice-based cropping systems, as opposed to those based on maize. Identifying appropriate technologies and improving the likelihood of farmers' adopting new legume fallow options require coverage of the regional diversity in terms of biophysical and socioeconomic environments, and targeting of specific environments. The research approach consisted of the following activities:

1. on-farm analysis of fallow-related rice production constraints,
2. selection of legumes for a range of biophysical conditions,
3. adaptation of crop management practices to maximize benefits from legume fallows,

4. determination of socioeconomic factors that influence farmer adoption of improved fallows, and
5. farmer-participatory on-farm evaluation of best-bet technologies.

Materials and methods

Experiments were conducted in four benchmark watersheds in three agroecological zones as well as on the research farm of the West Africa Rice Development Association at Mbé near Bouaké in Côte d'Ivoire. A characterization of the five experimental sites is given in Table 1.

Constraint analysis

Diagnostic field trials were conducted in three agroecological zones of Côte d'Ivoire in 1994 and 1995 (Gagnoa, bimodal forest; Touba, derived savanna; and Boundiali, Guinea savanna). Traditional extensive upland rice production systems (>6 yr fallow, rice as first crop after fallow) were compared with intensified cropping (<5 yr fallow, rice after >3 crop cycles) in 191 farmers' fields. Weed species composition and dry biomass, and rice grain yield were determined under farmers' management, and in three superimposed researcher-managed subplots (handweeding at 28, 56, and 84 d, 15 kg ha^{-1} mineral fertilizer N application, and a combination of both). Soil samples (0-20 cm), taken at the start of the cropping season, were anaerobically incubated in the laboratory (composite samples of five soils each in five replications) and extracted after 1 and 3 mo for soil exchangeable ammonium (2 *N* KCl) to determine potential soil N-supplying capacity in a 1-mo period (NH_4^+ N at 2 mo - NH_4^+ N at 1 mo of incubation). Yield gaps were attributed to weeds and N, based on yield response to researchers' management in intensified systems.

Table 1. Characterization of experimental sites.

	Boundiali	Bouaké	Touba	Gagnoa	Man
Location					
Village	Poundiou	Mbé	Ouanenou	Guissihio	Yotta
Longitude	9.5° N	7.8° N	8.3° N	6.1° N	7.3° N
Latitude	6.3° W	5.0° W	7.7° W	6.1° W	8.2° W
Climate					
Agroecological zone	Guinea	Derived	Derived	Humid forest	Humid forest
Growing period (d)	savanna	savanna	savanna		
Annual rainfall (mm)	210	250	240	270	270
Rainfall distribution	1100-1500	900-1100	1500-2000	1100-1500	1500-2000
	Monomodal	Bimodal	Monomodal	Bimodal	Monomodal
Soil					
Soil class	Alfisol	Inceptisol	Inceptisol	Alfisol	Ultisol
Texture class	Loamy sand	Clay loam	Loam	Sandy loam	Sandy clay
pH range (KCl)	5.0-6.0	5.2-6.0	5.5-6.3	5.0-5.5	3.8-5.2

Legume adaptation

Fifty-four legume accessions (39 species of 28 genera) were selected from germplasm collections at the Centro Internacional de Agricultura Tropical (CIAT), the International Rice Research Institute (IRRI), and the International Institute of Tropical Agriculture (IITA), and from local markets and wild-growing plants in West Africa. Legumes were grown during the dry season for 6 mo between two crops of upland rice. Four sites were used in Côte d' Ivoire: Gagnoa, bimodal forest, sandy loam Alfisol; Man, monomodal forest, sandy clay Ultisol; Bouaké, derived savanna, clay loam Inceptisol; and Boundiali, Guinea savanna, loamy sand Alfisol (each with three replications). In addition, 15 accessions were grown for 3 mo (April to July) at the research farm at Mbé near Bouaké, each in three replications of 1-m-wide and 60-m-long strips along a toposequence ranging from drought-prone upland over the hydromorphic valley fringe to a flood-prone rainfed lowland, to determine legume adaptation to hydrological conditions. Dry biomass and N content of weeds and legumes were determined at bimonthly intervals (site adaptation study), or at monthly intervals (hydrological adaptation study). The percentage of N derived from BNF (Ndfa) was determined by difference methods (Hauck and Weaver 1986), using "nonfixing" *Cassia occidentalis* and *Cassia tora* as reference plants. Farmers' reactions to fallow legumes were recorded during individual and group interviews at the sites in the forest and savanna zones.

Fallow management practices

Considerations related to improved management of legume fallows included (1) stage of the rice crop when the legume is sown (maximize legume establishment for dry season survival and soil cover, while minimizing competition with the associated upland rice crop), (2) how the fallow vegetation should be cleared in preparation for the succeeding rice crop, and (3) how to fit a range of technology packages into existing rice-based cropping systems. The timing of fallow establishment in relation to rice was examined at the research farm at Bouaké (derived savanna, loamy sand Alfisol) for the legumes *Tephrosia, Stylosanthes,* and *Calopogonium*, sown at 1, 28, 56, and 112 d after sowing rice (cv WAB 56-50), in three replications. At the same study site, the effect of removing, burning, mulching, or incorporating fallow residues before the rice crop on crop and weed growth was compared for the fallow species *Calopogonium, Canavalia, Centrosema, Mucuna, Pueraria, Vigna,* and the natural weed growth in three replications. In each experiment, rice, weed, and legume growth were recorded at monthly intervals.

"Best-bet" legume species and management practices for the major rice-based systems are being evaluated in seven countries of West Africa in the framework of a regional research network, involving rice scientists from 12 national programs (Rice Cropping Systems Task Force).

Results

Analysis of biophysical constraint

Diagnostic field trials were conducted to determine if long-term productivity of upland rice can be sustained at current levels of intensification. Land-use intensification increases total upland rice production in the short term but substantially reduces plot-level grain yield (Table 2). These intensification-induced yield losses were higher in the forest (41%, p=0.03) than in the derived savanna (31%, p=0.05) or the Guinea savanna zones (20%, ns) and appeared to be related mainly to increased weed infestation (+95%) and reduced soil N-supplying capacity (-57%). The relative importance of weeds and N varied by agroecological zone. Weeds seem to be the dominant factor responsible for yield loss in the forest (68% of the yield gap), although they appeared to play a lesser role in the savanna. Short fallow fields in the forest zone were dominated by broadleaf species (mainly *Chromolaena odorata*), whereas grasses (*Imperata, Digitaria, Hackelochloa,* and *Andropogon*) were dominant in the intensively cultivated fields in the savanna zones. Cropping intensification generally reduced soil N-supplying capacity. This reduction was greatest in forest soils. However, N supply explained only 28% of the yield gap in the forest, compared with 44% in the savanna. Changes in soil physical parameters and

Table 2. Effect of intensification in upland rice-based cropping systems on yield, weed biomass, and soil N supply, and relative share of weeds and N in observed yield gaps (on-farm trials/surveys, Côte d'Ivoire, 1994-95). Extensive system: >6 yr fallow (forest); rice first crop after fallow (savanna); intensive system: <5 yr fallow (forest); rice after >3 yr of cultivation (savanna).[a]

	Gagnoa Forest zone (Ultisol)			Touba Derived savanna (Inceptisol)			Boundiali Guinea savanna (Alfisol)			Across zones
	Extensive n=35		Intensive n=39	Extensive n=25		Intensive n=27	Extensive n=28		Intensive n=35	
OBSERVATION										
Yield (Mg ha^{-1})	1.68	*	0.98	1.47	ns	1.22	1.63	ns	1.31	-38%
Weed weight (g m^{-2})[b]	15.5	*	30.2	27.4	*	43.9	36.8	ns	39.4	+67%
N supply (mg kg^{-1})[c]	29	*	16	53	ns	44	28	ns	25	-28%
YIELD GAP[d]										
due to weeds (%)	68			34			44			51
due to N (%)	28			39			24			30
unaccounted (%)	4			27			32			19

[a] *=significantly different (5%), ns=not significantly different (5%).
[b] Weed biomass at 28 and 56 d after rice seeding.
[c] Net NH_4^+ N mineralization after 2 mo of anaerobic incubation ($N_{(2\ mo)} - N_{(1\ mo)}$).
[d] Based on grain yield differences between researcher-managed and farmer-managed plots under intensive land use.

increased pest pressure appear to have played a role in the observed yield decline (data not shown). This may be particularly true for the nearly continuously cultivated mechanized rice - cash crop rotations of the savanna zone, where neither weeds nor N supply could explain more than 30% of the yield gap. These results clearly suggest that long-term upland rice productivity cannot be sustained at current levels of intensification. This is particularly true for the forest zone where more than 70% of West Africa's upland rice is grown. Improving the quality of fallow vegetation by introducing leguminous cover crops may help stabilize upland rice-based systems by reducing the buildup of weed infestations that occurs during short periods of natural fallow and by improving soil N supply through BNF.

Legume adaptation

The screening and evaluation of legume species for their growth, weed suppression, and N accumulation in different agroecological zones, hydrologies (as determined by toposequence position), soil types, and farmers' production systems indicate that no single species performs satisfactorily (>30 kg N ha^{-1} and >50% reduction in weed biomass) in all environments (Table 3; Fig. 1, 2). The forage legume species *Stylosanthes guianensis*, *Canavalia ensiformis*, *Macroptilium latyroides*, and *Aeschynomene histrix* outperformed other legumes on acid P-fixing Ultisols in the humid forest zone in terms of biomass production (p=0.05) and weed suppression (p=0.01) (Table 3). In the savanna zone, where N supply and soil cover are the prime objectives of cover-legume use, *C. ensiformis* showed the highest biomass accumulation. *Crotalaria juncea* and *Mucuna* spp. provided good ground cover and suppressed weeds, and *Cajanus cajan* showed best dry season survival. *Aeschynomene histrix, C. ensiformis,* and *S.s guianensis* appeared well adapted across sites, producing good biomass and suppressing weeds. *Cajanus cajan, C. juncea,* and *C. retusa*, which did well at three sites, failed at the forest Ultisol site. The woody shrubs *C. juncea, A. afraspera*, and *Sesbania rostrata* accumulated the largest amounts of N in rainfed lowland ecosystems (>60 kg N ha^{-1}) and effectively suppressed weed growth (Fig. 3). The creeping annual species *Mucuna* and *Vigna* produced good growth relative to the natural weed growth (>1000 kg ha^{-1}) in all hydrological situations studied.

Farmers' reactions

The results on farmers' perceptions of legumes should be considered preliminary because sample size was small and interviews were limited to two sites in Côte d' Ivoire. The farmers who rely on cutting and burning to clear land, as in the forest zone of Côte d'Ivoire, considered creeping cover legumes such as *Calopogonium, Centrosema, Mucuna,* and *Pueraria* inappropriate for their farming system because they were difficult to cut. In addition, *Centrosema, Calopogonium,* and *Pueraria* are known to these farmers as weeds that are difficult to control in upland rice. Large-seeded legumes, such as

Table 3. Maximum dry biomass accumulation (Mg ha^{-1}) by some selected short-season fallow legumes and associated weeds during 6-mo off-season growth between two crops of upland rice (four WARDA key sites[a] in Côte d'Ivoire, 1995-96 dry season).[b]

Fallow vegetation		Bouaké		Man		Boundiali		Gagnoa	
		Legume	Weeds	Legume	Weeds	Legume	Weeds	Legume	Weeds
Aeschynomene histrix	F[c]	3.59 cd	0.47	2.90 bc	0.64	1.94 bcd	0.09	13.68 b	0.00
Arachis hypogaea	F/G	2.64 de	0.61	0.65	1.35	0.99 d	0.19	3.66 e	2.16
Cajanus cajan	G	6.69 b	0.21	0.00 f	1.76	1.39 cd	0.28	21.11 a	0.00
Calopogonum mucunoides	GM	4.55 bcd	0.41	0.17 e	0.89	0.65	0.10	3.94 e	0.92
Canavalia ensiformis	GM	9.54 a	0.43	4.09 a	0.60	2.90 a	0.17	11.03 b	0.80
Centrosema pubescens	F	3.02 cd	0.45	0.22 e	1.55	0.39 ef	0.13	3.59 e	0.29
Clitoria ternata	F	2.75 de	0.51	0.00 f	1.84	0.66 e	0.21	0.98 ef	3.36
Crotalaria anageroides	Gm	1.16 fg	1.01	0.38 de	1.19	1.29 d	0.25	19.87 ab	0.00
Crotalaria juncea	GM	8.36 ab	0.50	0.00 f	1.76	2.46 b	0.04	6.33 cd	1.30
Dolichos lablab	F/G	3.88 cd	0.56	0.11 ef	2.08	1.28 cde	0.14	3.42 e	1.10
Macroptilium latyroides	GM	3.63 cd	0.68	1.49 cd	1.04	1.81 c	0.05	2.31 3	2.08
Macrotyloma geocarpum	G	0.83 fg	0.97	2.00 c	1.43	0.69 e	0.36	0.83 ef	3.72
Mucuna cochinchinensis	GM	5.14 c	0.35	1.77 cd	1.14	2.98 a	0.07	11.71 c	0.00
Mucuna pruriens var. *utilis*	GM	4.15 cd	0.30	0.98 d	0.86	2.80 ab	0.04	12.74 bc	0.00
Pueraria phaseoloides	GM	1.62 ef	0.41	0.51 de	1.81	0.56 ef	0.17	4.93 de	0.00
Stylosanthes guianensis	F	5.11 cd	0.58	2.99 b	0.40	1.98 bcd	0.03	16.55 b	0.00
Tephrosia villosa	F	4.16 cd	0.28	1.50 cd	1.25	0.98 de	0.14	7.53 cd	0.34
Vigna unguiculata	G/F	1.92 ef	1.16	0.74 d	1.49	0.67 e	0.10	0.31 ef	2.34
Voandzeia subterranea	G	1.44 efg	1.39	0.42 de	1.28	1.27 d	0.22	0.38 f	2.95
Weedy fallow control	-	-	1.12	-	1.70	-	0.29	-	3.59
LSD *(0.05)*		1.29	0.70	0.54	0.81	0.49	0.08	3.58	1.56

[a]Mbé: derived savanna, Inceptisol; Man: monomodal forest, Ultisol; Boundiali: Guinea savanna, Alfisol; Gagnoa: bimodal forest, Alfisol.
[b]Values followed by the same letter within one column do not differ significantly by DMRT (5%).
[c]G = grain legume, F = forage legume, GM = green manure legume.

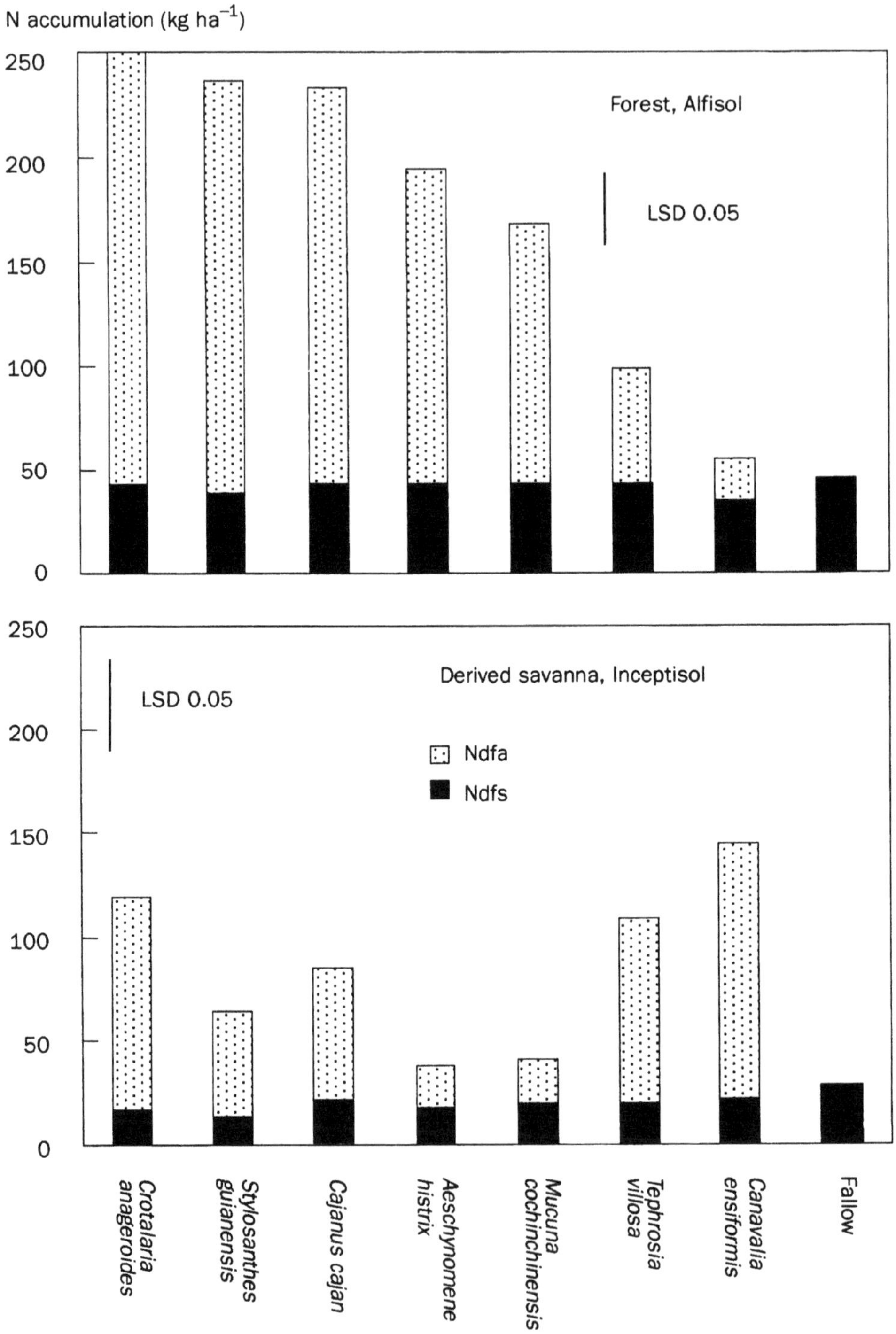

Figure 1. Nitrogen accumulation (N derived from soil [Ndfs] and N derived from atmosphere [Ndfa]) of selected cover legumes grown as short-season fallow during the 1995-96 dry season at two sites (Bouaké and Gagnoa) with favorable soil (Inceptisol, Alfisol) and hydrological conditions (bimodal rainfall pattern).

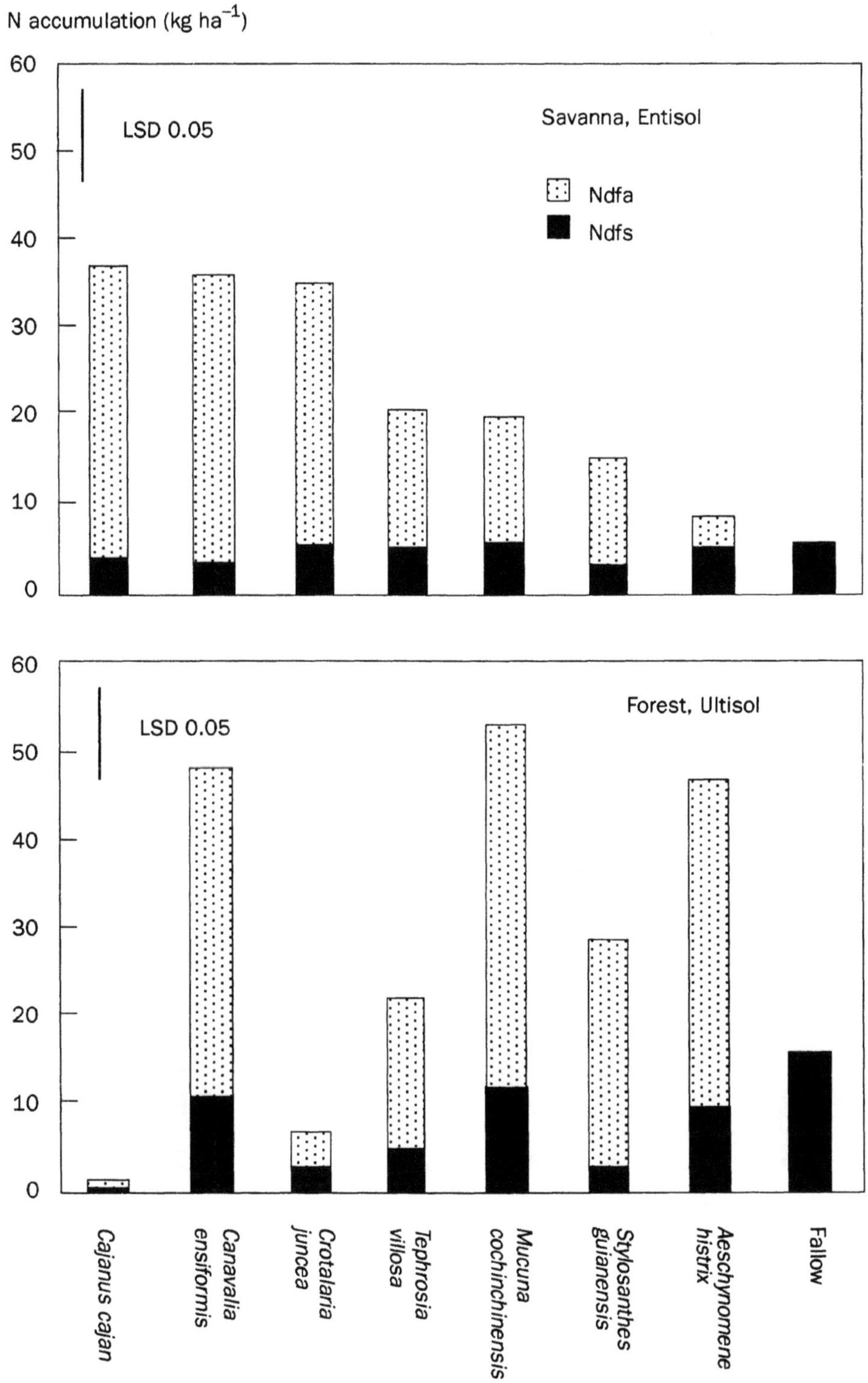

Figure 2. Nitrogen accumulation (N derived from soil [Ndfs] and N derived from atmosphere [Ndfa]) of selected cover legumes grown as short-season fallow during the 1995-96 dry season at two sites (Man and Boundiali) with unfavorable soil (Ultisol, Alfisol) and hydrological conditions (monomodal rainfall pattern).

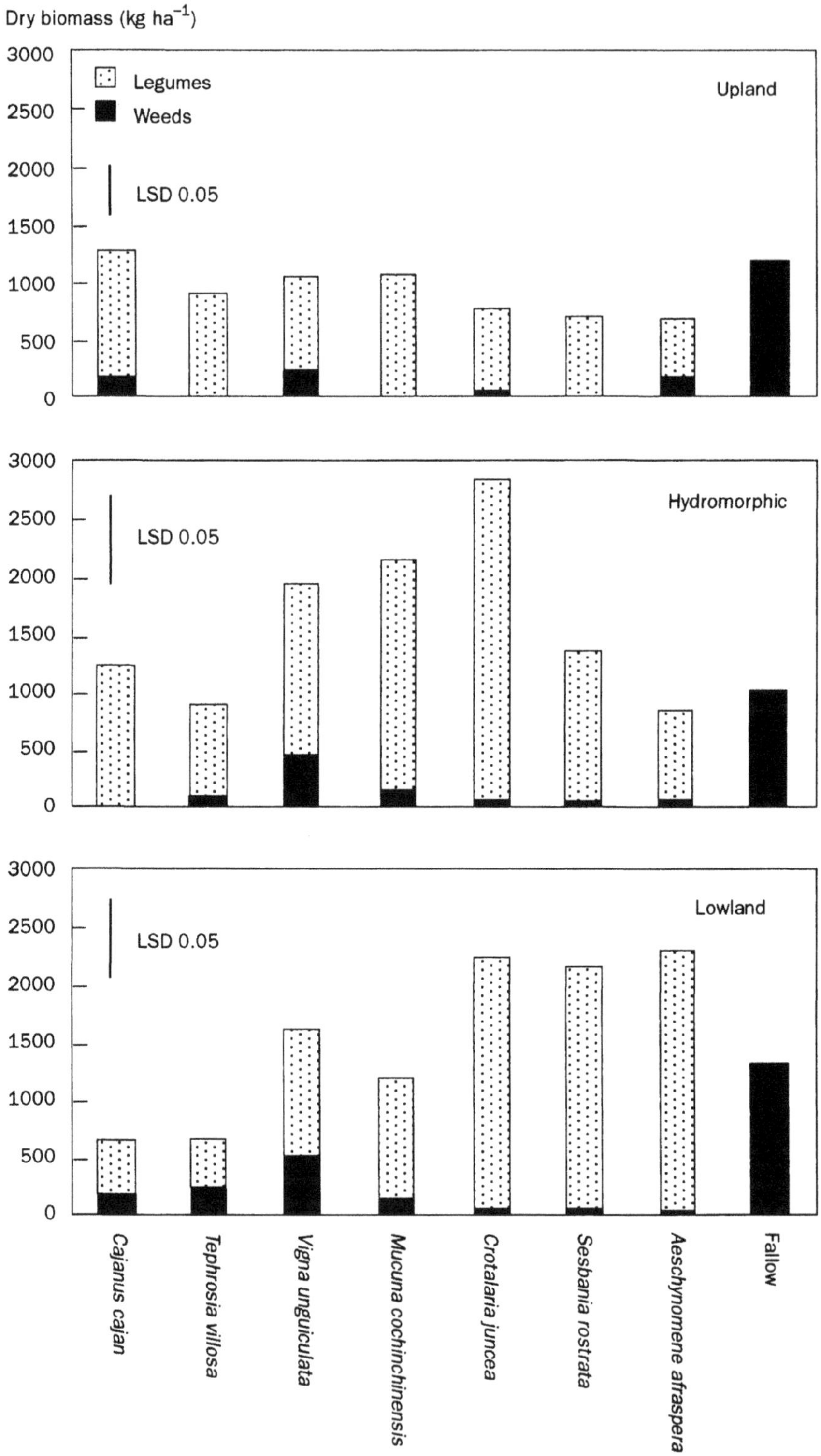

Figure 3. Dry biomass accumulation by selected 3-mo-old cover legumes and associated weed flora along a hydrological gradient (inland valley toposequence, Bouaké, April-June 1995).

Mucuna and *Canavalia,* were appreciated for their ability to suppress weeds, but were thought to take long to dry and difficult to burn once slashed. Farmers also mentioned that dense stands, mainly of *Mucuna,* harbored snakes. Erect, woody species (e.g., *Crotalaria, Cajanus*) could provide firewood and sticks for fencing (e.g., "agouti" bush rat control) and were considered more manageable by farmers who depend on slash-and-burn practices. On the other hand, in systems where farmers incorporate fallow vegetation, mainly manually (savanna upland rice-based systems and traditional or semi-improved lowlands), creeping plants were generally favored over erect ones (e.g., *Crotalaria* or flood-tolerant *Sesbania* spp.), because labor requirements for incorporation were perceived to be less. Farmers were interested in other possible uses of the legumes. Weed suppression by fallow vegetation was considered essential in both forest and savanna ecosystems, and high-quality forage during the dry season was considered desirable by most savanna rice farmers. A direct economic benefit from improved fallows would reportedly increase farmers' adoption of legume technology.

Fallow management practices

Across species, sowing legumes at the same time as rice provided best ground cover and highest N accumulation (39 kg ha^{-1}) in the fallow biomass at the onset of the dry season, but led to very low rice yields (0.3 Mg ha^{-1}) due to the interspecific competition (data not shown). Rice yields did not differ significantly when legumes were sown at 28, 56, or 112 d after rice (1.2 Mg ha^{-1}), but legume N accumulation at rice harvest increased with earliness of fallow establishment (6, 19, and 36 kg N ha^{-1} when seeded 84, 56, and 28 d after rice, respectively). The presence of legumes significantly reduced natural weed growth in rice at 56 and 84 d after seeding, compared with no-legume control plots. Modeling tools are being developed, as optimal fallow legume establishment dates are likely to vary with rice cultivar, legume species, and ecosystem.

A summary of the results of the fallow residue management study is given in Table 4. Across species, the legume fallow vegetation accumulated more than three times the N of the weedy fallow, while at the end of the fallow period the vegetation contained one-fifth of the weed biomass. Rice grain yield after legume fallow was significantly higher than that after a weedy fallow ($p=0.01$), and these tended to be highest when legume residues were incorporated. Across fallow management treatments, weed growth at 28 d after rice seeding was lower when fallow vegetation had been removed, compared with when it had been burned, mulched, or incorporated (32, 67, 57, and 76 g m^{-2}, respectively, $LSD_{0.05}$ 22).

Based on a preliminary analysis of the results on legume performance, site adaptation, and farmers' adoption criteria, we propose a number of "best-bet" legume scenarios for the major upland rice-based cropping systems in West Africa. Major rice-growing environments with their associated farming systems characteristics, their relative importance in Côte d'Ivoire, and some of the proposed fallow legume species and associated management practices are presented in Table 5.

Table 4. Effect of short-season fallow vegetation (Oct 1994-May 1995) and residue management on weed infestation and grain yield of upland rice (field experiment, Bouaké, 1995).

	Fallow vegetation		
	Weed fallow	Legume fallow[a]	LSD (0.05)
N accumulation (kg N ha^{-1}) before rice seeding	26	81	17
Mean weed biomas (g m^{-2}) across residue treatments	126	21	26
Rice grain yield (Mg ha^{-1})			
Residues removed	0.34	0.88	0.38
Residues burned	0.40	0.93	ns
Residues mulched	0.29	1.05	0.55
Residues incorporated	0.26	1.15	0.34
Mean yield across treatments	0.32	1.01	0.33
LSD (0.05)	ns	0.23	

[a]Mean of five legume species (*Calopogonium, Canavalia, Centrosema, Mucuna,* and *Pueraria*).

Table 5. Major upland rice-based farming systems in West Africa, their relative importance in Côte d'Ivoire, and suggested fallow legume species and associated management practices based on preliminary evaluation.

Agroecological zone	Systems characteristics	Share (%) of rice growing area in Côte d'Ivoire[a] (68% of total rice area)	Best-bet legume species	Associated management practices
Monomodal forest	Acid subsistence, slash-and-burn, broadcast	10	*Aeschynomene histrix* *Stylosanthes guianensis* *Tephrosia villosa*	Relay seeding
Bimodal forest	Neutral soils, subsistence, slash-and-burn, dibble	22	*Crotalaria anageroides* *Cajanus cajan* *Tephrosia villosa*	Post-rice seeding, burning, mulching
Forest	Valley fringes and hydromorphic areas	12	*Aeschynomene afraspera* *Crotalaria juncea*	Post-rice seeding, incorporation
Forest	Cash crop-based systems	< 1	*Mucuna pruriens* *Canavalia ensiformis*	Relay seeding, mulching, incorporation
Savanna	Subsistence/slash-and-burn	2	*Crotalaria juncea* *Cajanus cajan* *Tephrosia villosa*	Relay seeding burning
Savanna	Subsistence/Manual tillage	4	*Canavalia ensiformis* *Mucuna pruriens*	Relay seeding incorporation
Savanna	Animal traction/cash crop rotations	15	*Stylosanthes guianensis* *Aeschynomene histrix*	Relay seeding grazing, incorporation
Savanna	Valley fringes and hydromorphic areas, manual	2	*Calopogonium mucunoides* *Pueraria phaseoloides*	Post-rice seeding incorporation

[a]Adapted from Becker and Diallo (1992).

Discussion

Weeds, N, and labor are important biophysical and socioeconomic production constraints in upland rice rotations under shortened fallow. Important future problems in forest fallow and savanna cash crop rotations may include P deficiency (Sanchez and Salinas 1981) and soil physical degradation, including erosion (Pieri 1992). Legumes have potential to sustain productivity gains in intensified systems (Lathwell 1980, Becker et al 1995b, Peoples et al 1995). However, new fallow management options must consider the diversity of cropping systems, management practices, and farmers' production objectives. Our studies have shown that no single legume is adapted to the wide range of biophysical and socioeconomic rice production environments of West Africa. There is a clear need to exploit the existing genetic diversity and to select appropriate legumes for given environment and farmers' production objectives. Creeping legumes such as *Calopogonium, Centrosema, Mucuna*, and *Pueraria* have attracted much research attention (Agboola and Fayemi 1971, Akobundu 1993, Tarawali and Ogunbile 1995). However, farmers who rely on burning to clear land in the forest zone have not adopted such legumes because they do not fit their farming system. The use of creeping legumes that are fit for human consumption (e.g., *Mucuna* and *Canavalia*; Osei-Bonsu et al 1995) may partially overcome this problem. In mechanized systems in the savanna zone where soil erosion and soil physical degradation pose a serious threat to system sustainability (Pieri 1992), and where cattle have to be fed during the dry season (Hoefsloot et al 1993), forage legumes may provide an acceptable alternative to weedy fallows. Crop survival and good soil cover during the dry season (Lobo Burle et al 1992), however, may necessitate relay establishment of the legume crop (Balasubramanian and Blaise 1993). However, such practices may not be acceptable in mixed cropping systems where the cover legume would compete for land with associated noncereal crops (Milton 1989). Similarly, benefits from modified residue management practices (mulching and incorporation instead of residue burning) have to be evaluated in relation to their practical feasibility (establishment, farm equipment, termites, etc.). Further, the positive effects of legume fallows on populations of nematodes and insects and their natural enemies are likely to become increasingly important as rice-based systems intensify (Becker et al 1995a).

Labor is commonly the greatest constraint to production in many smallholder systems. In the traditional rice production systems, for example, returns to labor are approximately half the returns in improved systems using purchased inputs (Adesina 1994). Improvements in traditional systems must aim at increasing returns to labor. Ouattara (1994) reported that farmers in the savanna zone of Côte d'Ivoire spend 408-506 h ha^{-1} hand-weeding their upland rice crop, representing 40% of total labor investment in rice. Early attempts to introduce *Mucuna* to farmers in Nigeria were rejected because the legume involved extra labor (land preparation, sowing, and fallow management) and did not fit into the traditional mixed cropping (Faulkner 1934). Similarly, Rwandan farmers considered short-season fallows less attractive, partly because of additional labor requirements to plant and manage the fallow vegetation (Balasubramanian and Blaise 1993). Weed suppression is an essential factor to the viability of improved fallow management. A substantial reduction in labor requirements (e.g., for handweeding of upland rice) with legumes as compared with weedy fallow control was reported from several Rice Cropping Systems Task Force trials throughout the region, which may justify farmers' investment in an improved legume fallow. Likewise, the ability of *Mucuna* to suppress the perennial grass *Imperata cylindrica* in maize-based cropping systems (Versteeg and Koudopon 1990, Akobundu 1993) resulted in farmers' adoption of cover legume technology in Benin where agriculture is plagued by this weed. Further, widespread adoption of *Mucuna* in maize-based cropping systems of northern Honduras appears to be partly related to the reduction in labor requirements for the control of the grass weed *Rottboellia cochinchinensis* (Triomphe 1996).

In summary, legume fallows and associated management practices must be considered within the context of the cropping systems within which they are grown (including farmers' resource base and aspirations), not in isolation. Because farmers are looking for direct returns to their investments, legumes in most situations have to perform additional functions besides being an N source, in particular, reducing labor requirements or increasing returns to labor. This may be achieved by suppressing weeds and providing additional harvestable products such as food, fodder, or fuel. Improved productivity through reduced requirement for labor and maintenance of soil fertility, without additional cash investment, is likely to achieve progress in this direction. Particular solutions, as discussed in this paper, must consider the biophysical and socioeconomic specifics of prevailing systems in order to succeed. Ongoing studies within the framework of the Task Force activities are investigating the role of legume fallows on rice pests and their natural enemies, ex-ante and ex-post economic analyses, and farmer-participatory on-farm evaluation of "best-bet" options throughout the region.

Acknowledgements

The position of D. Johnson, Natural Resources Institute, and part of this research project were funded by the Overseas Development Administration, UK.

References

Adesina A A (1993) Farm-level profitability of crops in the rice-based systems of the humid forest zone of Côte d'Ivoire. Pages 30-32 in Annual report 1993. West Africa Rice Development Association, BP 2551, Bouaké, Côte d'Ivoire.

Adesina A A (1994) Farm-level determinants of chemical fertilizer use in rice production systems in Côte d'Ivoire. Pages 33-44 in Annual report 1994. West Africa Rice Development Association ,BP 2551, Bouaké, Côte d'Ivoire.

Adesina A A, Johnson D E, Heinrichs E A (1994) Determinants of insecticide and herbicide adoption in rice-based cropping systems. Pages 38-40 in Annual report 1993. West Africa Rice Development Association, BP 2551, Bouaké, Côte d'Ivoire.

Agboola A A (1994) A recipe for continuous arable crop production in the forest zone of western Nigeria. Pages 107-120 in Sanchez P A, van Houten H (Eds.) Alternatives to slash-and-burn agriculture. International Society of Soil Science and Mexico Society of Soil Science. INEGI, CNA, Mexico.

Agboola A A, Fayemi A A A (1971) Preliminary trials on the intercropping of maize with different tropical legumes in Western Nigeria. J. Agric. Sci. Camb. 77:219-225.

Akobundu I O (1993) Integrated weed management techniques to reduce soil degradation. Pages 278-294 in Proceedings of the First International Weed Control Congress, Melbourne, Australia.

Alegre J C, Cassel D K (1994) Soil physical dynamics under slash-and-burn systems. Pages 47-62 in Sanchez P A, van Houten H (Eds.) Alternatives to slash-and-burn agriculture. International Society of Soil Science and Mexico Society of Soil Science. INEGI, CNA, Mexico.

Balasubramanian V, Blaise N K A (1993) Short season fallow management for sustainable production in Africa. In Ragland J, Lal R (Eds.) Technologies for sustainable agriculture in the tropics. ASA-Special Publication No. 56. . American Society of Agronomy, Madison, Wisconsin.

Balasubramanian V, Sekayange L. (1992) Five years of research on improved fallow in the semi-arid highlands of Rwanda. Pages 405-422 in Mulongoy K, Gueye M, Spencer D S C (Eds.) Biological nitrogen fixation and sustainability of tropical agriculture. John Wiley & Sons, Chichester.

Becker L, Diallo R (1992) Characterization and classification of rice agroecosystems in Côte d'Ivoire. West Africa Rice Development Association, BP 2551, Bouaké, Côte d'Ivoire. 135 p.

Becker M, Assigbe P (1995) Rice-based cropping systems research in West Africa. In Proceedings of the International Colloquium "Quel avenir pour les riziculture en Afrique de l'Ouest?", 4-7 Apr 1995, CNRS, CIRAD, Bordeaux, France.

Becker M, Johnson D E, Heinrichs E A, Afun K, Russell-Smith A (1995a) Effect of cropping intensification on biotic and abiotic constraints in upland rice. In Proceedings of the International Colloquium "Quel avenir pour les ricicultures en Afrique de l'Ouest?". 4-7 Apr 1995, CNRS, CIRAD, Bordeaux, France.

Becker M, Ladha J K, Ali M (1995b) Green manure technology: potential usage, limitations: a case study for lowland rice. Plant Soil 174:181-194.

Bunch R (1990) The potential of intercropped green manures in Third World villager agriculture. Conference on the Socioeconomics of Organic Agriculture. IFOAM, Hamstead Marshall, UK.

Carangal V R, Rebancos Jr. E. T, Armada E C, Tengco P L (1994) Integration of food, forage, and green manure production systems. Pages 51-65 in Ladha J K, Garrity D P (Eds.) Green manure production systems for Asian ricelands. International Rice Research Institute, PO Box 933, Manila, Philippines.

Faulkner O T (1934) Some experiments with leguminous crops at Ibadan, Southern Nigeria, 1925-1933.

Gigou J J (1992) L'azote dans les systèmes de culture du nord et du centre de la Côte d'Ivoire. Pages 377-394 in Mulongoy K, Gueye M, Spencer D S C (Eds.) Biological nirogen fixation and sustainability of tropical agriculture, John Wiley & Sons, Chichester.

Hartmans E H (1981) Land development and management in tropical Africa. International Institute of Tropical Agriculture, PMB. 5320, Ibadan, Nigeria.

Hauck R D, Weaver R W (1986) Field measurement of nitrogen fixation and denitrification. SSSA Special Publication No. 18. Soil Science Society of America, Madison, Wisconsin. 115 p.

Heinrichs E H, Johnson D E, Afun J K, Robinson D J (1995) Rice pests of shifting cultivation in Côte d'Ivoire, West Africa. Pages 537-553 in Fragile lives in fragile ecosystems. International Rice Research Institute, P.O. Box 933, Philippines.

Hien V, Sedogo P M, Lompo F (1994) Etude des effets des jachères de courte durée sur la production et l'évolution des sols dans différents systèmes de culture du Burkina Faso. In Floret C, Serpenié G (Eds.) La jachère en Afrique de l'Ouest. Edition ORSTOM, Colloques et Seminaires, Paris, France.

Hoefsloot H, van der Pol F, Roeleveld L (1993) Jachères améliorées. Options pour le developpement des systèmes de production en Afrique de l'Ouest. KIT Press, Bulletin 333. Amsterdam, The Netherlands. 87 p.

IITA--International Institute for Tropical Agriculture (1993) Improvement of soil fertility and weed suppression through legume-based technologies. Collaborative Group on Maize-based Systems Research (COMBS). IITA Research Guide 48. International Institute for Tropical Agriculture, PMB. 5320, Ibadan, Nigeria.

Johnson D E, Adesina A A (1993) Farmers' perceptions of rice weeds and control methods in Côte d'Ivoire, West Africa. Pages 1143-1148 in Proceedings of the Brighton Crop Protection Conference, Brighton, UK.

Ladha J K, Garrity D P (1994) Green manure production systems for Asian ricelands. International Rice Research Institute, P.O. Box 933, Manila, Philippines. 195 p.

Lathwell D J (1980) Legume green manures: principles for management based on recent research. Tropical Soils Bulletin No. 90-01. Raleigh, NC. 30 p.

Le Roy X (1995) Le riz des villes et le riz des champs. In Procccdings of the International Colloquium "Quel avenir pour les ricicultures en Afrique de l'Ouest?", 4-7 Apr 1995, CNRS, CIRAD, Bordeaux, France.

Lobo Burle M, Suhet A R, Cravo M S, Bowen W, Bouldin D R, Lathwell D J (1992) Legume green manures: Dry-season survival on the effects on succeeding maize crops. CRSP Bulletin No. 92-04. Raleigh, NC. 35 p.

Milton F (1989) Velvet beans: an alternative to improve small farmers' agriculture. ILEIA Newsl. 5:8-9.

Nye P H, Greenland D J (1960) The soil under shifting cultivation. Technical Communication No. 51. Commonwealth Bureau of Soils, Farnham Royal, Bucks, England. 156 p.

Oldeman L R, Hakkeling R T A, Sambroek W G (1991) World map of the status of human-induced soil degradation. ISRIC and UNEP, Wageningen, The Netherlands.

Osei-Bonsu P, Buckles D, Soza F R, Asibuo J Y (1995) Traditional food uses of *Mucuna pruriens* and *Canavalia ensiformis* in Ghana. Internal document. Centro Internacional de Mejoramiento de Maiz y Trigo, Mexico. 5 p.

Ouattara A D (1994) Analyse *ex-ante* de la rentabilité financière et des contraintes à l'adoption des nouvelles technologies rizicoles dans la region de Korhogo. Thèse de Doctorat, Faculté des Sciences Economiques, Université d'Abidjan, Côte d'Ivoire. 320 p.

Peoples M B, Herridge D F, Ladha J K (1995) Biological nitrogen fixation: an efficient source of nitrogen for sustainable agricultural production? Plant Soil 174:3-28.

Peoples M B, Craswell E T (1992) Biological nitrogen fixation: investments, expectations and actual contribution to agriculture. Plant Soil 141:13-40.

Pieri C J M G (1992) Fertility of soils: a future for farming in the West African savannah. Springer-Verlag, Berlin, Germany. 348 p.

Roose E (1994) Capacité des jachères à restaurer la fertilité des sols pauvres en zone soudano-sahélienne d'Afrique occidentale. In Floret C, Serpenié G (Eds.) La jachère en Afrique de l'Ouest. Edition ORSTOM, Colloques et Seminaires, Paris, France.

de Rouw A (1994) Effect of fire on soil, rice, weeds, and forest regrowth in a rain forest zone (Côte d'Ivoire). Catena 22:133-152.

de Rouw A (1995) The fallow period as a weed-break in shifting cultivation (tropical wet forests). Agric Ecosyst. Environ. 54:31-43.

Sanchez PA, Salinas JG (1981) Low-input technology for managing Oxisols and Ultisols in tropical America. Adv. Agron. 34:280-406.

Tarawali G (1991) The residual effects of *Stylosanthes* fodder banks on maize yield at several locations in Nigeria. Trop. Grasslands 25:26-31.

Tarawali G, Ogunbile O A (1995) Legumes for sustainable food production in semi-arid savannahs. ILEIA Newsl. 11(4):18-23

Terry E R, Matlon P J, Adesina A A (1994) Enhancing productivity in the agricultural sector: the case of rice in sub-Saharan Africa. Paper presented at the UNDP Asia-Africa Forum, 12-16 Dec 1994, Bandung, Indonesia.

Triomphe B L (1996) Seasonal nitrogen dynamics and long-term changes in soil properties under the *Mucuna*/maize cropping system on the hillsides of Northern Honduras. Ph D thesis, Cornell University, USA. 217 p.

Versteeg M N, Koudopon V (1990) *Mucuna* helps control *Imperata* in southern Benin. West Africa Farming Syst. Network Bull. 7:7-8.

Wilson G F, Lal R, Okigbo B N (1982) Effects of cover crops on soil structure and on yield of subsequent arable crops grown under strip tillage on an eroded Alfisol. Soil Tillage Res. 2:233-250.

Yost R, Evans D (1986) Green manures and legume covers in the tropics. HITAHR Research Series No 055. University of Hawaii, USA.

Soil and crop management technologies for enhancing rice production under micronutrient constraints

P. SAVITHRI, R. PERUMAL, and R. NAGARAJAN
Department of Soil Science and Agricultural Chemistry, Tamil Nadu Agricultural University, Coimbatore, India

Key words: productivity decline, micronutrient deficiency/toxicity, soil amendment, foliar spray, multinutrient supply

Abstract
Micronutrient deficiency is considered as one of the major causes of the declining productivity trends observed in rice-growing countries. The submergence created for rice cultivation influences electrochemical and biochemical reactions, and alters pH, pCO_2, and the concentration of certain ions. This environment increases the availability of Fe and Mn with concomitant decrease in Zn and Cu. It is well known that Zn deficiency is predominant in lowland ecosystems. Sodic and upland soils and calcareous coarse-textured soils with low organic matter content suffer from Fe deficiency, besides Zn and Cu deficiencies. Rice cultivars do not experience deficiency of B and Mo. The acid soils and the low-lying, poorly drained alluvial and colluvial soils are prone to Fe toxicity. Experiments in different agroecological zones all over India showed that Zn doses to correct Zn deficiency varied from 2.5 to 22 kg ha^{-1}; 5.3 kg Zn ha^{-1} proved optimum and economical, with a maximum rice yield increase of 4.8 t ha^{-1}. In the lowland ecosystem, amending the soil with the required amount of Zn before transplanting was effective and easy to adopt, compared with repeated foliar sprays of 0.5% $ZnSO_4$ or use of Zn-enriched seedlings through seed soaking in 2-4% $ZnSO_4$ solution, fertilizing the nursery with Zn, or seedling root dipping in 2% ZnO slurry. Hepta as well as monohydrated $ZnSO_4$ were better than other sources of Zn (ZnO, $ZnCl_2$ and Zn frits). The Zn-blended diammonium phosphate (Zn-DAP), superphosphate, and nitrophosphates also proved effective. The Zn-enriched organic manures (farmyard manure, green leaf manure, and coir pith compost) were found advantageous for the direct and residual crops. Zinc fertilization with an optimal dose of 25 kg $ZnSO_4$ ha^{-1} once a year yielded high economic return. A differential response of rice up to a maximum increase in yield of 4.8 t ha^{-1} was observed with the foliar spray (1-2% $FeSO_4$ solution) or soil incorporation of Fe (50 kg $FeSO_4$ ha^{-1}) with bulky organic manure (12.5 t ha^{-1}). The application of 12.5 kg $CuSO_4$ ha^{-1} ameliorated Cu deficiency and significantly enhanced rice production. Management strategies such as liming and additional multinutrient supply (P, K, Mg, Zn, Cu, and B), besides improving drainage, enhanced the rice productivity of soils prone to Fe toxicity by correcting the multinutrient deficiency syndrome.

Introduction

In India, rice is grown predominantly in the Indo-Gangetic alluvial plains (Entisol and Inceptisol); river and irrigation command areas with Vertisols in the southern region; Alfisols of the eastern, central, and southern parts of the country; and, to a limited extent, in sandy soils of the coastal region. Rice is a component of cropping systems in 11 of the 20 agroecological regions of India (Seghal et al 1992). The gross area under rice increased from 30.8 to 42.7 million ha (38.7%). Rice production rose from 20.5 million t during 1950 to 74.3 million t in 1990, a 263% increase. This increase reflects the intensification of rice production over four decades (Fig. 1), which has enhanced the demand for secondary nutrients and micronutrients from soil and other sources.

Micronutrient deficiencies are becoming serious because of escalated nutrient demand from more intensive and exploitative agriculture, coupled with use of single-nutrient fertilizers and low amounts of organic manures. Phenotypical plasticity and physiological versatility of plants buffer nutrient deficiency and toxicity to some extent, thereby

V. Balasubramanian, J. K. Ladha and G. L. Denning (eds.), Resource Management in Rice Systems: Nutrients, 121–135.

moderating the effect on crop yields. The nutrient uptake from the soil is also greatly influenced by capacity, intensity, and kinetic factors (Neue 1994).

Soil characters such as soil reaction (pH), redox potential, texture, magnitude of calcium carbonate, and organic matter are important factors that influence micronutrient transformation and availability. Sorption and desorption kinetics are linked with mineral equilibria of the nutrients in the soil (Ponnamperuma 1972, Lindsay 1972, Jahiruddin and Cresser 1990). Problem soils with constraints such as coastal and inland salinity, sodicity, acidity, and cat clays exhibit micronutrient deficiency.

In the rice ecosystem, crop residue management by recycling wastes and manures is the prime factor that controls the long-term micronutrient balance of the soil. Micronutrient cycling in soils is closely associated with organic matter turnover because it is intricately related with trace elements. Organic matter decomposition also controls solubilization of primary and secondary minerals (Anand Swarup 1991). This paper discusses the micronutrient status of Indian soils, a diagnosis of the deficiencies, and the ameliorative technologies for the management of deficiency and toxicity of important micronutrients.

Micronutrient status in rice ecosystems

The concentration of nutrients and toxins in a flooded soil differs markedly from that in an upland nonflooded soil. A flooded rice soil is a complex of an aqueous phase, a solid phase, an interchangeable gaseous phase, and various flora and fauna. The main chemical changes brought about by flooding a soil have an impact on micronutrient supply; the decrease in redox potential due to the depletion of molecular oxygen leads to reduced Fe and Mn (Ponnamperuma 1972). Submergence for 10-12 wk increased the Fe^{2+} and Mn^{2+} concentration of soil solution, regardless of the soil type -- black, red, and lateritic (Katyal 1977), calcareous (Thind and Chahal 1983), sodic, neutral (Khind et al 1987), and acidic. The concentration of water-soluble Fe rarely exceeds 0.1 mg L^{-1} at submergence, and it rises to 600 mg L^{-1} within a few weeks after flooding. The concentration of water-soluble Zn and Cu decreases in the lowland soils, and Zn deficiency is a widespread nutritional disorder of wetland rice (Neue and Mamaril 1985, Neue and Lantin 1994). The reduction of SO_4 to H_2S in flooded soils further limits the availability of Cu and Zn, with a phenomenal reduction in the uptake of these nutrients.

The limiting micronutrients in the uplands and in coarse-textured lowland soils are Fe, followed by Zn and Cu. Submerged soils especially at low pH and acid sulfate soils exhibit higher Fe concentration leading to Fe toxicity and multinutrient deficiency (Benckiser et al 1982). In the coastal acid soils, rice suffers from B toxicity (Cayton 1985).

Analyses of more than 100,000 soil samples from different agroecological zones of India showed marked Zn deficiency (41.7%) and Fe deficiency (12%); deficiency of Mn and Cu was less conspicuous (Singh and Saha 1995) (Fig. 2).

The continuous rice cultivation over centuries in certain tracts depleted available Zn and Cu and increased soluble Fe and Mn. Old Cauvery Delta soils of Thanjavur District, Tamil Nadu, which have been cultivated for centuries, exhibit Zn deficiency in more than 80% of the area. Zinc availability is not a big problem in the new delta soils of the same district, where rice cultivation is recent (Table 1). The DTPA-Fe concentration is higher in these areas.

Diagnosis of micronutrient deficiency

Micronutrient deficiencies are usually diagnosed by visual symptoms and soil and plant analysis. Identification of micronutrient deficiency symptoms is complicated by the simultaneous occurrence of multinutrient deficiency, foliar discoloration due to disease or nematode infestation, etc. It is difficult for a farmer to identify Zn deficiency symptoms, which are described as Khaira in India, Hadda in Pakistan, Taya-Taya in Philippines, Akagare type II in Japan, and bronzing in the United States.

Soil analysis is a diagnostic tool to identify deficiency, sufficiency, or toxicity of a nutrient (Gupta et al 1994). Micronutrient availability varies with soil type, texture, agro-eco-region, and nature of the extractant. Selecting suitable soil tests with an extractant that simulates the environment of root-soil activity is essential. The generally used extractants are 0.1 N and 0.05 N HCl, DTPA, DTPA-ammonium acetate, and EDTA + $(NH_4)_2$ CO_3 (Trierweiler and Lindsay 1969, Lindsay and Norvell 1978). The DTPA method is universally used to estimate available Zn, Cu, Fe, and Mn. The critical levels have been established for different micronutrients and for divergent soils by the different cooperating centers of Indian Council of Agricultural Research-coordinated micronutrient scheme. For example, the DTPA-extracted Zn varied from 0.45 to 2.0 mg kg^{-1}(a fourfold variation) in Indian soils. The calcareous soils testing less than 6.96 mg kg^{-1} DTPA-Fe

Table 1. Available micronutrient status (mg kg^{-1}) of Cauvery Delta soils of Tamil Nadu, India (*Source:* Nagarajan 1983).

Micronutrient	Old Delta (745 samples) (Centuries of rice cultivation)	New Delta (400 samples) (Past five decades of rice cultivation)
DTPA-Zn		
Range	0.11-49.3	0.08-49.4
Mean	1.56	2.22
% soils deficient	80.4	47.0
DTPA-Cu		
Range	0.15-26.0	0.20-23.2
Mean	5.9	1.97
% soils deficient	18.8	14.2
DTPA-Fe		
Range	1.0-500	2.5-270
Mean	60.4	44.2
% soils deficient	3.3	Nil
DTPA-Mn		
Range	2.50-100	1.50-50.0
Mean	44.0	14.7
% soils deficient	0.30	Nil

Table 2. Critical limits for available Zn and Fe content of rice soils of India (*Source:* Takkar et al 1989).

Soil order	Extractant	Critical limit (mg kg^{-1})
	Andhra Pradesh	
Zinc		
Vertisol-CT[a]	Dithizone and NH_4OAc	1.0
Vertisol-FT[b]		2.0
Alfisol	DTPA	0.70
Inceptisol		0.6-1.0
	Bihar	
Calcifluvent	DTPA	0.78
Chromustert		0.90
Chromustert	EDTA-NH_4OAc	2.07
	Gujarat	
Inceptisol	DTPA	0.90
Entisol		
Inceptisol	0.1 N HCl	2.10
Entisol		
	Madhya Pradesh	
Inceptisol	DTPA	0.45
Alfisol		0.60
Vertisol		
Entisol		
	Punjab	
Typic Ustochrept	DTPA	0.80
	Tamil Nadu	
Alfisol	DTPA	2.0
Vertisol	EDTA + $(NH_4)_2CO_3$	1.1
	Bihar	
Iron		
Calciorthids	DTPA	6.95
Udifluvents		

[a]CT=coarse texture. [b]FT=fine texture.

responded to Fe application (Table 2). Deficiency of Cu, Mn, and B is rare in rice; hence, less research has been done on the evaluation of extractants and fixing of critical limits for these nutrients.

Foliar analysis is an effective and rapid method of diagnosing limiting micronutrients in rice plants. Analysis of index leaves of rice at active tillering stage can predict the micronutrient-supplying power of the soil better than preplant soil analysis. The results can, however, be used to correct deficiencies of succeeding crops.

Zinc threshold values (Table 3) for the rice plant reflect differences due to plant part tested, variety, age at sampling, and location. The values were invariably higher for the whole plant, compared with the third leaf. Neue (1994) gives the values for assessing the deficiency/toxicity/sufficiency status of the rice with reference to micronutrients.

Table 3. Threshold values (mg kg^{-1}) of micronutrients for rice plants in India (*Source:* Takkar et al 1989).

Micronutrient	Variety	Plant part/sampling stage	Critical limit
Zn	IR20	Third leaf (45 DAS)[a]	9.8
		Whole plant (45 DAS)	13.0
	Kanchana	Whole plant (50 DAS)	22.0
	Sita	Whole plant (50 DAS)	20.0
	Jaya	Third leaf (30 DAS)	12.0
		Plant top (30 DAS)	16.0
	Gorima	Whole plant (30 DAS)	21.7
		Plant top (30 DAS)	15.9
Cu		Whole plant (90 DAS)	4.1
Fe	Jaya	Top half (30 DAS)	85.7
	IR8	Leaves (55 DAS)	44.0

[a]DAS = days after seeding.

Amelioration of micronutrient deficiency

Zinc

Because lowland rice experiences widespread Zn deficiency in India, fertilization is common for Zn than for other micronutrients (Bansal and Nayyar 1989, Katyal 1985, Takkar et al 1989). Many sources of Zn can be applied by different methods. Well-tested application methods for Zn include broadcast/band application, foliar spray, soaking or dusting the seed in Zn solution/dust, nursery application, and dipping the roots in Zn suspension/slurry.

Soil application of Zn. Trials in different agroecological zones of India showed that soil application of 25-50 kg $ZnSO_4$ ha^{-1} is optimum for rice (Sadana and Takkar 1983, Rajagopalan and Palanisamy 1986). Increase in rice grain yield due to Zn fertilization varied from 0.14 to 6.5 t ha^{-1} (Table 4) with a fivefold variation (0.29 - 1.4 t ha^{-1}) in average response. Zn addition increased yield by 0.2-0.5 t ha^{-1} in 37% of the trials, less than 0.2 t ha^{-1} in 28%, and 0.5-1.0 t ha^{-1} in 24%. The wide variability in the response is natural, because it is mainly influenced by the inherent Zn status of the soil. In another series of trials, a high response of 0.3 to 0.8 t ha^{-1} was obtained in a Zn-deficient soil, compared with soils having medium Zn status (0.1-0.4 t ha^{-1}) (Takkar et al 1989).

Rice response to Zn fertilization rates varied with soil texture, available Zn status, and rice variety (Table 5). The optimum level seems to be 5.6 kg Zn ha^{-1} (25 kg $ZnSO_4$ ha^{-1}). The level of Zn for sodic soil was quite high--11.0 to 22.0 kg Zn ha^{-1} (Singh and Abrol 1985, Anand Swarup 1991, Channel 1992). Although different inorganic sources of Zn ($ZnSO_4 \cdot 7H_2O$, $ZnCl_2$, ZnO, Zn frits) are available, the hepta hydrated form is the cheapest and most commonly used Zn fertilizer. It outperformed all other sources in many experiments (Table 6) (Sajwan and Lindsay 1988, Chibba et al 1989, Deb 1990).

Table 4. Response of rice to Zn fertilization in India (*Source:* Takkar et al 1989).

State[a]	No. of trials	Response range ($t\ ha^{-1}$)				
		<0.2	0.2-0.5	0.5-1.0	>1.0	Average
Andhra Pradesh	536	232	146	81	77	1.48
Bihar	807	95	343	298	71	0.54
Gujarat	67	38	11	9	9	0.70
Haryana	75	13	35	8	19	0.58
Madhya Pradesh	27	14	8	4	1	0.50
Punjab	169	69	45	31	25	0.56
Tamil Nadu	176	64	91	17	4	0.32
Uttar Pradesh	18	1	7	9	1	0.29
All states	1875	526	686	457	207	0.62
% of total	100	28	37	24	11	–

[a]More than 50% of the sites gave grain yield increase of 0.2-1.0 $t\ ha^{-1}$ due to Zn fertilization.

Table 5. Response of rice varieties ($t\ ha^{-1}$) to Zn application (*Source:* Krishnasamy et al 1994).

Variety	Zn Level ($kg\ ha^{-1}$)[a]		
	0	5.6	11.2
ADT36	5.28	5.80	5.96
		(9.7)	(12.7)
IR20	3.29	3.88	4.42
		(17.1)	(33.4)
CO 41	2.78	2.18	3.57
		(7.0)	(28.1)
Jaya	3.00	3.52	3.95
		(9.8)	(23.4)
Kanchi	5.36	6.43	6.43
		(20.0)	(20.0)

[a]Figures in parentheses indicate percentage increase over control.

Table 6. Comparative efficacy of different sources of Zn on rice grain yield ($t\ ha^{-1}$).

Source	Hissar[a]	Hyderabad[b]	Ludhiana[c,d]
Control	6.14	4.90	1.7
$ZnSO_4$	6.89	5.70	4.5
ZnO	6.66	5.40	3.6
Zn frits	–	5.70	3.3
$ZnCO_3$	–	5.50	–
LSD (5%)	0.18	0.50	0.40

[a]*Source:* Gupta et al (1994). [b]*Sources:* ICAR-coordinated Scheme on Micronutrients, Annual Reports. [c]*Source:* Nayyar et al (1990). [d]Sodic soil.

Table 7. Effect of Zn-blended fertilizers on rice yield (Zn rate = 25 $kg\ ha^{-1}$ $ZnSO_4$).

Source	Punjab[a,b]	Bhavanisagar[c]	Coimbatore[d]
Control	4.80	2.94	5.19
$ZnSO_4$	6.30	4.09	6.11
Zincated super	5.90	3.56	-
Zincated DAP	-	-	6.23
LSD (5%)	-	0.05	0.04

[a]*Source:* Nayyar et al (1990). [b]Mean of 16 trials. [c]*Sources:* Annual Reports, Micronutrient Scheme (1983-84). [d]*Source:* Ilangovan (1986).

Mixing micronutrients with N, P, and K fertilizers will ensure uniform application and avoid separate application (Deb et al 1986, Ilangovan 1986, Mortvedt 1994). The data in Table 7 show the usefulness of Zn-urea, Zn-DAP, and Zn-superphosphate, compared with sole application of $ZnSO_4$.

Soil application vs other methods of Zn application. Basal Zn application may not correct Zn deficiency in some cases. For example, application of 25 kg $ZnSO_4$ ha^{-1} in two equal splits at transplanting and tillering proved to be on a par with the same level of Zn incorporated at transplanting, while two equal split applications at transplanting and panicle initiation was not effective (Table 8). In such cases, Zn-deficient crops can be saved by foliar sprays as an emergency treatment. Results reported by Sadana and Takkar (1983), Takkar et al (1989), and Sakal et al (1993). Data in Figure 3 show that foliar spray, root dipping, and seed soaking are either inferior to or just equal to soil application. The suspension made from 1 kg ZnO (2-4%) was sufficient to treat enough seedlings for planting 1 ha, and its yield was on a par with soil application. Bulk handling of seedlings for root dipping and protecting the treated seedlings from floodwater are tedious. Root dipping practiced in the Philippines (Castro 1977) is reported to be advantageous (Sadana and Takkar 1983).

Rice yield is more limited by Zn deficiency during rabi (winter) than during kharif (monsoon) season due to variation in temperature and radiation (Fig. 1). Residual effects of Zn fertilizers are substantial, as reflected in the response of subsequent crops (Chibba et al 1989, Devarajan 1989).

Although the addition of 25 kg $ZnSO_4$ ha^{-1} for every crop led to high total grain production of 6 crops compared with other frequencies of Zn application, Zn fertilization in alternate crops was cost-effective, with optimal maintenance supply of Zn in the soil (Fig. 2).

Organic manures and micronutrient availability. Organic manures supply Zn and other micronutrients to plants. They also mobilize the native Zn through chelation and complex formation with organic ligands. Fractionation of Zn in the Zn-enriched organic manures reflected an increase in the content of complexed Zn. The positive effect of organic manure and crop residues in supplying micronutrients to rice is well documented (Dravid

Table 8. Evaluation of methods and time of $ZnSO_4$ application[a] to rice (Sakal et al 1993).

Treatment	Grain yield (t ha^{-1})		Total Zn uptake (g ha^{-1})	
	Pusa	Dholi	Pusa	Dholi
Control	3.77	3.25	229	188
Basal (100%)	5.41	3.77	350	307
Basal (50%) + at tillering (50%)	4.63	4.03	384	332
Basal (50%) + at panicle initiation (50%)	4.18	3.56	325	316
LSD (5%)	0.19	0.14	32	91

[a]Level=25 kg $ZnSO_4$ ha^{-1}.

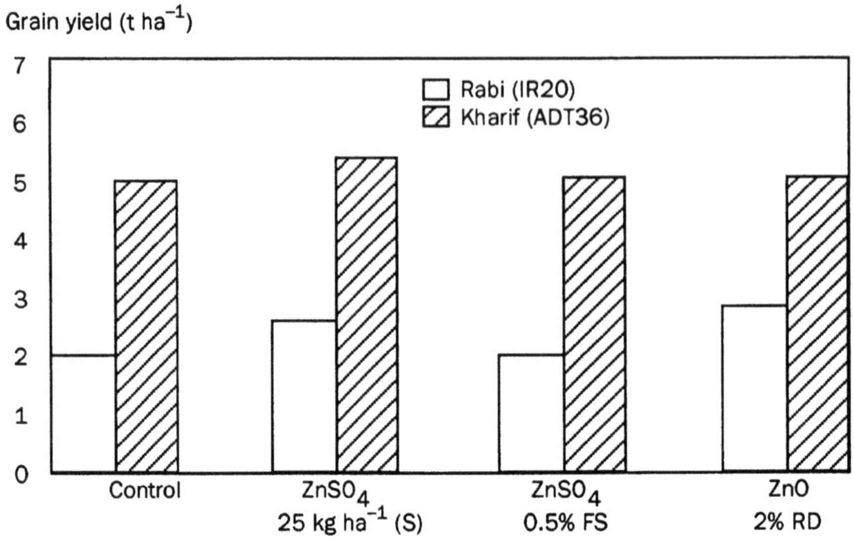

Figure 1. Efficacy of different methods of Zn application on grain yield (t ha^{-1}).

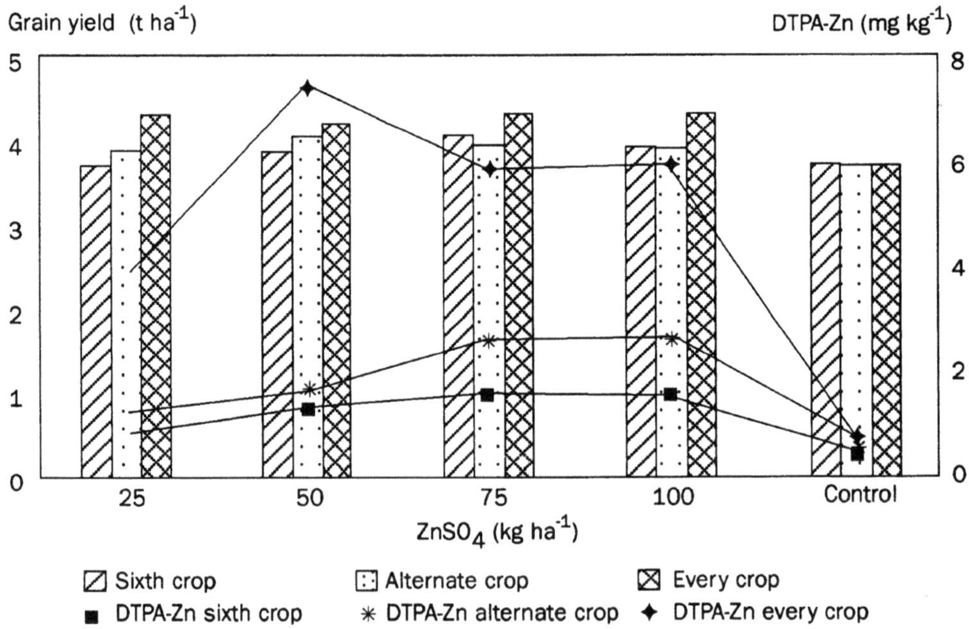

Figure 2. Frequency of Zn fertilization to rice - rice cropping system, grain yield, and DTPA-Zn (Devarajan 1989).

and Goswami 1987, Duraiswamy et al 1988). The effect of added organic manures was more pronounced in sodic soils than in other soils. The effect on Zn availability, uptake of Zn, and grain yield was more for dhaincha (*Sesbania aculeata*) than lignocellular coir pith (Channel 1992).

^{65}Zn studies confirmed the low availability and use efficiency of added soil-applied Zn (1-2%) (Rajarajan 1991). The soluble inorganic Zn, when added to soil, is readily adsorbed, fixed, and converted into less soluble compounds. Zn-enriched organic manures increase the availability of soil-applied Zn to rice (Sharma and Mitra 1990, Poongothai 1993). The use of Zn-enriched organic manures has to be encouraged due to their positive effect on yield and available Zn, with a 50% saving of Zn fertilizer (12.5 kg $ZnSO_4$ ha^{-1}) (Fig. 3). Use of synthetic Zn chelates is minimal due to their high cost.

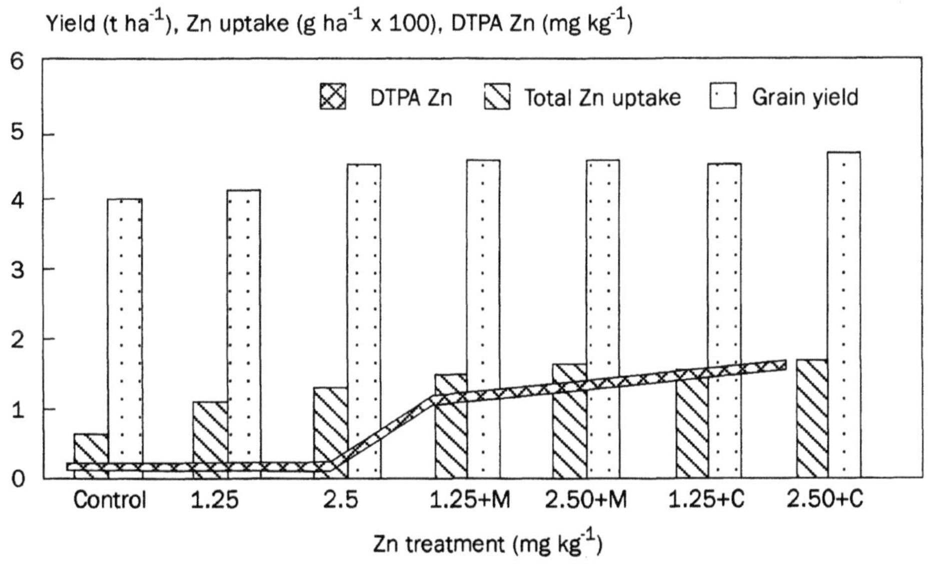

M - farmyard manure, C - composted coir pith.

Figure 3. Zinc-enriched organic manures for rice (Poongothai 1993).

Iron

Iron chlorosis in rice often occurs on coarse-textured soils with pH above 7.5, low organic matter content, low total Fe, and high $CaCO_3$. Iron deficiency is common on lands where water cannot be impounded for a longer period to create a reduced atmosphere. Moreover, the Fe nutrition of the rice crop is a complex one due to the involvement of nonspecific mechanisms (pH and exudation of organic acids [Suguira et al 1981]), and a specific one (Fe phytosiderophores). Correcting Fe chlorosis by soil application of Fe is difficult because applied Fe is readily converted into unavailable di/trivalent iron hydrated oxides (Lindsay 1972). Although soil application of $FeSO_4{\cdot}7\,H_2O$ at 10-20 kg Fe ha^{-1} (Fig. 4) helped correct Fe deficiency, the foliar spray of 1-2% $FeSO_4$ was more effective. The optimum rate varies with soil type, environment, and available Fe status, coupled with mode of application. Fifty-six trials all over India (Bihar, Gujarat, Punjab, and Tamil Nadu) showed a marked response--an increase of 0.2-4.4 t ha^{-1} grain yield (54 and 334%) over the untreated plots. The large increase observed for Fe fertilization reflects the serious nature of Fe deficiency in rice. The incorporation of organic manures, in addition to puddling of the soil, helped enhance the availability of native as well as applied Fe, and is often found advantageous in amending Fe-deficient soils (Fig. 4).

Combined Zn and Fe deficiency

In Tamil Nadu, direct-seeded, semidry rice is prone to Zn and Fe deficiencies. An increase of 16.5 and 14.5% over NPK control was realized in the direct-seeded, semidry rice from soil application of 25 kg $ZnSO_4$ and 50 kg $FeSO_4$, or foliar spray of 0.5% $ZnSO_4$ and 1.0% $FeSO_4$ at 15, 25, and 35 d after seeding, respectively, in Typic Haplustalf of Cauvery Delta (Krishnasamy 1990).

The best alternate strategy suggested for such nutrient stress conditions is the development of Zn- and Fe-efficient cultivars. Rice varieties identified as tolerant of Fe stress include CO 33 and CO 41. Rice varieties ADT36 and IET1722 are not recommended for Zn-stress situations; resistant varieties IR50 and ASD16 should be used (Velu et al 1982).

Mn, Cu, and B deficiencies

The sodic soil and highly coarse-textured soils in the rice ecosystem show low available Mn. Manganese fertilization studies conducted in northern India showed that soil application of 50 kg $MnSO_4$ ha^{-1} increased grain yield up to 1.8 t ha^{-1}. Nevertheless, a foliar spray of 1.0% $MnSO_4$ is equally efficient and cost-effective.

Little literature is available regarding the fertilization of Cu and B because the problem is not severe in rice. However, the continuous Zn fertilization of soils having marginal Cu status may cause Cu problem in the long run. Mostly copper sulfate ($CuSO_4.\ 5H_2O$) is applied at 12.5 kg ha^{-1} to correct Cu deficiency. Drenching the foliage with 0.2% $CuSO_4$ solution is practical because it is more cost effective. Scorching of the rice plant of sensitive varieties (e.g., IR20) may occur.

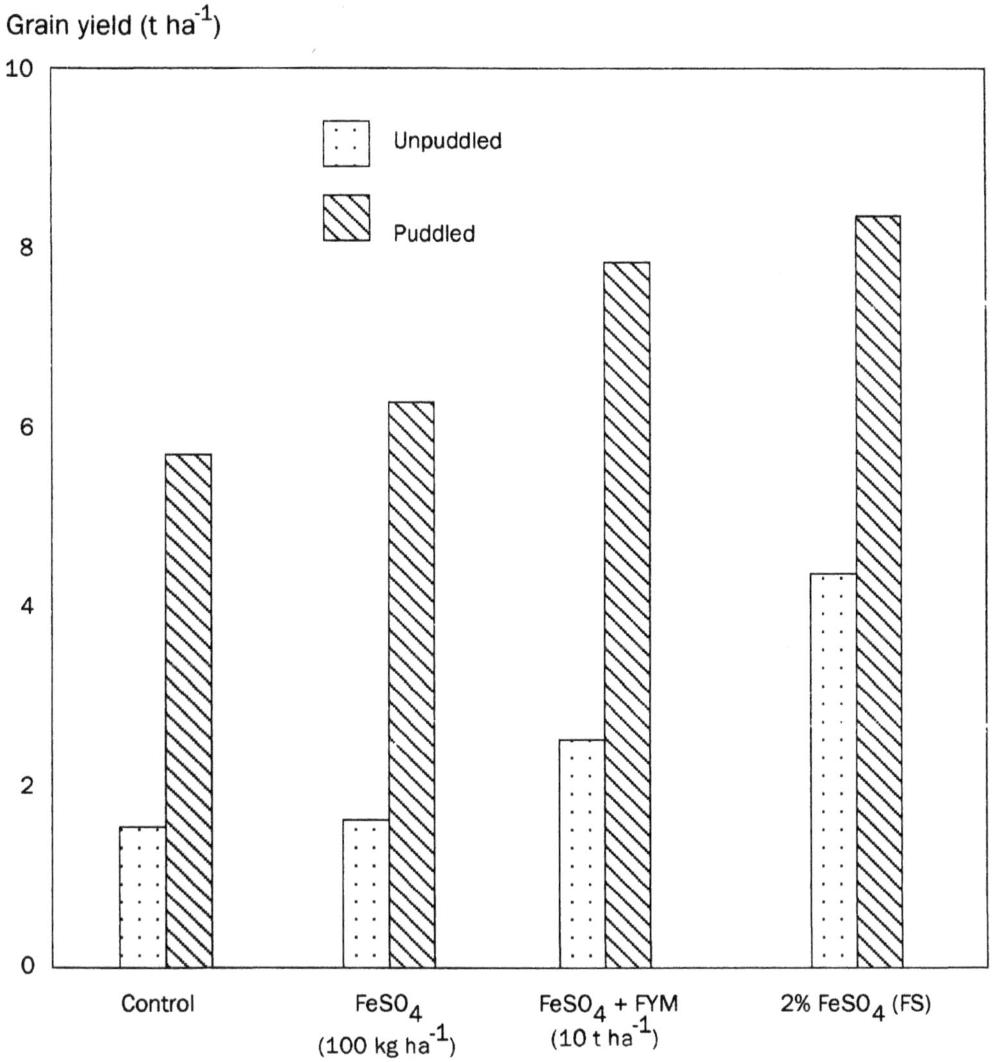

Figure 4. Iron fertilization of rice (Takkar et al 1989).

In calcareous, sodic, and excessively permeable soils in riverine flood plains, B deficiency may be a problem. Studies showed that applying 1.5-2.0 kg B ha^{-1} as borax will overcome B deficiency in such conditions.

Nutrient toxicity

Iron toxicity

Iron toxicity occurs in poorly drained areas: depression, swampy areas, and lowland valleys with lateral seepage and/or upwelling Fe-containing water (Sahu and Mitra 1992, Elsy et al 1994). The Ultisols, Vertisols, young acid sulfate soils, and acid Histosols show this disorder, which has been attributed to a concentration of high Fe^{2+} levels (30-500 mg kg^{-1}) and low pH (Makerim et al 1991, Konsten et al 1994). These soils are recognizable by a red brown oily scum on the surface of stagnant water, which is more pronounced at the lowest elevation (Benckiser et al 1982, Suresh 1996). Excessive uptake of Fe causes the physiological disorder called bronzing, reported to be distributed in 1 million ha of cultivated rice soils in Asia; it is reflected further in the reduced yield of rice. In modern rice varieties, excessive Fe applied alone, or combined Fe and Mn, may limit yields. Fe-toxic soils are characterized by a relatively low cation exchange capacity (10-15 mole $p^{+}kg^{-1}$), weak base saturation, low buffering capacity, and limited supply of available nutrients particularly P, K, Ca, Mg, Zn, and Cu (Ottow et al 1991). Further, Fe toxicity is aggravated by the low levels of P and K (Ismandaji and Ardjasa 1989, Singh et al 1992).

The excess Fe in the plant promotes sterility, which reduces rice grain yield by 20-80%. Iron toxicity can be overcome by liming, delayed planting, and late flooding. Rice yield can be enhanced threefold by draining the soil for 9 d 1 mo after planting. The water-soluble Fe content was halved by flooding and draining, and this reduction in soluble Fe substantially increased grain yield (Zini et al 1987, Singh et al 1993).

Table 9. Effect of liming and fertilization on yield and nutrition of rice variety ASD16 in Fe-toxic soils (*Source:* Suresh 1996).

	N : P_2O_5 : K_2O (T1) (125 : 50 : 50 kg ha^{-1})		T1 + $ZnSO_4$ 25.0 kg ha^{-1} + $CuSO_4$ 12.5 kg ha^{-1}		FS (0.5% $ZnSO_4$ + 0.2% $CuSO_4$ + 1% DAP + 1% MOP)		
	L0	L1	L0	L1	L0	L1	LSD(5%)
			Yield (t ha^{-1})				
Grain	5.35	6.47	5.45	7.92	5.93	7.79	1.015
Straw	5.50	6.32	6.56	8.21	7.25	9.01	-
			Total uptake of nutrients (kg ha^{-1})				
N	72.5	91.9	81.6	119.0	101.0	124.0	6.010
P	12.9	19.3	15.8	27.3	17.3	28.2	0.713
K	69.0	96.5	74.7	118.0	96.6	140.0	4.070
Ca	9.9	21.5	14.0	31.1	14.3	27.7	1.017
Fe	10.9	4.3	10.2	4.8	12.7	6.4	0.509
Mn	2.2	1.3	2.4	1.5	2.7	1.7	0.902
Zn (g ha^{-1})	154	156	205	207	234	261	8.13
Cu (g ha^{-1})	67	62	86	82	92	89	ns

L0 = control; L1= lime 9.6 t ha^{-1}; FS= foliar spray at tillering, active tillering, and panicle initiation.

The multinutrient stress of rice grown in Fe-toxic soils can be improved by balanced fertilization with macronutrients, secondary nutrients, and micronutrients. In a trial in sandy clay loam soil (Aquic Hapludalf with pH 4.7 and DTPA-Fe 311 mg kg^{-1}), liming increased grain yield by 1.6 t ha^{-1} and straw yield by 1.3 t ha^{-1}. Foliar spraying of multinutrients (0.5 % $ZnSO_4$ + 0.2 % $CuSO_4$ + 1.0% DAP + 1% MOP) in addition to soil application of major nutrients improved crop nutrition and grain yield (Table 9). In addition, use of varietal tolerance appears to be an attractive strategy. The tolerance of IR36, IR42, IR46, IR52 and Suakoka for Fe toxicity was established by Ponnamperuma et al (1981). Screening rice varieties is important for specific environments with Fe toxicity. Twelve varieties were evaluated in an Aquic Hapludalf soil with excess Fe; varieties TPS1, ASD16, ASD18, TKM9, and CO 41 were not affected much, whereas ADT36 was highly susceptible to iron toxicity (Suresh 1996).

Conclusion

The information presently available clearly demonstrates the importance of Zn deficiency in rice soils of India. The need for Zn and Fe fertilization for direct-seeded semidry and upland rice is growing. The problems will be aggravated by the intensification of rice production to meet the growing demand. Rice response to Zn, Fe, Cu, and Mn fertilization is impressive. Micronutrient research in India has developed various management strategies by focusing on level, source, mode, and frequency of micronutrient fertilization. Organic manures fortified with micronutrients are highly useful. A number of efficient cultivars have been developed for certain micronutrient problems.

Future research strategy includes

- generating data base for micronutrient status of rice soils;
- developing models for predicting crop response and rate of depletion of micronutrient reserves in soils;
- understanding the mechanisms of bioagents-agents (blue-green algae and VAM) on micronutrient availability;
- increasing micronutrient use efficiency with innovative technologies such as enriching or fortifying available organic manures; and
- coordinating micronutrient research with breeders, soil scientists, physiologists, biotechnologists, pathologists, and microbiologists, besides animal human nutrition scientists.

References

Anand Swarup (1991) Effect of gypsum, green manure, FYM and zinc fertilization on the zinc, iron and manganese nutrition of wetland rice on a sodic soil. J. Indian Soc. Soil Sci. 39:530-536.

Bansal R L, Nayyar V K (1989) Effect of zinc fertilization on rice grown on Typic Ustrochrepts. Int. Rice Res.Newsl. 14 : 24-25.

Benckiser G, Ottow J C G, Santiago S, Watanabe I (1982) Physico-chemical characterization of iron toxic soils in some Asian countries. IRRI Res. Pap. Ser. 85. 11p.

Castro R V (1977) Zinc deficiency in rice: a review of research at the International Rice Research Institute. IRRI Res. Pap. Ser. 9. 18 p.
Cayton M T C (1985) Boron toxicity in rice IRRI Res. Pap. Ser. 113. 10 p.
Channel H T (1992) Studies on the effect of amendments on the transformation of applied zinc in sodic soil in rice-pulse cropping system. Ph D thesis, Tamil Nadu Agricultural University, Coimbatore.
Chibba M et al (1989) Direct and residual effect of some zinc carriers in rice, wheat rotation. J. Indian Soc. Soil Sci. 37:585-587.
Deb D L (1990) Effect of zinc sources on yield and utilization of zinc in rice - wheat sequence. J. Nuclear Agric. Biol. 19:236-241.
Deb D L et al (1986) Radioisotope aided micronutrient research for increasing fertilizer use efficiency. Fert. News. 31:21-29.
Devarajan R (1989) Direct, residual and cumulative effect of applied Zn in rice. In Krishnasamy R et al. (Eds.) Twenty-five years of micronutrient research in soils and crops of Tamil Nadu. Tamil Nadu Agricultural University, Coimbatore.
Dravid M S, Goswami N N (1987) Effect of FYM, P and Zn on dry matter yield and uptake of P, K, Zn, Fe by rice under non-saline and saline soil conditions. J. Nuclear Agric. Biol. 16:201-205.
Duraiswamy P et al (1988) Effect of amendments and Zn on the availability, content and uptake of Zn and Fe by rice Bhavani in sodic soil. Madras Agric. J. p 119-124.
Elsy C R et al (1994) Varietal response of rice to iron toxicity. Oryza 31:67-68.
Gupta V K et al (1994) Twenty-five years of micronutrient research in soils and crops of Haryana Agriculture University, Hisar. 99 p.
Ilangovan R (1986) Evaluation of Zn-DAP in lowland rice. MS (Ag) thesis, Tamil Nadu Agricultural University, Coimbatore.
Ismandaji M, Ardjasa W S (1989) Effect of fertilization on Fe toxicity in lowland rice. Paper presented at a seminar at the Bogor Research Institute for Food Crops, Bogor, Indonesia.
Jahiruddin M, Cresser M S (1990) Effect of submergence on zinc, iron, manganese and copper availability in Bangladesh soils. Bangladesh J. Agric. Sci. 17(1):85-91.
Katyal J C (1977) Influence of organic matter on the chemical and electrochemical properties of some flooded soils. Soil Biol. Biochem. 9:259-266.
Katyal J C (1985) Research achievements of All India Coordinated Scheme of micronutrients in soils and plants. Fert. News 30(4):67-80.
Khind C S, Jugsujinda A, Lindau C W, Patrick W H Jr. (1987) Effect of sesbania straw in a flooded soil on soil pH, redox potential and water soluble nutrients. Int. Rice. Res. Newsl. 12(3):42-43.
Konsten C J M et al. (1994) Effects of flooding on pH of rice producing acid sulphate soils of Indonesia. Soil Sci. Soc. Am. J. 58:871-883.
Krishnasamy R (1990) Zinc and Fe nutrition of direct sown semi-dry rice In Krishnasamy R et al. (Eds.) Twenty-five years of micronutrient research in soils and crops of Tamil Nadu. Tamil Nadu Agricultural University, Coimbatore.
Krishnasamy R et al (1994) Twenty-five years of micronutrient research in soils and crops of Tamil Nadu. Tamil Nadu Agricultural University, Coimbatore. 192 p.
Lindsay W L (1972) Inorganic phase equilibria of micronutrients in soils. In Mortvedt J J, Giordano P M, Lindsay W L (Eds.) Micronutrients in agriculture. Soil Science Society of America, Madison, Wisconsin. 41 p.
Lindsay W L, Norvell W A (1978) Development of a DTPA soil test for Zn, Fe, Mn and Cu. Soil Sci. Soc. Am. J. 42:421-428.
Makerim A K et al (1991) An overview of major nutritional constraints to rice production. In Deturck P, F.N. Poonamperuma (Eds.) Rice production of acid soils of the tropics. Institute of Fundamental Studies, Kandy, Sri Lanka.
Mortvedt J J (1994) Needs for controlled-availability micronutrient fertilizers. Fert. Res. 38:213-221.
Nagarajan R (1983) Zinc fertilization and nutrition in rice soils. Ph D dissertation, Tamil Nadu Agricultural University, Coimbatore.
Nayyar V K et al (1990) Micronutrients in soils and crops of Punjab. 148 p.
Neue H V (1994) Variability in rice to chemical stresses of problem soils and their method of identification. In Senadhira D. (Ed.) Rice and problem soils in South and South east Asia. p 115-144.
Neue H V, Mamaril C P (1985) Zinc sulfur and other micronutrients in wet land soils Pages 307-320 in wet land soils : Characterization, classification and utilization. International Rice Research Institute, P.O. Box 933, Manila, Philippines.
Neue H V, Lantin K S (1994) Micronutrient toxicities and deficiencies in rice. In Xeo A R, Flowers T J (Eds.) Monogr. Theor. Appl. Genet. 21:175-200.
Ottow J C G et al (1991) Strategies to alleviate iron toxicity of wet land rice on acid sulphate soils In Detrula P, Ponnamperuma F N (Eds.) Rice production on acid soils of the tropics. Institute of Fundamental Studies, Kandy, Sri Lanka, 186 p.
Ponnamperuma F N (1972) The chemistry of submerged soils. Adv. Agron. 24:29-96.

Ponnamperuma F N et al (1981) Adverse soils tolerance. Paper presented at the Internal Program Review, International Rice Research Institute, Los Baños, Laguna, Philippines.
Poongothai S (1993) Zinc-enriched organic manures application in rice - groundnut based cropping system. Annual progress report of All India Coordinated Scheme on micro and secondary nutrients and pollutant elements in soils and plants. Tamil Nadu Agricultural University, Coimbatore.
Rajagopalan S, Palanisamy S (1986) Effect of nursery bed nutrient management and seed treatment on rice grain yield. Int. Rice Res. Newsl. 11:26.
Rajarajan A (1991) Radiotracer studies on Zn nutrition of rice (*Oryza sativa* L.) Ph D thesis, Tamil Nadu Agricultural University, Coimbatore.
Sadana V S, Takkar P N (1983) Methods of zinc application to rice in sodic soil. Int. Rice Res. Newsl. 8(2):21.
Sahu S K, Mitra G N (1992) Iron-potassium interaction of nutrient balance in rice. J. Pot. Res. 8:311-319.
Sajwan K S, Lindsay W L (1988) Response of flooded rice to various sources of zinc. J. Agric. Sci. 11:197-198.
Sakal R et al (1993) Evaluation of methods and time of Zn application to rice. J. Indian Soc. Soil Sci. 41:195-196.
Seghal et al (1992) Agroecological regions of India. Technical Bulletin NBSS and LUP Publ. No. 24. NBSS and LUP, Nagpur, India. 130 p.
Sharma A R, Mitra B N (1990) Complementary effect of organic, bioagents and mineral fertilizers in rice based cropping system. Fert. News. 35:43-51.
Singh M V, Abrol I P (1985) Direct and residual effect of zinc application on the yield and chemical compositions of rice - wheat crops in an alkali soil. Fert. Res. 8:179-191.
Singh M V, Saha J K (1995) 26th progress report of All India Coordinated Research Project of Micro and Secondary Nutrients and Pollutant Elements in Soils and Plants. IISS, Bhopal. 145 p.
Singh B P et al (1992) Characteristics of Fe toxic soils and affected plants and their correction in acid Haplaquent of Meghalaya. Int. Rice Res. Newsl. 17:18.
Singh B P et al (1993) Response of rice to phosphorus and potassium grown on iron toxic Haplaquent under mid altitude of Meghalaya. Ann. Agric. Res. 14:100-102.
Suguira Y et al (1981) Structure, properties and transport mechanism of iron (III) complex of mugineic acid, a possible phytosiderophore. J. Am. Chem. Soc. 103:6979-6982.
Suresh S (1996) Nutrition of rice and Banana in soils prone to iron toxicity in the high rain fall zone of Tamil Nadu. Ph D dissertation, Tamil Nadu Agricultural University, Coimbatore.
Takkar P N et al (1989) Twenty years of coordinated research on micronutrients in soils and plants. Bull. No. 1. Indian Institute of Soil Science, Bhopal. 114 p.
Thind H S, Chahal D S (1983) Iron equilibria in submerged soils as influenced by green manuring and iron application. J. Agric. Sci. (Camb.) 101:207-221.
Trierweiler J F, Lindsay W L (1969) EDTA- ammonium carbonate soil test for zinc. Soil Sci. Soc. Am. Proc. 33:49-53.
Velu G et al (1982) Varietal susceptibility of rice to Zn deficiency. Tamil Nadu Agric. Univ. Newsl. 12:11.
Zini Z et al (1987) Alleviation of iron toxicity in lowland rice grown on newly opened paddy field. Sukarami Research Institute for Food Crops Technology Transfer Booklet. p 14-15.

Strategies for developing and promoting innovative fertilizer formulations for efficient crop production: industry's view

OLUF CHR. BØCKMAN[1], MARIT LÆGREID[1], and SOLVEIG NYVOLD[2]

[1]*Norsk Hydro ASA, The Research Centre, N-3901 Porsgrunn, Norway*

[2]*Norsk Hydro ASA, N-0240 Oslo, Norway*

Key words: variable nutrient needs, location-specific fertilizer formulations, product development strategy, integrated fertilizer production-distribution-sale, marine fertilization, fertilizer-related pollution

Abstract

The predicted increase in human population implies that agricultural yields must be sustainably increased and crops adequately supplied with plant nutrients. Raw materials for fertilizers and soil amendments differ in quality. The use of local resources should increase in importance, but global scarcities are not expected within the next generation. Because of variations in climate, soil fertility, and crop requirements, the need for nutrients varies. The need for external supplies is usually limited to a few elements: N, P, K and sometimes S, Mg, Zn, and B. However, variations in needs are so extensive that a single or a few formulations cannot adequately cover all the requirements and conditions. Quantifying local needs and fitting formulations to these is one of the challenges that must be addressed in nutrient management. A further challenge is the development of appropriate systems and physical product characteristics to suit special conditions, e.g., soils with a low cation exchange capacity, or areas with high rainfall intensity. Ways to increase the plant uptake of applied nutrients to minimize losses and to maximize the crop's ability to resist stresses such as drought and diseases through appropriate nutrition must be found. However, chemical reactivity and economic considerations restrict the options available to the formulator. Production, packaging, transport and distribution costs, and farm-level realities also set practical limits to the range of formulations that can be supplied. Further, marketing and promotion must be tailored to the social and economic realities of farmers. A strategy for addressing all these issues implies a close cooperation among fertilizer producers, distributors, financial institutions, state and local authorities, plant breeders, soil and crop scientists, advisors, and finally--the most important persons in food production --the farmers themselves. Although appropriate product formulation is important, the main strategic challenge is to integrate all the participants in food production into a harmonious working relationship which is profitable to all.

Introduction

World human population is increasing, although the growth rate is declining. The United Nations median projections (Table 1) illustrate the seriousness of the situation. Even more relevant is regional population density. Figure 1 gives the number of people per hectare of arable land in the world's five main regions. Table 1, taken together with Figure 1, indicates that the main challenges are in Asia and Africa. The challenge for world agriculture and the farmers is to

- produce and market sufficient food for all;
- improve dietary variation and quality for the world's multitudes; and
- maintain and, preferably, enhance soil productivity, water quality and availability, and conservation of scarce resources.

There are diverging opinions on the feasibility of attaining these goals. The World Watch Institute presents arguments that support the pessimistic view, and there are indeed

V. Balasubramanian, J. K. Ladha and G. L. Denning (eds.), Resource Management in Rice Systems: Nutrients, 137–150.

Table 1. Population growth, 1950-2025. UN median projections (*Source:* UN 1995).

Region	Population in millions				
	1950	1990	2000	2010	2025
World	2520	5280	6160	7030	8290
Africa	220	630	830	1070	1500
Asia	1400	3190	3740	4260	4960
Latin America and the Caribbean	166	440	520	600	710
Northern America	170	280	310	330	370
Europe	550	720	730	730	720
Oceania	13	26	31	35	41

numerous problems ahead (Brown 1995, Gardner 1996). Unfortunately, a pessimistic view of the future can hinder the development and adoption of appropriate policies and technologies by encouraging despondency and inaction due to the belief that a food crisis is inevitable. In contrast, the FAO (Alexandratos 1995) presents well-supported arguments for the technical possibility of meeting the challenges for world agriculture (see preceding paragraph), *provided that the appropriate policies are adopted.*

Currently, food production is not at its maximum level, technically speaking:

- the European Union and United States of America have deliberately reduced the areas under cultivation to avoid cereal overproduction;
- production in the former Eastern Europe has suffered as a result of recent political changes and should have potential for higher productivity; and
- low prices have discouraged intensive food production.

In large areas such as China, for example, great efforts are being made to produce as much food as present technology permits, but on a global scale the principal factors limiting production are socioeconomic and political rather than technical. One of the key issues that must be addressed in a high-yield sustainable agriculture is the availability of plant nutrients to crops. Fertilizer companies have, therefore, an important role to play in the struggle toward sustainable global food security, but they are only one group of participants among many in this great endeavor.

Plant nutrients in a high-yield, sustainable agricultural economy

It has been known for about 150 yr that a crop's access to inorganic nutrients is one of the major factors limiting yield, the others being water availability, temperature, insect pests, and diseases. Concentrated plant nutrients, in the form of mineral fertilizers, enable farmers to control nutrient availability and achieve yields as large, and produce quality as high, as the other factors restricting yield and quality permit. The use of mineral fertilizers is based on the experience of 150 yr.

Present and predicted global fertilizer use is given in Table 2. Market demand for food and fertilizer varies with economic conditions and changes in line with developments in

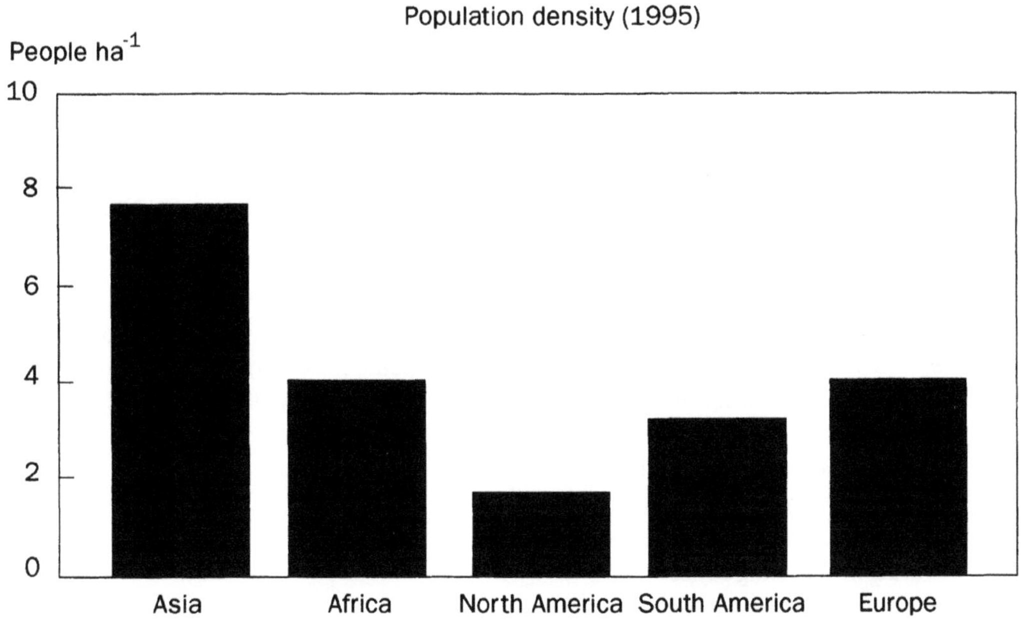

Figure 1. Population density per arable hectare (Sutton 1995).

the efficiency of using plant nutrient sources: crop residues, manures, and fertilizers. Forecasts of fertilizer requirements beyond the next few years are thus speculative, but it is likely that demand will remain strong and probably continue to grow by 2-3 % annually in the coming decade, mainly in developing nations. This global trend of fertilizer demand masks important details. In many OECD nations, fertilizer use has stagnated and declined for various reasons:

- cultivated areas have been reduced;
- fertilizer use has become more efficient; and
- utilization of farm nutrient resources (e.g., crop residues and manures) has improved.

Large regions such as much of sub-Saharan Africa make very meager use of fertilizers, so that cultivated soils slowly lose their fertility (van Reuler and Prins 1993). Unevenness in crop fertilization is high; for example, cash crops are often better fertilized than cereals, and some important food crops (e.g., cassava) are hardly fertilized at all. The use of fertilizers throughout the farming community varies due to availability and economic factors. Experience and information regarding requirements, and optimal fertilizer use for a given locality and crop, may be inadequate. Local conditions limit fertilizer efficiency, e.g., frequent heavy rains or droughts, soils with low capacity for retaining nutrients (low cation exchange capacity) (Weischet and Caviedes 1993).

However, the need for increased agricultural production should gradually manifest itself in enhanced demand and higher prices for food products. This should, in time, increase fertilizer use, notably in those nations where nutrient availability to farmers forms a principal restraint on productivity. As suppliers of plant nutrients, fertilizer com-

Table 2. Forecasts of world fertilizer consumption (10^6 t of nutrients) (*Source*: IFA 1996).

Nutrient[a]	1995	2000
N	77.8	88.6
P	13.0	14.8
K	17.0	19.3

[a]Nutrient amounts are given as the elements P and K, not as oxides P_2O_5 and K_2O.

panies have an important role in agriculture. The basic strategy must be to satisfy customer at all times, as that is the only key to a sustainable commercial operation. To achieve this, the producer has to operate under conditions that vary widely:

- competition in the market
- desired product composition
- desired physical form and quality
- desired package type
- transportation and delivery system
- extension services
- agronomic knowledge
- application methods
- environmental considerations
- legal, fiscal, and economic factors, e.g. payment structure

Development strategies of fertilizer companies differ, reflecting company and regional traditions and conditions. This presentation is based on the worldwide experiences of a major fertilizer company, Norsk Hydro ASA, but should reflect the general realities of the business.

Product composition and form

To produce fertilizers, (a) inert atmospheric N_2 must be fixed--i.e., converted to forms that crops can use; (b) water solubility of rock phosphate must be increased by treatment with acids, and the byproducts (calcium salts) utilized or disposed of; and (c) soluble potassium and other nutrient minerals must be mined, purified, and processed into usable forms.

Lime (including dolomite) and gypsum can be considered the simplest fertilizers. Lime is mainly used for correcting soil pH, but calcium is also an essential plant nutrient. Lime for agriculture, where needed, is applied in large amounts (e.g., 5 t ha^{-1} applied every fifth year is typical for Norway), and is used after simple processing of local resources. However, production of fertilizers with N, P, and K is not a cottage industry. The basic principles of fertilizer manufacture may appear simple, but in practice many technical difficulties require special knowledge and experience. Research and development costs are, therefore, a prerequisite to resolve many challenges and to sustain success in the market. For safety, economy, and efficiency, fertilizer plants have to be large with a complicated structure. The majority of world fertilizer plants are old, often rather small, and serve local or regional markets. However, a new factory for complex fertilizer (with N, P, and K) starting from natural gas and minerals may make about 2 x 10^6 t of product per year and cost some US$700 million to build. Capital, manpower, and other costs associated with the process represent the major part of fertilizer costs.

A wide variety of processes are available (Engelstad 1985). Environmental considerations are important in the selection of processes, notably for P fertilizers. Nitrophosphate processes produce hardly any solid waste, while sulfuric acid-based factories discharge large amounts of gypsum waste containing Ca and S, though both are plant nutrients. This is one of the numerous instances where environmental costs and concerns enter into the strategic evaluations of fertilizer companies. Once the process has been selected, the range of products that can be produced is limited by the type of process chosen. Hence, a careful analysis of the present and future needs of the intended market, as a basis for process selection, forms part of any strategy.

These needs vary, as exemplified in Table 3. The need for N is universal, but soils vary considerably in their reserves of other nutrients. If these are not replenished, yields will eventually suffer. Up to now, too much emphasis has been placed on straight N fertilization in many regions. In the future, we expect to see an increasing requirement for complete fertilizers. This can be done by producing compounds where all particles have the same composition, or blends made by mixing particles differing in composition. In addition, fertilizer solutions (blends dissolved in water) are used increasingly. All these product types differ locally in price, transport costs, precision in use, and need for special equipment. No clear-cut recommendation can be made as to which of these will be the product of the future; all have their merits and disadvantages. However, it is our view that there will be increasing emphasis on product quality, both chemical and physical.

Safety considerations place considerable restrictions on what can be made and transported. Certain combinations (e.g., products rich in both NH_4NO_3 and KCl, notably when they also contain Cu^{2+}) are thermally unstable and exhibit self-sustaining decomposition when the heat source is removed. Such safety considerations are given legal force through official regulations, which classify dangerous goods that are subject to transport or marketing restrictions. Further, slow solid-state reactions between fertilizer components can cause particle disintegration. These topics also arise in product development strategies. Liquid fertilizers have inherent restrictions on composition because the precipitation of insolubles (e.g., gypsum) must be avoided. Some compounds are deliquescent and liquefy when exposed to humid air. Furthermore, segregation can easily occur in blends if the components differ too much in shape, size, and density. Most fertilizers are solids, and their physical strength is an important aspect of quality. Particles must be able to withstand transport: they must not cake or degenerate into powder; they must not absorb water during transport, yet should dissolve rapidly after application. Paying attention to these physical aspects is an essential part of product development, and of promotion. Deficiencies in physical quality are immediately visible to the customer.

There are, thus, a multitude of chemical and physical factors that formulators must consider in their planning, production, marketing, and development work. Anticipation of agricultural needs and product development is a permanent feature of fertilizer company strategy, as described in more detail elsewhere (Kongshaug 1995).

Table 3. Examples of national fertilizer recommendations (kg ha^{-1}) (*Sources:* IFA 1992, FAO/IFA/IFDC 1992, 1994).

Crop	N	P	K	Country
Wheat	60-120	13-26	0-25	China
Rice	90-125	11-20	37-42	India
Maize	120-180	20-26	25-50	Indonesia
Soybean	45-60	20-26	50-100	China

Logistics

Logistics are very important in fertilizer production and marketing strategies. A modern fertilizer plant produces large volumes of products, and transport costs can make or break a project. It is much more expensive to transport products on land, than in bulk on large ocean-going ships and even on barges. Fertilizer plants and larger terminals must, therefore, be located near harbors if they are to be competitive. It is also more expensive to package and handle sacks than to handle bulk products, and distribution costs and complexity increase if the number of product grades increase. In consequence, the development of improved, more appropriate, and cheaper distribution methods and networks forms one of the major challenges for fertilizer suppliers, as does the development of a short list of product grades suitable for the range of needs encountered in the field. Cooperation among customers, local distributors, transport companies, and producers is essential here.

It is necessary to stress this point in presenting strategies for developing and promoting fertilizer formulations for efficient crop production, because emphasis is often placed on product development per se rather than on the product as perceived by the customer. In this context, packaging and logistics form an important part of the total picture.

Crop and soil need for nutrients

A principal concern in fertilizer production and marketing is that farmers should profit both in the short and long term through fertilizer use. This implies that fertilizer application must be balanced, with the needs and removals of all plant nutrients attended to. Unbalanced fertilization generally implies that excessive amounts of N are applied compared with the input of P, K, S, Ca, Mg, and micronutrients. The result is low N efficiency and low yields. In Egypt, the emphasis has long been on N fertilization: the applications in 1990 in 10^3 t were N = 793, P = 72, and K = 17.5. Demonstrations of more balanced fertilization, based on soil testing and plant analysis, increased yields by about 20 %, but N applications could be somewhat reduced and farmers' profits increased (El-Fouly and Fawzi 1996). This experience illustrates the importance of crop- and location-specific recommendations, and the inclusion of the appropriate products in fertilizer marketing strategies.

Although in principle it is necessary to supply a complete range of nutrients, the main research challenge is represented by the correct and efficient supply of N because of the complexity of the N cycle. This has been described in numerous reviews, e.g., Ramos

(1996). A crop obtains many of its nutrients from crop residues and other forms of soil organic matter through mineralization. At the same time, applied fertilizers not only supply the plants but also fertilize the soil. However, soil processes are not amenable to precise quantification; thus, there is an element of imprecision in fertilizer recommendations, which represents a major challenge to agronomists and to fertilizer producers. One method of improving the situation is through colorimetric leaf analysis, which indicates when crops need more N. This is exemplified by the use of chlorophyll meters for N management in lowland rice (Johnkutty and Palaniappan 1996), but the scatter in crop response to predicted N supply is still substantial. Although such techniques are promising, the mainstay of fertilizer recommendations remains field trials guided by soil analysis and local agricultural knowledge. Participation in such trials is an important part of fertilizer formulation and marketing strategy.

Many mathematical models, e.g., of crop growth and nutrient utilization, are available. However, knowledge about nutrient transformation in soil and losses from soil remains incomplete. This is especially noticeable in the tropics where crops, soils, and climates can differ markedly from those in temperate regions. Much information is available on tropical soils and fertilizer use (Ahn 1993). Still, our knowledge of N loss processes, such as denitrification to N_2 and N_2O, is inadequate for tropical conditions. Under semiarid conditions, N is hardly lost by leaching and denitrification, so that recoveries are generally high. Under humid conditions, where rainfall exceeds evapotranspiration, the opposite generally holds (van Duivenboden et al 1996). However, although such generalized experiences are useful, the lack of site-specific data hampers the rational development of fertilizer products and recommendations. Close cooperation with local scientists and attention to the farmer's experiences and traditions are required for a successful development strategy for fertilizer use.

Another problem is presented by fertilizers for highly weathered soils in hot humid climates where cation exchange capacity is low. Such soils have low fertility and present a notable challenge to fertilizer formulators, but progress depends on improved insight into the relevant processes in such soils (Weischet and Caviedes 1993).

Although the emphasis is on N, P, and K, the importance of other nutrients should not be ignored. The rapidity of development of sulfur deficiencies in Europe, after environmental measures had markedly reduced sulfur dioxide emissions from power and industrial plants, provides a salutary lesson (Ceccotti 1996). There is also room in product development for investigating a possible role for nontraditional nutrients, such as silicon, used on a large scale in Japanese rice fertilization (Ahn 1993).

Customer service and support

The present lack of data on optimal fertilizer application rates and timing is a main constraint to efficient fertilizer use in many developing nations. The fertilizer industry seeks to alleviate this constraint through cooperation with national universities and agricultural research institutes so that improved advice can be given to farmers. Establishing field experiments and demonstration plots and disseminating research results to farmers as well as to extension agents are an important part of this activity. This activity needs intensification in the future. There is also a need for methodological development, e.g., in soil and leaf analysis, suitable for field work under tropical conditions.

Product commitment also forms part of the producer-distribution-farmer relationship, so that the customer can rely on his supplier to deliver the product that research and trials have demonstrated to be the appropriate one.

Fertilizer application

The span of technologies used for fertilizer application is widening, from hand-spreading in small farms to automated spreading of solids or liquid, according to computer-generated field fertility maps, on tractors guided by the global positioning system (GPS). The latter technology is developed rapidly; it promises to provide more detailed control of nutrient application than is possible with traditional methods, with savings in the amounts applied and enhanced nutrient efficiency. We may see a shift, at least in some regions, toward contract application, making the fertilizer producer and/or distributor also a partner in the application.

Part of development strategy is to include the application aspect in work on products and the development of services adapted to local conditions. Although impressive developments in application equipment have been made utilizing very advanced technology, there is also a need for simpler, serviceable systems. Free-flowing, dust-free fertilizers facilitate spreading, whether applied mechanically or manually; hence, the importance of physical characteristics in product development.

Environmental issues

Environmental concerns play an increasing role in the design of fertilizer products and services. Fertilizer production issues are well understood and codified as best available techniques (BAT; EFMA 1995), supported by environmental and safety audits and occupational health services. The principal issues are the effects of nutrient losses. However, nutrient losses during fertilizer production are considerably less than 1% of the nutrients handled. Thus, the main environmental concerns associated with fertilizers are not at the production site, but on the farms. The principal issues with plant nutrient management in farms are

- soil fertility and nutrient balances
- nutrient efficiencies and losses
- gaseous emissions: NH_3, NO, and N_2O
- leaching and surface runoff: NO_3^-, SO_4^{2-}, and PO_4^{3-}
- accumulation of undesirable elements: Cd
- utilization of farm nutrient resources: manures and crop residues
- effects of fertilizers on crop resistance to insect pests and diseases
- recycling or use of packaging materials

- sustainability of raw material supply
- should N be supplied by fertilizer or through biological N_2 fixation ?

These matters are discussed at length elsewhere (Bockman et al 1990), hence only those aspects most relevant to this topic will be included here. Fertilizers are essentially a means of maintaining soil fertility. Harvested nutrients and unavoidable losses should be replenished so that soil fertility does not diminish. However, soil fertility depends not only on nutrient status but also on other factors, e.g., soil organic matter. Frequent cultivation diminishes a soil's carbon content. Crops differ in their ability to contribute organic matter to soil, grass being one of the most useful crops in this respect. Proper fertilization contributes to production of root mass and crop wastes that provide organic matter to the soil. We expect this aspect of selection of rotations and of fertilizer use to increase in importance in the future.

Nutrient efficiency is a very complex issue, partly because nutrients enter into and are released from soil nutrient reserves in ways difficult to quantify. Reports on nutrient efficiencies vary greatly, from very high to very low. Under favorable conditions, winter cereals can take up 86% of applied N (Jenkinson 1982). However, crop uptake in the year of application is usually considerably lower. Some of the applied N goes to maintain soil N reserves, and some is lost. The overall N efficiency in agriculture (N output compared with N inputs) is generally low; for example, it is only 20% in Western European agriculture (Bøckman et al. 1990). However, much of the apparent inefficiency is due to N losses from farm animals and their manure, where volatilization of NH_3 is a major factor (ECETOC 1994). Most of the applied P and K enter soil reserves through chemical reactions with soil constituents, and only a minor part is utilized by the crop in the year of application.

Increased nutrient efficiency remains a prominent target in the development of products and services in fertilizer industry. Despite much work on fertilizer development, losses remain high in some cases. One reason is that losses are often indirect rather than from the fertilizer itself: NO_3^- in leachate can derive directly from the applied fertilizer, but also from mineralization of roots and other forms of soil organic matter (Addiscott et al 1991). Much effort has been spent on developing specialty fertilizers with delayed or controlled nutrient release so that direct nutrient leaching can (in principle) be diminished. Whatever the problem to be addressed, economic considerations dominate. Numerous fertilizer products have been marketed

- with special coatings that delay dissolution;
- where the N is chemically bonded and made gradually available to plants, e.g., oxamide and various polymers; and
- where N transformation in the soil is retarded by the addition of enzyme inhibitors.

Products incorporating these lines of development have been launched, but market acceptance has been limited. All extra processes increase production costs. The general experience of the fertilizer industry is that even small to moderate price increases form a barrier to product acceptability, even where increased efficiency decreases the amounts that should be applied. Ammonia losses are diminished if the fertilizer is placed at some depth in the field. Deep placement of urea is facilitated if the urea granules are large. In flooded rice cultivation on clay soils, farmers can save 30-50% of their present urea use through deep placement of supergranules without sacrificing yield. For a variety of reasons, commercial development of this product has so far been slow (Savant and Stangel 1990). This experience is a deterrent to product development but also a challenge to analyze the causes of product success or failure and to learn from past experiences. Farmers will accept a new product only if they derive noticeable economic benefits and/or the product is easier to use than traditional material.

So far, the most successful approach to managing efficiency and environmental issues with fertilizer use has not been to make "magic bullets," but to develop sensible nutrient combinations and recommendations for users, based on a thorough understanding of soil processes, crop physiology, and well-designed field trials. This approach has a prominent place in any strategy for future plant nutrient management. It is complex, slow, and laborious, but experiences indicate that it is a viable strategy which places emphasis on development of customer service.

In practice, farmers have considerable amounts of farm-generated nutrients that must be returned to the soil--crop residues and animal manures. The rate of nutrient release from these tends to be slow and difficult to predict with desired accuracy. A special case is N supply through biological N_2 fixation. Legumes obtain a variable, but often generous, amount of their N through symbiosis with N_2-fixing bacteria. Efforts have long been made to study such symbiotic systems in the hope that N_2-fixing symbiosis can be extended to other crops, e.g., rice. The target still seems distant but not impossible (Ladha and Reddy 1995). Extensive use of N_2-fixing crops without fertiliztion should be similar in its effects to unbalanced straight N fertilization, and it will not be a sustainable practice. Hence, N_2-fixing crops cannot be expected to eliminate the need for fertilizers, although the preferred composition will change, presumably to something similar to the present recommendations for soybean (Table 3).

Fertilizers and crop resistance to stress

The relationship between a crop's resistance to stress and its nutrition is complex. Adequate nutrition enhances a crop's ability to resist insect and disease attacks and to compensate damage through growth. However, unbalanced nutrient supply, notably N in excess of crop requirements, can make crops more attractive to insects as a superior N source and diminish their resistance to fungal damage. Nitrogen fertilizers acidify soils, which can enhance or diminish crop susceptibility to root diseases.

In principle, therefore, it should be possible to design a fertilizer recommendation that represents optimal fertilizer use for crop resistance. Much work has been done in this direction, but the target has proved elusive, as perhaps is to be expected because of the many problems and variables in the field. It is known that an adequate supply of K is especially important for disease resistance (Perrenoud 1990) and vigorous research into the role of micronutrients is being carried out (Engelhardt 1989). However, it appears unlikely that a magic composition that greatly enhances crop resistance will be found.

Fertilization is only one of the tools available to farmers for crop protection, together with appropriate crop rotation and the use of biological and chemical agents. The use of pesticides is contentious, but appears unavoidable in the foreseeable future. However, much research has been carried out on defining integrated pest management practices, where pesticide use is reduced to a minimum. It is not yet clear to what extent this requires modifications in fertilizer recommendations, and, in what direction. Much development work still has to be done.

Water deficiency is a common stress factor in plant production. Fertilization plays a role in water management through ensuring adequate root development and water economy in crop nutrient uptake. This issue is best tackled by means of studies reflecting local conditions.

Sustainability issues: the natural resource base

Fertilizer production implies that mineral raw materials are converted and consumed. Hence, the concern: when will the availability of nonrenewable resources become a restraint on an agriculture dependent on external inputs? Is agriculture based on the use of nonrenewable resources sustainable? The principal concern is with fossil fuels, mainly with petroleum (for traction and transport) and natural gas (for NH_3 production). Current known reserves and consumption are listed in Table 4. Present consumption of petroleum and natural gas occurs at a rate that causes concern, both on account of resource management and because of the possibility of the undesirable climatic effects of the CO_2 formed when fossil fuels are burned. The data in Table 4 indicate that we will run out of oil and gas within a few decades. However, known reserves do not adequately reflect the amounts actually present on this planet. The total resources of fossil fuels that can be utilized are unknown, but these must be much larger than the known reserves. The rise in known reserves and the general constancy of prices, despite heavy consumption during the last decade, indicate that it is possible to identify new reserves. Hence, no real scarcity is foreseen for the next few decades, although prices may temporarily rise abruptly if supplies are disrupted due to political disturbances, as has occurred in the past. In the more distant future, as reserves are depleted, prices will rise, and much of the present use of fossil fuel, notably in the transport sector, will then diminish and the economics of world trade in food and fertilizers may change. When petroleum and gas become more expensive, other energy sources become competitive, and past experience demonstrates that all energy forms (e.g., electricity, coal) can also be used in fertilizer production. The use of other energy sources will help alleviate scarcity situations when such situations eventually develop on a permanent basis.

Much the same can be said about phosphate rock resources. Phosphorus is not a scarce element. It is the 11th most common element on this planet. In 1994, world phosphate reserves that can be utilized at present phosphate prices were 11 x 10^{12} t phosphate rock, compared with an annual production of 0.13 x 10^{12} t rock. If the price rises from about US\$40 t^{-1} to US\$100 t^{-1}, workable reserves triple to 34 x 10^{12} t (U.S. Bureau of Mines 1995, Llewellyn 1994). At present, only the richest sources can be worked. As these are depleted, prices will rise and other sources should become competitive, including the rec-

Table 4. Oil and gas: proved reserves and current consumption (in 10^9 t oil or equivalent) (*Source:* BP 1996).

Particular	Oil		Natural gas	
	1985	1995	1985	1995
Reserves	96.4	138.3	90.6	128.3
Consumption	2.8	3.2	1.5	1.9

lamation of P in sewage sludge. Many known sources cannot be worked today because they are too small, their quality is too poor, or they are too inaccessible compared with mines close to harbors. However, local sources of raw materials have attractions that may compensate for higher costs: reduction of imports, regional development, and security of supply. In the longer term, increasing development of local resources seems a likely prospect. But a general raw material scarcity seems to be a long way off, although some sources will become depleted and trade patterns may shift. Potassium, magnesium, and sulfur are so abundantly available that a global resource crisis is impossible, although local needs may eventually have to be filled through imports. In conclusion, raw material supplies do not appear to be a limiting factor within the period under consideration, and will not be so for at least 20-30 yr.

The farming economy

There are also other concerns. Farming is a risky business and crops may fail. Use of inputs enhances the farmer's productivity, but also his economic vulnerability. Society has long recognized the need to protect the farmer in such cases: the laws of Hamurabi in ancient Babylon (in present Iraq) about 3,750 yr ago stipulated the farmer's right to a moratorium on debts in cases of crop failure (Saggs 1965). Hence, private and government financial institutes also have a role to play in developing strategies for marketing fertilizer to a sustainable, highly productive farming sector.

New opportunities

Fish production in ponds has a long tradition in Asia. The possibility of extending fish production, for instance, to some of the coastal areas where marine productivity now is severely limited by nutrients, exists. Thus, fertilization of marine areas emerges as a new opportunity for increasing food production. Fertilizing marine areas will require new and perhaps unusual formulations. It is desirable, for health reasons, to increase the availability of fish and crustaceans, and Norsk Hydro is an active participant in a large European program (Maricult) to explore and develop this possibility (Hoell 1995).

Conclusion

In conclusion, strategies for developing and promoting innovative fertilizer formulations for efficient crop production can be summarized in one word: cooperation. There are many participants in the quest for an efficient and productive agriculture. Success can be achieved only if all participants cooperate profitably. New improved fertilizer products and services are being gradually developed and marketed at present as in the past. But the complexities of crop nutrition are such that fertilizers in the future are likely to be more diverse and complex than present grades. The challenge facing fertilizer formulators is to provide this diversity cheaply, reliably, and efficiently.

Acknowledgement

The authors thank Rune Ingels, Gunnar Kongshaug, Torstein Obrestad, and Jarle Ryan for their helpful comments.

References

Addiscott T M, Whitmore A P, Powlson DS (1991) Farming, fertilizers and the nitrate problem. CAB International, Wallingford.

Ahn P M (1993) Tropical soils and fertilizer use. Longman Scientific & Technical, Harlow.

Alexandratos N (Ed) (1995) World agriculture: towards 2010. FAO Rome/J. Wiley Chichester.

Brown L R (1995) Nature's limits. In State of the world 1995. World Watch Institute, Washington D.C., p 3-20.

BP--British Petroleum (1996) Statistical review of world energ. London.

Bøckman O C, Kaarstad O, Lie O H, Richards I (1990) Agriculture and fertilizers. Norsk Hydro, Oslo.

Ceccotti S P (1996) Plant nutrient sulfur--a review of nutrient balance, environmental impact and fertilizers. Fert. Res. 43:117-125.

ECETOC--European Centre for Ecotoxicology and Toxicology of Chemicals (1994) Ammonia emissions to air in Western Europe. Technical Report No. 62. Brussels.

EFMA--European Fertilizer Manufacturers Association (1995) Best available techniques for pollution prevention and control in the European fertilizer industry. Booklet 1-8. Brussels.

El-Fouly M M, Fawzi A F A (1996) Higher and better yields with less environmental pollution in Egypt through balanced fertilizer use. Fert. Res. 43:1-4.

Engelhardt A W (Ed) (1989) Soilborne plant pathogens: management of diseases with macro- and microelements. APS Press, St. Paul MN.

Engelstad O P (Ed.) (1985) Fertilizer technology and use. 3d ed. Soil Science Society of America, Madison WI.

FAO/IFA/IFDC--Food and Agriculture Organization/International Fertilizer Industry Association/International Fertilizer Development Center) (1992, 1994) Fertilizer use by crop. Vol. 1 & 2. FAO, Rome.

Gardner G (1996) Preserving agricultural resources. Pages 78-94 in State of the world 1996. World Watch Institute, Washington D.C.

Hoell E (1995) Marine cultivation--a new area for fertilization. Pages 46-52 Proceedings of the 45th Annual Meeting, Fertilizer Industry Round Table, Glenn Arm MD.

IFA--International Fertilizer Industry Association (1992) World Fertilizer use manual. Paris.

IFA--International Fertilizer Industry Association (1996) Fertilizer demand forecasts. A/96/76B. Paris.

Jenkinson D S (1982) The nitrogen cycle in long-term field experiments. Phil Trans Royal Soc. London B 296:563-571.

Johnkutty I, Palaniappan S P (1996) Use of a chlorophyll meter for nitrogen management in lowland rice. Fert. Res. 45:21-24.

Kongshaug G (1995) Fertilizers for the future. The Fertilizer Society, Peterborough. Proc. no 374.

Ladha J K, Reddy P M (1995) Extension of nitrogen fixation to rice--necessity and possibilities. Geojournal 35:363-372.

Llewellyn T O (1994) Phosphate rock 1993. U.S. Bureau of Mines, Washington D.C.

Perrenoud S (1990) Potassium and plant health. 2d ed. International Potash Institute, Bern.
Ramos C (1996) Effect of agricultural practices on the nitrogen losses to the environment. Fert. Res. 43:183-189.
Saggs H W F (1965) Everyday life in Babylonia and Assyria. Dorset Press New York. 140 p.
Savant N K, Stangel P J (1990) Deep placement of urea supergranules in transplanted rice: principles and practices. Fert. Res, 25:1-83.
Sutton C D (1995). Future farming practice in Asia. Report prepared for Norsk Hydro.
UN—United Nations (1995) World population prospects. 1994 revision. United Nations, New York.
U.S. Bureau of Mines (1995) Mineral commodity summaries 1995. Phosphate Rock. Washington D.C. p 124-125.
Van Duivenboden N de Wit C T, van Keulen H (1996) Nitrogen, phosphorus and potassium relations in five major cereals reviewed in respect to fertilizer recommendations using simulation modelling. Fert. Res. 44:37-49.
Van Reuler H, Prins W H (Eds.) (1993) The role of plant nutrients for sustainable food crop production in sub-Saharan Africa. Dutch Association of Fertilizer Producers (UPK), Leidschendam.
Weischet W, Caviedes C N (1993). The persisting ecological constraints of tropical agriculture. Longman Scientific & Technical, Harlow.

PART II

Farmer adoption of nutrient management technologies: constraints and opportunities

Farmer adoption of improved nitrogen management technologies in rice farming: technical constraints and opportunities for improvement

V. BALASUBRAMANIAN
International Rice Research Institute, P.O. Box 933, 1099 Manila, Philippines

Key words: crop management, water control, plant population, weed pressure, nitrogen use efficiency, balanced fertilization, integrated nutrient management, soil constraints, climatic constraints, on-farm technology adoption

Abstract
Generally, agricultural research is conducted under the controlled conditions of research stations which, in many cases, do not represent farmers' circumstances. Thus, several technologies do not perform well in farmers' fields and so are not adopted by farmers. Many technical factors and management practices constrain the adoption of improved technologies including efficient N management techniques: (a) poor water control; (b) low plant population; (c) partial nutrient application; (d) insufficient weed control; (e) untimely sowing, transplanting, weeding, and/or harvesting; and (f) poor postharvest processing. Unless farmers improve many, if not all, of these conditions, they cannot reap the full benefits of modern rice varieties and related production technologies. In addition, soil-related constraints such as high spatial and temporal variability in nutrient status, nutrient imbalance, poor drainage, soil degradation (salinity, alkalinity, acidity), and subsoil compaction lessen the effectiveness of nutrient management techniques. Similarly, climatic factors such as variable and unpredictable rainfall, drought, flood, low radiation, and extremes in temperature affect crop growth and nutrient use in rice systems. Most of these factors are beyond farmers' control. Stress-tolerant rice varieties and improved management practices are being developed and evaluated on rice farms to tackle soil and climatic constraints, especially in rainfed lowland rice systems. Two of the management strategies rely on better N placement techniques and use of controlled-release fertilizers to improve nutrient use efficiency in rainfed lowlands. Some tools for site-specific N management in rice include the chlorophyll meter and the leaf color chart which are being evaluated on farmers' fields. Another strategy is integrated nutrient management to enhance crop nutrition and minimize fertilizer costs. Adequate farmer training is needed to receive, process, and effectively exploit improved rice varieties and related information, knowledge, and technologies.

Introduction

Agricultural scientists have been generating new research findings (knowledge and technologies) that work well under the controlled, ideal conditions of research stations, but not so well in farmers' fields for obvious reasons. Several technical, socioeconomic, institutional, and policy-related factors constrain the full exploitation of improved nutrient management technologies on rice farms. Pandey (1998) discusses the socioeconomic, institutional, and policy-related constraints to farmer adoption of nutrient management technologies. This paper critically reviews the crop management, soil, and climatic factors that influence the applicability and performance of improved N management techniques in farmers' ricefields.

Crop management factors

Most farmers' rice crops suffer from one or more of the following conditions, listed in order of decreasing importance:

V. Balasubramanian, J. K. Ladha and G. L. Denning (eds.), Resource Management in Rice Systems: Nutrients, 153–165.

1. poor water control, especially during the monsoon;
2. less-than-optimum plant population due to inadequate land preparation and leveling, poor seed quality, use of aged seedlings at transplanting, contract transplanting on area basis, pattern and depth of transplanting and direct seeding, biotic and abiotic stresses on direct-seeded rice, etc.;
3. partial nutrient application;
4. insufficient weed control;
5. uncertainty in the timing of sowing, transplanting, weed control, and/or harvest operations; and
6. poor postharvest processing.

All of these conditions have certain impact on crop growth and the effectiveness of improved N management practices under farm-level conditions. Unless farmers address and alleviate these management constraints, it is impossible to fully exploit the potential of modern rice varieties and related production technologies.

Poor water control

Water control is the most important management practice that determines the efficiency of all other production inputs (nutrients, herbicides, pesticides, farm machines, etc.) in rice farming. Farmers in many areas are not able to control water in their ricefields for many reasons.

Water-related constraints in rainfed lowland rice ecosystem. The onset and distribution of rainfall are highly variable between and within monsoon seasons. Both dry spells and floods could occur at different times of the same season, and farmers experience severe yield losses if dry spells and floods occur at critical crop growth stages. Moreover, floods do not allow the timely application of N to a growing crop. These are some of the major reasons rainfed farmers do not use adequate amounts of purchased inputs such as fertilizers to increase crop yields. Rice varieties tolerant of abiotic stresses (specifically drought and flood) and nutrient management practices such as deep placement and use of controlled-release fertilizers are being developed and evaluated under rainfed conditions to alleviate some of these constraints (Zeigler and Puckridge 1995). Long-term climate data can be processed to predict with certain probability the dates of receiving adequate rainfall for crop establishment in direct-seeded rice systems (My et al 1995).

Water-related problems in irrigated rice ecosystems. Many irrigation systems do not have adequate provision to drain excess water when necessary. Poor drainage is highly damaging to crops and, in the long run, degrades soils with the eventual development of salinity and alkalinity. In several places, irrigation infrastructures are not maintained properly either by public (irrigation) authorities or by farmers themselves due to scarcity of funds or breakdown of organization. This situation is aggravated by deforestation-induced deterioration of upper watersheds, leading to excessive silting of rivers, dams,

Table 1. Rice area, yield, and total production of IR8 at the Mbarali Rice Farm in Tanzania, 1980-81 to 1988-89 (*Source*: Mbarali Rice Farms Ltd - Food Strategy Unit 1989, as cited in Tran 1996).

Year	Area (000 ha)	Yield (kg ha^{-1})	Production (000 t)
1980-81	2.65	8150	21.57
1981-82	2.90	7900	22.89
1982-83	2.84	7500	21.33
1983-84	2.88	7000	20.17
1984-85	2.89	5500	15.91
1985-86	2.69	4000	10.63
1986-87	2.68	3900	10.18
1987-88	2.79	3500	8.92
1988-89	2.25	4000	9.01

lakes, and canals. As a result, irrigation authorities cannot maintain the timing, frequency, and quantity of water supply to farmers, and those at the tail end of canal systems suffer most due to uncertain water supply. In many irrigated areas, the systems have deteriorated so badly that they are not much different from rainfed ecosystems in terms of water-related constraints to crop production. In an irrigation system (1,200 ha with two crops per year) in the Kou Valley of Burkina Faso, mean rice yields declined from 6.7 t ha^{-1} in 1970 to 4.0 t ha^{-1} in 1994, mainly due to poor water control and crop management (inappropriate water management, poor land leveling, disregard for cultural calendar, irregular use of inputs) (FAO 1994). Similarly, in the Mbarali Rice Irrigation Scheme (2,600 ha) of Tanzania, the average yield of rice variety IR8 decreased from 8.2 t ha^{-1} in 1980-81 to 4.0 t ha^{-1} in 1988-89 (Table 1), mostly due to deterioration of irrigation infrastructure, red rice infestation, and use of poor-quality seeds (Tran 1996).

Because of uncertain water supply, farmers tend to flood their fields with excess water whenever they get an opportunity to do so, regardless of the recommended water level for the particular growth stage. This practice not only wastes water but also leads to leaching of soluble nutrients in light-textured soils which may pollute the groundwater. Nitrogen topdressing is least efficient when water levels are high (De Datta 1986), and a large part of the applied N could be lost in runoff when water overflows from one field to another (Takamura et al 1976, Padmaja and Koshy 1978). Foliar spraying of urea solution may help nourish the plants under such difficult conditions of flood or overflowing floodwater. Single application of controlled-release fertilizers at planting or when moisture regime is favorable is being tried for rice crops under variable floodwater conditions.

Effect of good water control on rice yields. Tandon (1989) emphasized the importance of integrated water-N management for securing high rice yields and N use efficiency. Rice yield was optimized with the use of 'researcher's best split' N application: incorporating basal applied urea into the soil with no standing water, followed by post-plant topdressing of N onto 50 mm of standing water at panicle initiation (De Datta 1986). Some farmers in Indonesia drain fields before topdressing urea, but this practice is not common elsewhere. Good water control and fertilizer use resulted in 155-261% increase in grain yields of rice in the Tananarive Plains of Madagascar (Table 2) (Celton and Marquette 1972). In 1992-

Table 2. Effect of water control and fertilizer use on rice yield in the Tananarive plains of Madagascar, 1971-72 (*Source*: Celton and Marquette 1972).

Variety	Water control	Fertilizer use	Yield (kg ha^{-1})	Relative yield (%)
1632	Good	Yes	5147	361
	Poor	Yes	4166	292
	Good	No	3625	254
	Poor	No	1426	100
1285	Good	Yes	5661	255
	Poor	Yes	3736	168
	Good	No	4041	182
	Poor	No	2219	100

Table 3. Effect of water control and systems of rice cultivation on rice yield (kg ha^{-1}) in farmers' fields, Central Highlands of Madagascar, 1993 (*Source*: ODR 1993).

Water control	Research recommended system	Intensive system, 'SRI'
Good	4000	5000
Poor	2000	2000

Table 4. Mean rice yields (kg ha^{-1}) of wet-seeded (WSR) and transplanted rice (TPR) under different water regimes imposed in station trials in the Philippines (*Source*: Bhuiyan et al 1994).[a]

Water regime	TPR	WSR	Difference
Central Luzon State University Farm (silty clay loam soil), UPRIIS, 1989-90			
Continuous submergence (5-7 cm)	6632	6685	53
Two-cm irrigation in alternate days (continuous, near saturation)	6376	6601	225
Two-cm irrigation at 4-d intervals (continuous, near saturation)	4169	5396	1227
Five-cm irrigation at 7-d intervals (with appearance of small cracks)	3378	3908	530
Maligaya, UPRIIS, 1990-91 DS			
Continuous submergence (5-7 cm)	7357	7585	228ns
Saturated soil throughout	6743	7338	595**
Mild vegetative stress (no watering for10 d starting from saturated soil at 30 DAS or 9 DT)	6297	7007	710**
Severe vegetative stress (same as 3, but no watering for 20 d)	5303	6075	772**
Mild reproductive stress (no watering for 10 d starting from saturated soil at 50 DAS or 29 DT)	6038	6397	359*
Severe reproductive stress (same as 5, but no watering for 20 d)	4204	5210	1005**

[a]ns=nonsignificant; * and ** significant at 5 and 1%, respectively.

93, similar increases in rice yields due to good water management were obtained in farmers' fields in the Central Highlands of Madagascar (Table 3). Under increasing water stress in controlled experiments, wet-seeded rice produced higher grain yields than transplanted rice in Nueva Ecija, Philippines (Table 4) (Bhuiyan et al 1994).

Low plant population

It is necessary to maintain optimum plant population to fully exploit the potential of high-yielding varieties. Plant population is affected by the extent of land preparation and leveling, seed quality, crop establishment method, seedling age, plant spacing, and pattern and depth of transplanting or direct seeding.

Land preparation and leveling. In well-prepared and leveled ricefields, it is easier to achieve uniform seeding and good water control, both of which result in better and uniform seedling emergence. The development of laser technology for precision leveling and water management has improved crop stand and productivity in direct-seeded rice in temperate countries (Hill et al 1991). Such technologies are being tested in tropical lowland rice to achieve uniform crop stand and to improve rice yield and efficiency of irrigation water.

Seed quality. Quality seeds assure good crop establishment at low seeding rates, minimize weed contamination, lessen insect pest and disease incidence, result in good grain filling, and lead to high grain yields, especially in direct-seeded rice. Use of good-quality seeds has been shown to contribute 15% higher yield than use of seeds recycled by farmers (Malabanan 1993). Improved methods of on-farm seed storage are necessary to preserve the quality and viability of farmer-harvested seeds.

Transplanting. Transplanting is one of the most time-consuming and laborious operations in rice farming. No successful and cost-effective mechanical transplanter has been developed to ease this task. Thus, most farmers transplant rice at random with variable plant spacing. In many areas, labor for transplanting rice seedlings is contracted on an area basis. Contract transplanting results in large spacing between plant hills and poor anchorage of seedlings, which lead to less-than-optimum plant population. In Haryana and Punjab, India, rice yield is often limited by low planting density, which could be as low as 50% of the recommended plant population. The number of seedlings per hill has to be increased as seedling age increases, to optimize the number of tillers and panicles per unit area. The pattern of transplanting with alternating narrow and wide spacing and fertilizer N placement have been shown to increase grain yield and N use efficiency (Daftardar and Savant 1995).

Most fertilizer trials in research stations are conducted on row-transplanted rice, which is not commonly used by farmers. Thus, many of the research findings and fertilizer recommendations emanating from research stations are not directly applicable to farmers' conditions.

Direct seeding. Many farmers broadcast dry or pregerminated rice seeds on well-prepared soil to save on labor for crop establishment. Weeds are controlled by selective herbicides.

Table 5. Effect of crop establishment and N management on grain yield of wet-seeded rice in farmers' fields, Nueva Ecija, Philippines, 1996 dry season (mean of 10 farms).

Crop establishment method	Seed rate (kg ha^{-1})	N (kg ha^{-1})	N splits (no.)	Grain yield (kg ha^{-1})
Broadcasting	150	0	--	2909 c
Broadcasting	150	150	3	5789 a
Broadcasting	150	150	2	5157 b
Row seeding	75	150	2	5978 a

In Malaysia, a modified blower is used to broadcast rice seeds uniformly over the field. In surface broadcasting, seeds are exposed to inclement weather and attack by birds and rats, and the resulting plant population is often low and not uniformly distributed over the entire field. A new method called 'anaerobic broadcast seeding' involves broadcasting the seeds immediately after puddling so that they will be covered by a thin (about 5 mm) layer of settling mud. Seeds buried by the mud are less vulnerable to damage by birds and other organisms, disturbance from heavy rains or flood, desiccation by water stress and/or dry weather, and lodging due to poor anchorage (Yamauchi et al 1994). All these factors improve the chances of uniform crop establishment. Rice varieties tolerant of low oxygen conditions during germination will perform better under anaerobic seeding.

Row wet-seeding on or below the soil surface requires 50% less seeds, assures good crop establishment in rows, and facilitates better weed control, compared with broadcast-sown rice. At the same level of inputs, this method gave higher grain yields than broadcast-sown rice in farmers' fields in Nueva Ecija, Philippines, during the 1995 dry season (Table 5) (Balasubramanian et al 1996). Further testing continues to confirm the yield advantage of row wet-seeding over broadcasting.

Drilling dry seeds in rows in well-prepared dry soil, either alone or with fertilizers, is effective in securing adequate plant population in dry-seeded upland or semidry flooded rice (Panda et al 1988). The seed drill-cum-fertilizer applicator is being evaluated for dry seeding in flood-prone areas of eastern India.

Partial nutrient application

Among the commonly applied nutrients to rice crops (N, P, K, S, Zn, etc.), N gives the most dramatic visual effect on crop growth and greening of foliage. When there are limitations of cash and differences in availability and price of single fertilizers due to differential subsidy levels (policy aberration), farmers tend to buy and use mostly N fertilizers. In such circumstances, soils supply for some time many nutrients not found in applied fertilizers when yield levels are moderate. However, prolonged selective application of nutrients in intensive systems leads to soil mining of elements not supplied from external sources. Soil phosphorus (P) will be depleted with repeated applications of N alone, and severe P deficiency will limit crop yields, especially in high P-fixing soils (Tandon 1989, Balasubramanian et al 1995). Potassium (K) becomes a limiting element after a few years of intensive cultivation with high doses of N and P application. Progressive farmers in the Mekong River Delta of Vietnam apply K in split doses to obtain high yields in direct-seeded rice. Deficiencies of S and Zn occur in isolated areas, depending on soil type and fertilization history (Balasubramanian et al 1995).

Deficiency of any one nutrient will impair the crop uptake and utilization of other nutrient(s). Rice response to N application is adversely affected by the deficiency of P, S, K, and/or Zn, in order of decreasing severity. Timely correction of deficiencies of all nutrients is vital to sustain high crop yields. This calls for balanced fertilization of rice crops through integrated nutrient management--judicious and combined use of all nutrient sources (soil, biologically fixed N_2, crop residue, compost, fertilizer). Management practices should be devised to fully exploit the deep soil N reserves (Kundu et al 1996) and prerice accumulated soil nitrate (George et al 1994) in certain cases. Organic manures and crop residues, if applied regularly, will replenish the soil reserve of essential nutrients. This is the reason for the better performance of crops under combined organic manures and fertilizers than either alone. Utilization of crop residues and green manures may pose some disease problems on rice or other cereals in a cropping sequence, which will require special attention.

Poor weed control

Weeds compete with rice plants for all growth factors including nutrients. Timely and adequate weed control is a prerequisite for obtaining high grain yields and efficient nutrient use. Selective herbicides are available now to control weeds in direct-seeded and transplanted rice. However, continuous and widespread use of herbicides may lead to the development of herbicide-resistant weeds and create environmental problems. Therefore, research is geared to develop alternative weed management strategies that are safe to use with least adverse effect on the environment. These strategies include allelopathy, weed-suppressing rice varieties, and biocontrol.

Other crop husbandry practices

Insufficient land preparation and leveling of fields lead to poor crop establishment and water control, which in turn lower grain yield. Untimely sowing, transplanting, weeding, and/or harvesting will diminish crop yields and reduce the efficiency of applied nutrients. Planting at optimum dates is necessary to expose the rice crop to maximum solar energy at the reproductive and ripening periods (De Datta 1986). Adequate protection of plants against insect pests and diseases is important to obtain high yields. Quality of harvested grains will deteriorate with poor postharvest handling.

Soil conditions

Certain soil processes such as chemical degradation (e.g., salinity, alkalinity, acidity, acid sulfate soil), physical deterioration (e.g., subsoil compaction, breakdown of soil structure), and substantial reductions in populations of beneficial soil organisms will impede crop growth and decrease grain yields. For example, high pH increases ammonia volatilization (Kissel et al 1988), as in saline-alkaline soils, especially when urea is applied on soil surface and not incorporated into the soil, as practiced in Punjab and Haryana, India. Subsoil compaction leads to formation of hardpan in some soils due to repeated shallow plowing and puddling (Kundu et al 1996). This condition will restrict root growth and nutrient uptake.

Nitrogen leaching losses are high in permeable, coarse-textured soils (Vlek et al 1980, Singh et al 1991), especially under conditions of inefficient N application methods (single N dose at early crop growth stage, partial nutrient application, etc.) and high levels of irrigation water or high rainfall, again exemplified in some parts of western India, northeast Thailand, Cambodia, and Laos. Denitrification losses are favored by poor drainage, conditions favoring movement of accumulated nitrate into reduced soil layer, and high organic matter content in soils (Freney et al 1990). Runoff losses of applied N will be high on sloping lands and poorly leveled fields and under conditions of overflowing floodwater. Estimated runoff losses could be as high as 66-70% of applied N (Takamura et al 1976, Padmaja and Koshy 1978).

Tolerant varieties and appropriate management practices must be used to obtain satisfactory yields in saline and alkaline soils (Mishra and Tyagi 1986). Periodic deep plowing will break the subsoil hardpan, increase the penetration of roots into larger volumes of soil, and improve nutrient uptake (Kundu et al 1996). Appropriate soil amendments should be used to restore and maintain the population of beneficial soil organisms.

Split N application is the most widely recommended and used practice to increase the efficiency of N fertilizers. The number and timing of splits and the relative amount of N applied in each split vary with soil and crop conditions (Table 6). Methods of N application, such as basal incorporation, subsoil banding, deep placement, and co-situs placement with seed, and use of modified urea such as urea supergranules/pellets/briquettes/tablets and neem-, sulfur-, or polyolefin-coated fertilizers have been shown to be highly efficient in different soil and water conditions (Cao et al 1984; O'Brien et al 1985; Wetselaar 1985; INSFFER 1985, 1986; De Datta and Buresh 1989; Kumar et al 1989; Schnier et al 1990; Savant et al 1992; Kanno and Shoji 1994). Foliar spraying of N and other nutrients is given as a supplement to soil application of fertilizers to correct deficiencies quickly in the same growing season. It is an effective method for saline, drought-affected, and waterlogged areas where topdressing of N is difficult (Tandon 1989).

High spatial and temporal variability of soil N status also affects the efficiency of applied nutrients, especially N. Appropriate decision tools (soil test kit, chlorophyll meter, leaf color chart, pH paper strip, phosphorus sorption paper strip, crop model, etc.) are now available to vary fertilizer rates in relation to dynamic soil nutrient supply (Furuya 1987; Peterson et al 1993; Turner and Jund 1991, 1994; Peng et al 1996). Nitrogen management tools are being evaluated and adapted to different soil and crop conditions in farmers' fields, in collaboration with national research and extension organizations. Preliminary results are discussed in Balasubramanian et al (1998). Extension agents and farmers require adequate training to effectively use these tools to manage soils for higher and sustainable crop production.

Table 6. Suggested strategy (splits, timing, and amounts) for efficient N application to rice under different soil, crop, and pest conditions (various sources).

Condition[a]	Splits[b] (no.)	Timing of splits
Less permeable vs porous soils with low soil N		
SD (100-110 d) variety	3-4[c]	0, 28, 49 (PI) / 0, 14, 28, 49 (PI)
MD (130-140 d) variety	4-5[c]	0, 21, 42, 63 (PI) / 0 and every 14-21d to PI
LD (> 150 d) variety	5-6[c]	0, and every 14-21 d to PI
Less permeable vs porous soils with high soil N		
SD (100-110 d) variety	2-3[c]	21 and 42 (PI) / 14, 28, 42-49 (PI)
MD (130-140 d) variety	3-4[c]	21, 42, 63 (PI) / 14, 28, 42, 56-63 (PI)
LD (> 150 d) variety	4-5[c]	21, 42, 63, 77 (PI) / 14 and every 14-21 d to PI
Aged seedlings	Same	Same, but more N as basal and less at topdressing
Danger of bacterial leaf blight	4,5,6[d]	Every 14 to 21 d, based on disease severity

[a]SD = short duration, MD = medium duration, LD = long duration
[b]% of N to apply at each split: 2 splits: 33 & 67 or 50-50;
3 splits: 33-33-33 or 20-40-40 or 50-25-25;
4 splits: 25-25-25-25 or 20-30-30-20;
5 splits: 20-20-20-20-20 or 10-20-30-30-10;
6 splits: 15-15-15-20-20-15,
[c]Lower number for clayey soils, higher number for sandy soils.
[d]For SD, MD, and LD varieties, respectively.

Climatic conditions

Climatic conditions that affect the efficient use of N fertilizers are (a) variable and unpredictable onset and distribution of rainfall; (b) in-season drought and/or flood; (c) low radiation due to high cloud cover, especially during the monsoon; and (d) high and low temperatures at the reproductive phase, leading to heat or cold injury. High temperature (McGarry et al 1987, Burch and Fox 1989, Cai et al 1992) and high windspeed (Fillery et al 1986) accelerate the volatilization of ammonia from N fertilizers applied on soil surface or floodwater. Mineralization of organic N is limited by low temperature in the highlands (Balasubramanian et al 1995).

Research needs

Introducing a green manure crop in the rice - wheat system has been reported to increase the incidence of bacterial leaf blight and stem borer on rice (R. Nagarajan, Directorate of Wheat Research, Karnal, India, 1996, pers. commun.), probably due to excess N application to rice after green manure. The effect of prerice green manure or grain legumes on incidence of insect pests and diseases on rice and other cereals in a cropping sequence has to be clarified by further research.

In western India, rice straw bundles are allowed to decompose in situ in a corner of the wheat fields in rice - wheat system. Rice straw bundles form an ideal habitat for insects and disease organisms that infect the wheat crop. Fast biodecomposition methods are needed to quickly compost such crop residues and kill the pathogens, to minimize the carryover of insect pests and diseases from one crop to another.

Conclusions

Numerous technical and management factors constrain the performance of improved nutrient management technologies in farmers' fields. The following major crop management practices affect the efficiency of nutrient use in most rice farms of Asia:

- poor water control, especially during the monsoon;
- less-than-optimum plant population due to inadequate land preparation and leveling, low seed quality, use of aged seedlings at transplanting, contract transplanting on area basis, pattern and depth of transplanting and direct seeding, biotic and abiotic stresses on direct-seeded rice, etc.;
- partial nutrient application;
- insufficient weed control;
- uncertainty in the timing of sowing, transplanting, weed control, and/or harvest operations; and
- poor postharvest processing.

It is important that farmers improve many, if not all, of these conditions to reap the full benefits of modern rice varieties and related crop management technologies.

Several soil and climatic constraints also adversely affect the efficient use of nutrients in farmers' fields. Soil-related constraints include high spatial and temporal variability in soil nutrient status, nutrient imbalance, poor drainage, soil degradation (salinity, alkalinity, acidity), and subsoil compaction. Climatic factors such as variable and unpredictable rainfall, drought, flood, low radiation, and extremes in temperature affect crop growth and nutrient management in rice systems. Most of these factors are beyond the farmers' control.

Stress-tolerant rice varieties and improved management practices are being developed and evaluated on rice farms to tackle soil and climatic constraints. Good land preparation, use of quality seeds for direct sowing or young seedlings for transplanting, close spacing and arrangement of plants in rows will ensure optimum plant population. Judicious and combined use of all available nutrient sources (integrated nutrient management [INM]) is essential to reduce fertilizer cost, increase nutrient use efficiency, minimize pollution, and sustain productivity and resource quality. The following INM practices can enhance the N nutrition of rice crops :

- Adopting need-based, variable-rate fertilizer application to supplement the indigenous N supply;
- Tapping deep soil N reserves and prerice soil nitrate accumulations;

- Adjusting fertilizer N rates when organic manures are used with fertilizers (available nutrient contents of organic manures should be included in calculating nutrients added); and
- Using refined fertilizer recommendations for cropping systems, such as reducing N dose from 120 to 80 kg ha^{-1} for rice following a green manure crop as advocated in Haryana, India (Tandon 1989). Similar refined fertilizer recommendations are needed for other cropping systems.

In addition, better methods of N application such as basal incorporation, subsoil banding along seed rows, deep placement, and co-situs placement with seeds, etc., plus the use of controlled-release fertilizers will increase both rice yields and efficiency of applied N.

Portable decision aids are available to improve nutrient management through need-based nutrient application to individual ricefields. They include soil test kit, chlorophyll meter, leaf color chart, pH paper strip, phosphorus sorption paper strip, pocket penetrometer, soil thermometer, soil moisture tester, grain moisture meter, and others. Adequate training is necessary for extension agents and farmers to effectively use these tools for better soil and crop management.

References

Balasubramanian V, Rabeson R, Razafinjara L, Ratsimandresy J (1995). Rice soil constraints and fertility management in the highlands of Madagascar. Pages 313-324 in Fragile lives in fragile ecosystems. International Rice Research Institute, P.O. Box 933, Manila, Philippines.

Balasubramanian V, Morales A C, Cruz R T, Abdul Rachman S (1996) Farmers' evaluation of knowledge-intensive nitrogen management technologies to sustain rice yields under irrigated conditions in Asia. Paper presented at the Expert Consultation on Technological Evolution and Impact for Sustainable Rice Production in Asia and the Pacific, 29-31 Oct 1996, FAO Regional Office, Bangkok, Thailand.

Balasubramanian V, Morales A C, Cruz R T, Abdul Rachman S (1998) On-farm adaptation of knowledge-intensive nitrogen management technologies for rice systems. Nutr. Cycles Agroecosyst. (in press)

Bhuiyan S I, Sattar M A, Tabbal D F (1994) Wet-seeded rice: water use efficiency and productivity, and constraints to wider adoption. Paper presented at the Workshop on constraints, opportunities, and innovations for wet seeded rice, 31 May-3 Jun 1994, Bangkok, Thailand. 19 p.

Burch J A, Fox R H (1989) The effect of temperature and initial soil moisture content on the volatilization of ammonia from surface applied urea. Soil Sci. 147: 311-318.

Cai G X, Peng G H, Wang X Z, Zhu J W, Zhu Z L (1992) Ammonia volatilization from urea applied to acid paddy soil in southern China and its control. Pedosphere 2: 345-354.

Cao Z H, De Datta S K, Fillery I R P (1984) Nitrogen-15 balance and residual effects of urea-N in wetland ricefields as affected by deep placement techniques. Soil Sci. Soc. Am. J. 48: 203-208.

Celton J, Marquette J (1972) Plaine de Tananarive: Resultats des essais agronomiques de la campaigne 1971-72. Document No. 359. IRAM/IRAT, Antananarivo, Madagascar. 14 p.

Daftardar S Y, Savant N K (1995) Evaluation of environment-friendly fertilizer management for rainfed lowland rice on tribal farmers' fields in India. Pages 173-186 in Fragile lives in fragile ecosystems. P.O. Box 933, Manila, Philippines: International Rice Research Institute.

De Datta S K (1986) Improving N fertilizer efficiency in lowland rice in tropical Asia. Fert. Res. 9: 171-186.

De Datta S K, Buresh R J (1989) Integrated nitrogen management in irrigated rice. Adv. Soil Sci. 40: 143-169.

FAO--Food and Agriculture Organization (1994) Special programme for food security in Burkina Faso: exploratory mission report, 8-25 Nov 1994. 23 p.

Fillery I R P, Simpson R, De Datta S K (1986) Contribution of ammonia volatilization to total nitrogen loss after application of urea to wetland rice fields. Fert. Res. 8: 195-202.

Freney J R, Trevitt A C F, De Datta S K, Obcemea W N, Real J G (1990) The interdependence of ammonia volatilization and denitrification as nitrogen loss process in flooded rice fields in Philippines. Biol. Fert. Soils 9: 31-36.

Furuya S (1987) Growth diagnosis of rice plants by means of leaf color. Jpn. Agric. Res. Q. 20: 147-153.

George T, Ladha J K, Garrity D P, Buresh R J (1994) Legumes as nitrate catch crops during the dry-to-wet transition in lowland rice cropping systems. Agron. J. 86: 267-273.

Hill J E, Bayer D E, Bocchi S, Clampett W S (1991) Direct seeded rice in the temperate climates of Australia, Italy, and the United States. Pages 91-102 in Direct seeded flooded rice in the tropics. International Rice Research Institute, P.O. Box 933, Manila, Philippines:

INSFFER--International Network on Soil Fertility and Fertilizer Evaluation for Rice (1985) Summary reports on the fifth international trial on nitrogen fertilizer efficiency in irrigated wetland rice (1981-84). International Rice Research Institute, P.O. Box 933, Manila, Philippines.

INSFFER--International Network on Soil Fertility and Fertilizer Evaluation for Rice (1986) Progress report on the first international trial in comparison of hand and machine applied prilled urea and urea supergranules in lowland rice (1984-85). International Rice Research Institute, P.O. Box 933, Manila, Philippines.

Kanno H, Shoji S (1994) A new fertilizer placement method--co-situs placement using controlled availability fertilizers. Pages 492-493 in Transactions of the 15th World Congress of Soil Science, Acapulco, Mexico, 10-16 Jul 1994. Vol 5b. International Society of Soil Science, Vienna, Austria.

Kissel D E, Cabrera M L, Ferguson R B (1988) Reaction of ammonia and urea hydrolysis products with soil. Soil Sci. Soc. Am. J. 52: 1739-1796.

Kumar V, Shrotriya G C, Kaore S V (Eds.) (1989) Soil fertility and fertilizer use. Vol. III. Urea supergranules for increasing nitrogen use efficiency. Indian Farmers Fertilizer Cooperative, New Delhi, India. 143 p.

Kundu D K, Ladha J K, Lapitan-de Guzman E (1996) Tillage depth influence on soil nitrogen distribution and availability in a rice lowland. Soil Sci. Soc. Am. J. 60: 1153-1159.

Malabanan F M (1993) Association of seed vigor with field performance of high-yielding and traditional rice (*Oryza sativa* L.) cultivars. Ph D dissertation, University of the Philippines Los Baños, Laguna, Philippines.

McGarry S J, O'Toole P, Morgan M A (1987) Effect of soil temperature and moisture content on ammonia volatilization from urea-treated pasture and tillage soils. Irish J. Agric. Res. 26: 173-182.

Mishra B, Tyagi N K (1986) Highlights of research on crops and varieties for salt affected soils. Central Soil Salinity Research Institute, Karnal, India. 28 p.

My T V, Tuong T P, Xuan V T, Nghiep N T (1995) Dry seeding rice for increased cropping intensity in Long An Province, Vietnam. Pages 111-122 in Denning G L, Xuan V T (Eds.) Vietnam and IRRI: a partnership in rice research. International Rice Research Institute, P.O. Box 933, Manila, Philippines.

O'Brien D T, Irawan, Prawirasumantri J (1985) Economic evaluation of alternative methods of urea deep placement in rice. Special Publication SP-6. International Fertilizer Development Center, Muscle Shoals, Alabama, USA. p. 51-54.

ODR--Operation developpement recherche (1993) Ministere de la vulgarisation agricole. Antananarivo, Madagascar.

Padmaja P, Koshy M M (1978) A note on the runoff losses of nutrients in water logged rice soil. J. Indian Soc. Soil Sci. 26: 74-75.

Panda D, Samantaray R N, Patnaik S (1988) Management of fertilizer nitrogen for direct sown, rainfed lowland rice (*Oryza sativa* L.). J. Agric. Sci. UK 110: 475-479.

Pandey S (1998) Farmer adoption of nutrient management technologies: Economic and institutional constraints and opportunities. Nutr. Cycles Agroecosyst. (in press)

Peng S, Garcia F V, Laza R C, Sanico A L, Visperas R M, Cassman K G (1996) Increased N-use efficiency using a chlorophyll meter on high-yielding irrigated rice. Field Crops Res. 47: 243-252.

Peterson T A, Blackmer T M, Francis D D, Schepers J S (1993) Using a chlorophyll meter to improve N management. A Nebguide in Soil Resource Management: D-13, Fertility. Cooperative Extension, Institute of Agriculture and Natural Resources, University of Nebraska, Lincoln, USA:

Savant N K, Ongkingco P S, Garcia F D, Dhane S S, Khadse R R, Chavan S A, Rao KS (1992) Agronomic performance of urea briquette applicator in transplanted rice. Fert. Res. 32: 139-142.

Schnier H F, Dingkuhn M, De Datta S K, Marqueses E P, Farolina J E (1990) Nitrogen-15 balance in transplanted and direct seeded flooded rice as affected by different methods of urea application. Biol. Fert. Soils 10: 89-96.

Singh B, Singh Y, Khind C C, Meelu O P (1991) Leaching losses of urea-N applied to permeable soils under lowland rice. Fert. Res. 28: 179-184.

Takamura Y, Tabuchi T, Sujuki S, Hangal Y, Uneno T, Kubuta H (1976) The fate and balance sheet of fertilizer nitrogen and phosphorus: 1. Applied to rice paddy field in the Kasumigaura Basin. J. Sci. Soil Manure, Japan 47: 398-405.

Tandon H L S (1989). Urea supergranules for increasing nitrogen efficiency in rice--an overview. Pages 10-22 in Kumar V, Shrotriya G C, Kaore S V (Eds.) Soil fertility and fertilizer use. Vol. III. Urea supergranules for increasing nitrogen use efficiency. Indian Farmers Fertilizer Cooperative, New Delhi, India.

Tran D V (1996) Evolution of rice yield in Asia and the Pacific. Paper presented at the Expert Consultation on Technological Evolution and Impact for Sustainable Rice Production in Asia and the Pacific, 29-31 Oct 1996, FAO Regional Office, Bangkok, Thailand. 18 p.

Turner F T, Jund M F (1991) Chlorophyll meter to predict nitrogen topdress requirement for semidwarf rice. Agron. J. 83: 926-928.

Turner F T, Jund M F (1994) Assessing the nitrogen requirements of rice crops with a chlorophyll meter method. Aust. J. Exp. Agric. 34: 1001-1005.

Vlek PLG, Byrnes B H, Crasswell E T (1980) Effect of urea placement on leaching losses of nitrogen from flooded rice soils. Plant Soil 54: 441-449.

Wetselaar R (1985) Deep-point placed urea in a flooded soil: a mechanistic view. In: Special Publication SP-6. International Fertilizer Development Center, Muscle Shoals, Alabama, USA. p.7-14.

Yamauchi M, Aragones D V, Casayuron P R, Winn T, Borlagdan P C, Quick G R, Aguilar A M, Cruz R T, Sta. Cruz P C, Asis C A (1994) Rice anaerobic direct seeding in the tropics. Paper presented at the workshop on constraints, opportunities, and innovations for wet seeded rice, 31 May - 3 Jun 1994, Bangkok, Thailand. p 19.

Zeigler R S, Puckridge D W (1995) Improving sustainable productivity in rice-based rainfed lowland systems of South and Southeast Asia. GeoJournal 35: 307-324.

Adoption of nutrient management technologies for rice production: economic and institutional constraints and opportunities

S. PANDEY
International Rice Research Institute, P.O. Box 933, Manila, Philippines

Key words: Green Revolution, technology adoption model, rainfed rice system, irrigated rice system, nutrient management options, targeted interventions, evolution of rice systems

Abstract
When semidwarf, fertilizer-responsive varieties became available in the late 1960s, a range of policy and institutional changes were introduced to exploit the potential of these varieties. As a result, adoption of both high-yielding varieties and chemical fertilizers increased rapidly, especially in irrigated areas. A review of how these policy and institutional changes have influenced adoption of chemical fertilizers is presented using a microeconomic model of technology adoption. It is argued that fine-tuning of these policy and institutional innovations will continue to be important in further increasing rice yields and farmers' incomes, especially in rainfed areas where fertilizer use is still very low. In the more intensive irrigated areas, where chemical fertilizer use is already high, a change in the paradigm--from that of encouraging higher input use to that of achieving increased input use efficiency--is suggested. Based on a conceptual model of the likely evolution of the nature of rice production systems in Asia, a somewhat targeted approach to the design of technological, policy, and institutional interventions for improving farmers' nutrient management practices is recommended.

Introduction

The impressive success of the Green Revolution (GR) technology in increasing food production and improving food security of the poor is well documented (Anderson et al 1985). The GR technology, which was based on a combination of high-yielding varieties, irrigation, fertilizers, and other complementary inputs, was rapidly adopted by farmers in areas with access to these inputs. The public sector encouraged the adoption of the technology through a range of policy and institutional interventions.

Despite the achievements of the past several decades, the demand for rice will continue to outstrip production into the future due to the expansion of population (Hossain and Fischer 1995). The growth rate of rice production, on the other hand, has slowed down considerably in recent years. A long-term decline in the price of rice has reduced the profitability of rice farming and, thus, the incentives for growing rice. In addition to the technological breakthrough in raising the yield frontier and encouraging input intensification in areas where substantial yield gaps exist, appropriate crop management technologies to increase the efficient use of inputs and to reduce production cost of rice are needed to increase the supply and keep the price of rice at a level affordable to low-income consumers.

Inorganic fertilizers are one of the major inputs in rice farming. Traditionally, farmers relied on organic sources of nutrients such as crop residues and animal manures. The nutrient management (NM) technology refers to other ways of utilizing these sources of nutrients to increase rice yields and profitability.

V. Balasubramanian, J. K. Ladha and G. L. Denning (eds.), Resource Management in Rice Systems: Nutrients, 167–180.

This paper analyzes the economic factors determining the adoption of NM technologies in rice production and identifies intervention points to enhance the profitability and sustainability of rice production through better NM. The patterns of fertilizer use in rice production are first examined, and a microeconomic model of adoption of NM technology is presented. The changes in the nature of NM technology demanded in the course of the GR are subsequently discussed. A typology of rice production systems is presented and its relevance for research on development of NM technology is discussed. Finally, some suggestions on the design of policy and institutional interventions to enhance the adoption of NM technologies are made.

Fertilizer use in rice production

Fertilizer use in agriculture increased rapidly in the wake of the GR. Global fertilizer use increased from about 47 million t of nutrients in the mid-1960s to about 146 million t by the end of the 1980s. In Asia, which accounts for about 47% of the global fertilizer use, the growth rate of fertilizer use is around 7.3% per annum. Although countries such as Korea, Japan, and China have a very high rate of application of nutrients per unit area, Bangladesh, Pakistan, Thailand, and India have application rates below 100 kg of nutrients per hectare (Fig. 1) (Pingali et al 1995).

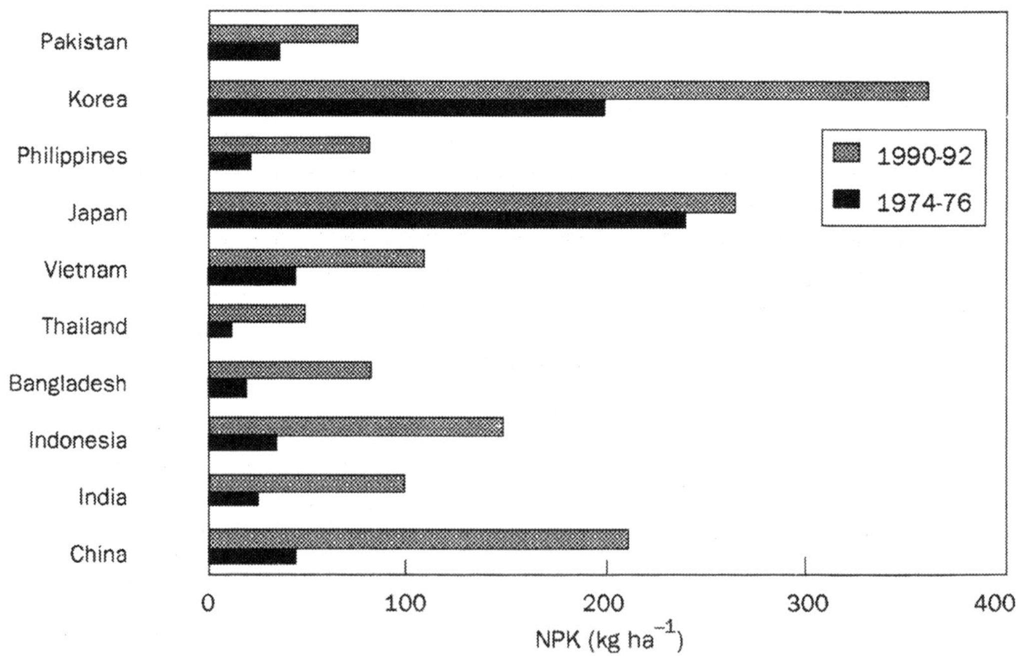

Figure 1. Nutrient use in major rice-producing countries (Pingali et al 1995).

At the aggregate level, the rate of chemical fertilizer application appears to be determined mainly by irrigation. The application rate is higher in countries where rice is produced under fully irrigated conditions than in countries with limited irrigation coverage. Even within a country with low aggregate level of application, the application rates in irrigated areas approach those in the more advanced East Asian countries. In contrast, rainfed areas lag far behind. For example, NPK application in India is more than 150 kg ha^{-1} in the irrigated areas, but less than 50 kg ha^{-1} in the rainfed areas (Fig. 2).

Overall, considerable economic potential exists for increasing fertilizer use in countries with low application rates. Production function estimates indicate that the marginal productivity of fertilizer at the current level of use is considerably higher than its real price, especially in regions where rice is grown mainly under rainfed conditions (Table 1). Rice production could be further increased in these areas by encouraging higher application rates. In the more intensive rice areas where application rate is already high, the marginal productivity is close to the real price of fertilizers, indicating that it would not be economical for farmers to apply more fertilizers. Given that fertilizer cost accounts for almost a quarter of the production cost in these intensive areas, farmers could perhaps benefit more by increasing fertilizer use efficiency rather than by further increasing the quantity. Such an approach seems reasonable, especially in the light of environmental concerns and reduction or elimination of price subsidies on fertilizers.

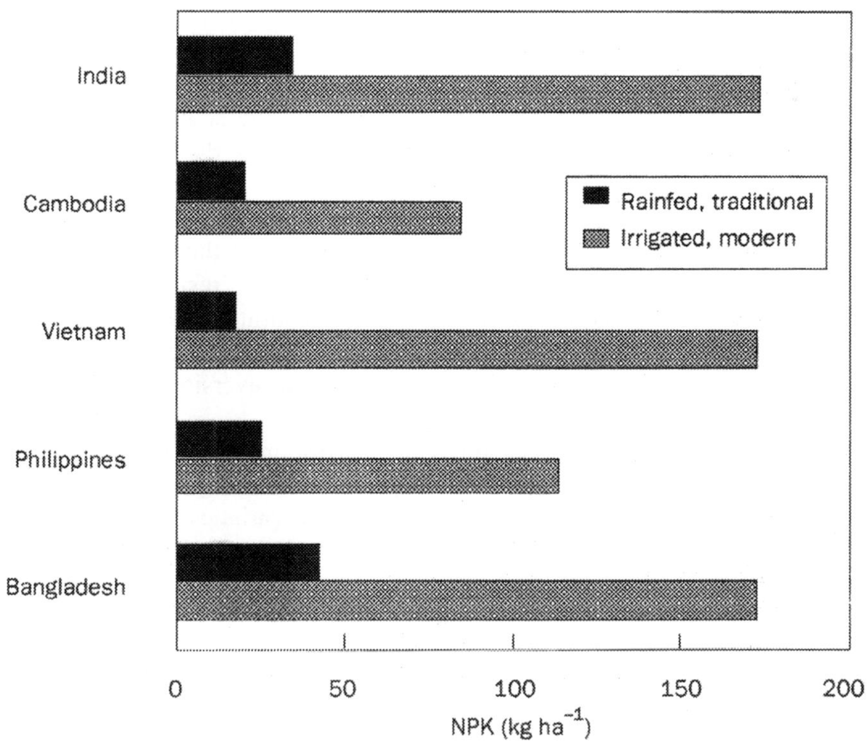

Figure 2. Estimates of fertilizer use, by rice ecosystem (Pingali et al 1995).

Table 1. Marginal productivity of fertilizer use in different rice production environments, selected Asian countries.

Country/season/environment	Mean NPK (kg ha^{-1})	Standard deviation (SD)	Marginal productivity at mean NPK	Marginal productivity at mean + 1 S.D. NPK	Fertilizer-rice price ratio
Bangladesh (1987 wet season)					2.30
Irrigated	86.54	141.30	10.99	8.79	
Rainfed	76.94	40.37	20.98	14.61	
Thailand (1987 wet season)					3.79
Irrigated	71.67	41.91	9.88	8.94	
Rainfed	16.15	26.62	8.85	8.47	
Philippines					4.08
1985 wet season	84.20	24.46	15.32	13.90	
1986 dry season	133.64	58.05	8.26	3.33	
Indonesia (1988 dry season)					1.70
Irrigated	176.38	53.62	2.67	-6.14	

A microeconomic model of NM technology adoption

For divisible inputs such as chemical fertilizers, adoption can be measured by the extent of use (Feder et al 1985). To study factors determining adoption, it would be useful to consider adoption as consisting of two steps. There must first be a latent demand for the technology. The potential for a technology to increase farmers' welfare generates a latent demand for it. In the second step, this latent demand is translated into an actual demand and adoption. Different sets of factors determine the latent and the actual demand. Of course, actual demand cannot occur without the latent demand, but the existence of a latent demand is not a sufficient condition for adoption.

The latent demand for a technology depends on the arguments of the utility function of a farmer. If farmers are assumed to be profit maximizers, all factors that determine the level of profit influence the latent demand. The characteristics of the response function, the price of the input under consideration, the price of other substitute or complementary inputs, and the price of output are the main factors which determine the profitability. If the farmer is risk-averse, the level of risk and the degree of risk aversion also become relevant.

The characteristics of a technology are captured by the response function, which is defined as the quantitative relationship between the output and the input in question, with the levels of all other inputs being held constant. Modern rice varieties are considered more responsive to fertilizers in the sense that for a given level of nutrient application, the modern varieties yield more than the traditional varieties (Fig. 3). The fertilizer responsiveness of modern varieties is one of the major factors that led to a rapid increase in yield and level of fertilizer use in the wake of the GR.

In addition to the nature of the response function, the profit-maximizing level of fertilizer application depends also on the price factors. The yield-maximizing level of fertilizer is applied only if the cost of fertilizer is zero. The lower the ratio of the price of fertilizers to the price of rice, the higher will be the optimal quantity demanded. Thus, a

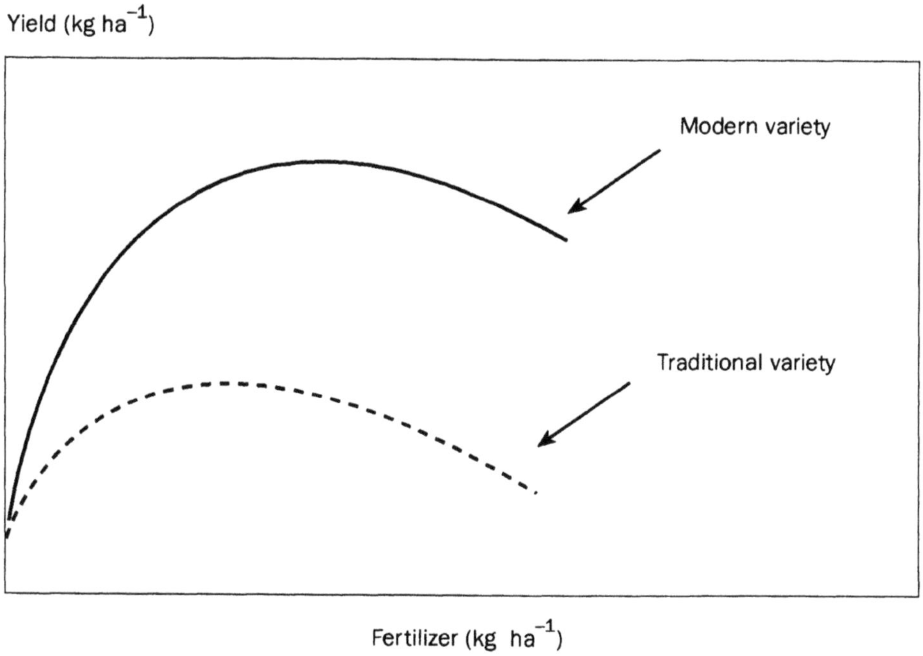

Figure 3. Hypothesized yield response curves for modern and traditional varieties.

farmer can be considered to be moving up the response curve as the fertilizer-rice price ratio falls. The difference between the rates of fertilizer application in, say, Japan and India could be partly explained by the difference in this ratio. Although the price ratio has declined over time in both countries, the N-rice price ratio in India is more than double that of Japan (Pingali et al 1995).

Risk considerations can lead to a level of use which is different from the profit-maximizing value (Anderson et al 1977). The quantity of inputs used (especially purchased inputs such as chemical fertilizers) by a risk-averse farmer can be lower than that used by a risk-neutral farmer if risk increases with an increase in input (Feder et al 1985). Empirical evidence shows that reductions in optimal N use due to moderate risk aversion range from 0 to 30% of the risk-neutral level (Roumasset et al 1989) in irrigated areas of the Philippines. District-level data from eastern India indicate that the average level of fertilizer application is negatively correlated with the magnitude of yield risk, suggesting that in more variable environments, farmers apply smaller average quantities of fertilizers (Hossain and Laborte 1994). Similar results have been obtained from a microlevel study of fertilizer application in rainfed rice in Central Luzon, Philippines (Pandey et al 1997). The importance of risk considerations on the level of fertilizer application depends also on the opportunities for diffusing income risk by participation in the land, labor, and capital markets (Anderson and Hazell 1994). Where such markets are effective in diffusing risk, farmers are likely to display an essentially risk-neutral behavior.

The resource endowments may be considered to constrain the choices available to farmers so that the optimal level of adoption under resource constraint is different from that of the utility-maximizing value. Differences in resource endowments among farmers can lead to differential degrees and patterns of adoption. Resource endowments are usually defined to include access to conventional inputs such as land, labor, and capital. The effect of differential access to these inputs on technology adoption and, hence, on farmer welfare was the subject of much of the debate on the GR technology. Earlier literature indicated that technology adoption was greater and occurred at a faster pace among farmers who had better access to these inputs (Griffin 1974). Small farmers, farmers with insecure tenancy, and farmers with limited access to capital and markets were hypothesized not to adopt the GR technologies. The more recent evidence of this, however, indicates that all categories of farmers ultimately adopted divisible components of the GR technologies, especially the high-yielding varieties, regardless of access to these conventional resources (Anderson et al 1985, Dalrymple 1986, David and Otsuka 1994). What seems to matter more is not the access to these conventional resources per se but the differential quality of resource base, especially that of the human capital. For simple technologies such as the replacement of traditional varieties by modern varieties without any major change in farm management, the importance of human capital may, however, be less important than for management-intensive technologies such as improved NM. Recent empirical evidence from areas where high levels of inputs are being applied indicates that the differences in technical efficiency across farms are due mainly to differences in farmer skills and access to information (Ali and Byerlee 1991).

Technologies requiring large investments are economically profitable only if the scale of operation is above some critical value. Such technologies are not scale-neutral and are less likely to be adopted by small farmers. The tractor is considered a typical example of indivisible technology, even though the services of tractors may be made divisible through the operation of the rental market. Improved seeds and chemical fertilizers, on the other hand, are divisible, hence there are no scale advantages associated with their use. However, even for seemingly divisible technologies, access to information and cost of learning may impart indivisibility. Efforts in acquiring and learning to use improved NM technologies may not be economically justifiable for farmers with a small scale of operation. The cost of learning, in addition to the friction caused by risk aversion and liquidity constraint, can also generate a sequential pattern of adoption whereby more profitable components of NM technologies are first adopted (Ryan and Subramanian 1975, Byerlee and Hesse de Polanco 1986). Farmers approach economically optimal level of input use over time as they learn more about the technology (Antle and Crissman 1990).

Given that the latent demand exists, whether or not a technology is adopted depends on the physical availability of inputs, farmers' purchasing capacity, and the cost of acquiring information about technology and learning how to use it. The public sector can play an important role in determining the level of latent demand by manipulating the price policy and in translating the latent demand into actual demand through institutional interventions such as the provision of inputs, credit, extension, and education. In the early periods of the GR, Asian governments actively encouraged the adoption of improved technologies through policy and institutional interventions (Mellor and Ahmed 1989). Price subsidy was an important element of price policy, which is believed to have encouraged adoption of fertilizers. Stabilization of output price is another aspect of price policy used to reduce risk caused by wide swings in the price of output. Although stabilization policies based on food procurement and distribution monopolies have their costs, price stabilization policies in India and Pakistan played a critical role in encouraging the adoption of the GR technologies (Desai 1989, Byerlee 1994).

Another major role played by the public sector in enhancing the adoption of NM technologies is through investment in irrigation (Rosegrant and Svendsen 1992). By dramatically increasing the marginal productivity of fertilizers, the availability of irrigation made fertilizer use highly desirable. The latent demand thus created was subsequently translated into actual demand by the active involvement of the public sector in the distribution of improved seeds and fertilizers. In the early stages of technological change when the private sector was unable to provide agricultural inputs due to poorly developed rural infrastructure and marketing systems, the public sector played a very important role in supplying these inputs to farmers. Government-backed institutional credit programs may have similarly helped the adoption where cash constraint was binding, although such interventions had other undesirable effects (Adams et al 1984).

From increasing input use to increasing input use efficiency

Broadly speaking, the GR phase of technological change in agriculture can be characterized as an intensification of inputs where farmers used more inputs to exploit the opportunities provided by the availability of input-responsive, high-yielding varieties (Byerlee 1994). The public sector played an important role in encouraging such input intensification through a range of policy and institutional interventions, as discussed in the preceding section. Most of the areas with high levels of input use can now be considered to be in the post-GR phase where no further gains from increasing input use are likely to be realized. This situation occurs when no allocative inefficiencies exist. Instead, opportunities may exist for increasing the input use efficiency as farmers attempt to obtain a higher level of output for the same level of input or produce the same level of output with a lower input. The increase in input use efficiency (or technical efficiency) can be represented by an upward shift in the production function. A shift from production function Y_0 to Y_1 thus represents an increase in input use efficiency[1] (Fig. 4).

Increase in technical efficiency requires substitution of input by knowledge. Instead of just applying more nutrients, technical efficiency is improved by matching the supply of nutrients with the plant demand. Such matching would require knowledge about the crop status and the N-supplying capacity of the soil at different points in time throughout the crop growth cycle. Precision in the depth of application, improved timing, and application based on specific factors such as soil or plant tests would require additional time investment to learn, acquire, and use such knowledge-intensive technologies. The adoption of technologies that improve technical efficiency will occur only when, for a given output, the cost of input saved is greater than the cost of acquiring and using knowledge-intensive technologies.

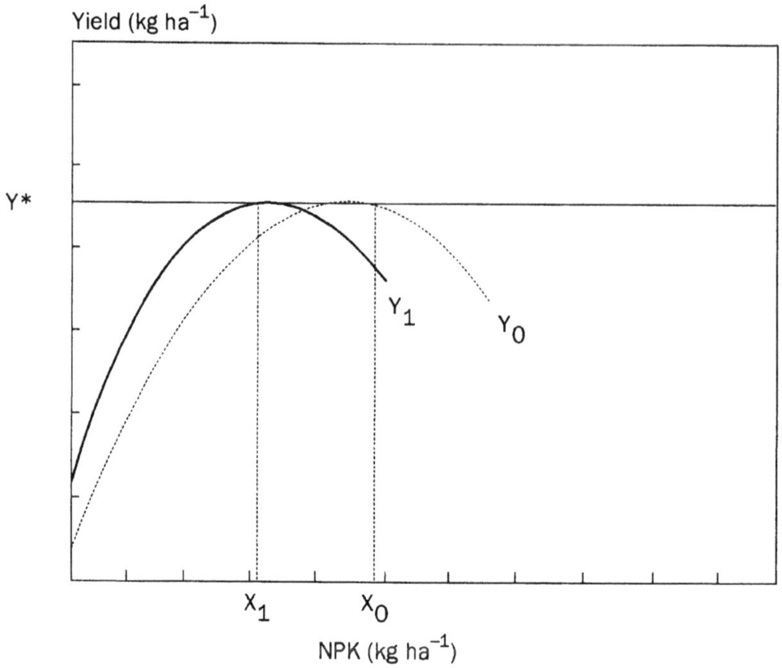

Figure 4. Fertilizer response functions at different levels of efficiency.

Adaptive research and NM technology

Improvement in nutrient use efficiency requires conditional recommendations, as opposed to blanket recommendations used during the input expansion phase. As discussed by Byerlee (1994), adaptive research, extension, and farmer education assumed a greater importance during the post-GR stage where input use efficiency is of major concern.

Adaptive research seeks to develop recommendations targeted to specific environmental conditions. Because environmental conditions can vary even within a small area depending on soil type, rainfall pattern, and management regime, methods and approaches that can take into account these factors simultaneously are needed to conduct adaptive research. Obviously, much of such research would have to be conducted in farmers' fields and could potentially benefit from farmer participation, although operational problems remain (Bentley 1994). Detailed characterization of socioeconomic and environmental conditions at a disaggregated level could facilitate identification of more or less homogeneous domains where a particular set of NM recommendations may be appropriate. Many of the methods and tools developed for farming systems research (Simmonds 1985) are hence relevant for adaptive research.

What would be the best mechanism for transmitting conditional recommendations generated from adaptive research? Because these recommendations are complex and may contain different kinds of if-then rules, the extension system designed to deliver simple messages and blanket recommendations are not appropriate. As compared with high-yielding varieties which spread from farmer to farmer, this mode of transmission is unlikely to be successful for complex information. Research in the context of integrated pest management indicates that farmers are unlikely to adopt such technologies unless they understand some basic principles involved (Goodell 1984, Waibel 1986). Thus, the role of extension would be not only to transmit the recommendations but also to impart knowledge to farmers so that they can monitor the crop and field conditions better and make appropriate adjustments in NM where necessary. For effective absorption and assimilation, the extension messages need to be packaged such that they fit well in the farmers' knowledge systems. Extension systems in most developing countries have severe capacity constraints and are unlikely to be able to fulfill these additional tasks without substantial investment in upgrading.

Economics of adoption of organic fertilizers

Farmers have traditionally relied on organic sources, such as farmyard manure, compost, straw, and green manure crops, to maintain and enhance soil fertility. A closer integration of livestock with cropping and the low opportunity costs of land and labor in traditional systems have made the use of organic nutrients economically viable.

To the extent that nutrients provided by organic sources are perfect substitutes for nutrients from inorganic sources, the effective prices of nutrients from these two sources determine their use. A long-term decline in the real price of inorganic fertilizers has made their use economically more attractive to farmers. Organic fertilizers require land, labor, and other inputs for their production and application. The effective price of nutrients from organic sources has been found to be higher than that of nutrients from inorganic sources (Rosegrant and Roumasset 1987; Garrity and Flinn 1987, Ali and Narciso 1994).

Agronomic evidence indicates that organic fertilizers can improve soil characteristics (Singh et al 1991), and this can result in additional yield gains in rainfed lowland systems. Possible long-term adverse effects of inorganic fertilizers on soil properties and the environment can be ameliorated by the use of organic sources that match nutrient-release patterns with crop demand (Becker et al 1994). These long-term and environmental benefits are not priced in the market, and organic fertilizers may socially be more beneficial than they appear in terms of the short-term benefit-cost criterion. Available information on this is, however, very limited (Ali and Narciso 1994). Further research on quantifying long-term benefits under farmers' conditions in a range of environments and management practices is needed.

Where organic fertilizers do, in fact, improve soil quality and enhance sustainability, it does not necessarily mean that they should be produced in situ. They could be produced ex situ by other producers and sold to users. Where the opportunity costs of labor and land are rising, in situ production of organic fertilizers may not be economical. Recent trends in vegetable production in peri-urban areas in the Philippines using organic manure imported from poultry farms and other sources attest to this. Often, the discussions on organic fertilizers are confused by the mixing up of issues related to their production and the economics of their use.

Production system characteristics and NM technology

At a broader level of generalization, the nature of agricultural production systems can be considered to be determined by population density and stage of economic growth (Pandey 1996). Population density is a major factor determining land use intensity (Boserup 1965, 1981). An increase in population pressure, when population density is low initially, results in an expansion of agricultural area, as returns to labor (which is a relatively scarce factor) are maximized by land-extensive strategy. Economic growth affects the agricultural sector in two ways: first, it alters the income-induced demand for staples, and second, it draws resources away from agriculture. When incomes are low, most of the income is spent on staples; but as incomes increase, demand for nonstaple commodities increases. Such shifts in consumption patterns provide an incentive for diversification of agricultural production systems. With economic growth, the agricultural sector must increasingly compete for inputs with expanding secondary and tertiary sectors. This competition creates a demand for agricultural technologies that save on inputs such as land, labor, and water, which are becoming increasingly expensive. The combined effect of population density and level of economic growth (as indicated by income level) on the nature of agricultural production systems is stylized in Figure 5.

When both income levels and population density are low, agricultural production systems tend to be land-extensive and subsistence-oriented. At an early stage of economic growth, production strategy is subsistence-oriented due to the small size of the nonfarm sector and limited infrastructure development (Type I). An increase in population density leads to the evolution of land- and labor-intensive production systems (Type II) that are subsistence-oriented, unless access to export market provides incentives for commercialization. At higher levels of income, production systems tend to be initially diversified (Type III), although further economic growth may result in somewhat specialized and mechanized production systems (Type IV).

		Population pressure: High	Population pressure: Low
Income level	High	* Small farms * Diversified * Commercial Type III	* Large farms * Mechanized * Commercial * Specialized (?) Type IV
Income level	Low	* Small farms * Subsistence-oriented * Labor-intensive Type II	* Large farms * Subsistence-oriented Type I

Figure 5. Effect of population density and income level on the nature of rice production systems.

The nature of the NM technology demanded by farmers will differ in different regions (or countries), depending on the predominant nature of the production system. In Type I systems, the demand for improved NM technologies that increase yield and returns to land will be limited because it is a relatively abundant factor of production. Such technologies will be adopted only if they also help save labor, which often is the more limiting input. The stimulation of demand for yield-increasing technologies (including NM technologies) in these regions will initially require policy reforms (such as expansion of export market, improved market infrastructure) that will increase the profitability of foodgrain production. Such reforms can be expected to lead these regions along the path of input expansion where the demand for inorganic fertilizers will continue to increase. Type I systems are found mainly in sparsely populated regions of Laos, Myanmar, and Vietnam.

In Type II systems, a high labor-land ratio will generate demand for NM technologies that will increase yield. Yield-increasing technologies will be demanded even if they are labor-intensive. For example, labor-intensive NM technologies such as green manuring and composting are used mostly by farmers in Type II systems. The demand for inorganic chemical fertilizers tends to be low, however, due to low-level income and subsistence nature of production systems. These type II systems demand NM technologies that help recycle the nutrients by utilizing crop and livestock byproducts. The rainfed rice-growing areas in eastern India and Bangladesh are examples of Type II systems. Rice production in these environments suffers from a host of biotic and abiotic stresses.

The rice production systems of Type III demand NM technology that fits well with diversified activities at the farm level. In most cases, this will mean increased reliance on inorganic fertilizers as commercialization improves farmers' purchasing power. As high levels of nutrients are applied to both the subsistence and cash crops, NM technologies that increase input use efficiency become increasingly in demand. Often, NM for cash crops occupies the center stage due to the overwhelming importance of these crops in the household economy. Green manuring and other organic sources of fertilizers are generally not economical due to high opportunity cost of land and labor. The diversified rice production systems of Thailand, Indonesia, and parts of the Philippines are examples of Type III systems.

In Type IV systems, nutrient use efficiency is a major concern in determining the profitability of rice farming. In addition, environmental impact of high levels of nutrient use is important. Farmers and policymakers thus increasingly demand NM technologies that help reduce the quantity of inorganic fertilizers applied. In addition, consumers may express preference for products grown with low chemical inputs by their willingness to pay a price premium for such products. These factors may ultimately lead to increasing supplementation of inorganic nutrients by organic sources. NM technologies become more sophisticated, and specialized agencies for providing NM advice may evolve. These kinds of technologies are likely to be demanded by farmers in intensive rice production systems in irrigated areas of Malaysia, Japan, Taiwan, and Korea.

Implications for research and public policy

The preceding discussion indicates that different types of NM technology and policy interventions are required, depending on the characteristics of rice production systems. In intensive rice production areas that are in the post-GR stage, the emphasis has shifted from input expansion to input use efficiency. NM technologies that help make more efficient use of nutrients are needed in these areas. In addition to raising the profitability of rice production, such technologies can help reduce the negative environmental effects of high level of nutrient application.

The adoption of more efficient NM technology, of course, depends on whether or not the cost of fertilizer saved is greater than the additional cost of acquiring and using efficient NM technology. The real price of chemical fertilizers has declined over time due to technical changes in the fertilizer industry (Tomich et al 1995). If the relative price of fertilizer remains low in the future, the economic incentives for adopting efficient NM technologies may not exist. The challenge to researchers is to develop efficient NM technologies that not only save on fertilizer but also reduce the cost of using the technology. Because labor cost is generally high in intensive rice production areas, efficient NM technologies that require increased labor use are unlikely to be acceptable to farmers. Since the adoption process tends to follow a sequential pattern, initial emphasis should perhaps be on developing NM technologies in the form of simple rules that easily fit in the local knowledge systems and, hence, are likely to be retained. This points to the need to develop a better understanding of farmers' knowledge systems in relation to NM.

Public policy and institutional arrangements can play a critical role in facilitating the adoption of efficient NM technologies. Subsidies on fertilizer prices tend to dampen the demand for efficient NM technologies. Although there is a general trend toward liberalizing input and output markets for a variety of reasons, fertilizers are still subsidized in many countries. More rapid progress is needed in reducing and ultimately removing subsidies on chemical fertilizers. Such policies could also be justified on the grounds of reducing adverse environmental effects of high levels of fertilizer use.

New institutional arrangements for adaptive research and dissemination of efficient NM technologies are also needed to reduce the cost of adoption. The research-extension approach suitable during the GR period is less appropriate for achieving a higher input use efficiency. A decentralized institutional setup for conducting adaptive research for specific locations needs to be developed. Many of the features of the farming systems research are relevant in this context, and the approach may need to be revived. The extension system similarly requires new orientation and strengthening to be effective in transferring efficient NM technologies. Given the current capacity constraint of most extension agencies, other nontraditional conduits such as nongovernment organizations, the private sector, and mass media can play an important role.

In rainfed rice areas where nutrient use is still very low, the spread of fertilizer-responsive modern varieties is limited due to high levels of abiotic stresses. Development of improved varieties adapted to these environments will induce the demand for nutrients. Due to a close integration of rice and livestock production, improved recycling of nutrients through the use of crop and livestock residues may have a bigger role in these environments. The demand for inorganic fertilizers is likely to increase with an increase in income and commercialization of production systems. Public policies and institutional arrangements used during the GR phase of the 1970s may still have a role to play in facilitating the adoption of NM technologies in these areas. However, direct intervention of the public sector in fertilizer procurement and distribution may not be necessary because private sectors are now generally well developed in most areas. Since the use of inorganic fertilizers in rainfed environments is unlikely to be as high as in intensive irrigated areas, the demand for efficient NM technology in these environments will be somewhat limited. The essential features of interventions to encourage adoption of such technologies will, however, be similar to those in the post-GR areas.

References

Adams D, Graham D, von Pischke J (1984) Undermining rural development with cheap credit. Westview Press, Boulder, Colorado.

Ali M, Byerlee D (1991) Economic efficiency of small farmers in a changing world: a survey of recent evidence. J. Int. Dev. 3(1):1-27.

Ali M, Narciso J H (1994) Economic evaluation and farmers' perception of green manure use in rice-based farming systems. Pages 173-195 in Ladha J K, Garrity D P (Eds.) Green manure production systems for Asian ricelands. Selected papers from the International Rice Research Conference. International Rice Research Institute, P.O. Box 933, Manila, Philippines.

Anderson J R, Dillon J L, Hardaker B (1977) Agricultural decision analysis. Iowa State University Press, Iowa.

Anderson J R, Hazell P B R (1994) Risk considerations in the design and transfer of agricultural technology. Pages 321-339 in Anderson J A (Ed.) Agricultural technology: policy issues for international community. CAB International, UK.

Anderson J R, Herdt R W, Scobie G M (1985) The contribution of agricultural research to world agriculture. Am. J. Agric. Econ. 67(5):1080-1084.

Antle J M, Crissman C C (1990) Risk, efficiency, and the adoption of modern crop varieties: evidence from the Philippines Econ. Dev. Cult. Change 38(3):517-537.

Becker M, Ladha J K, Ottow J C G (1994) Nitrogen losses and lowland rice yield as affected by residue nitrogen release. Soil Sci. Am. J. 58:1660-1665.

Bentley J W (1994) Facts, fantasies, and failures of farmer participatory research. Agriculture and human values. J. Agric. Food Hum. Values Soc. 11(2-3):140-150.

Boserup E (1965) The conditions of agricultural growth. Allen and Unwin, London.

Boserup E (1981) Population and technological change: a study of long-term trends. University of Chicago Press, Chicago.

Byerlee D (1994) Technology transfer systems for improved crop management: lessons for the future. Pages 208-230 Anderson J R (Ed.) Agricultural technology: policy issues for the international community. CAB International, UK.

Byerlee D, Hesse de Polanco E (1986) Farmers' stepwise adoption of technological packages: evidence from the Mexican Altiplano. Am. J. Agric. Econ. 68(3):519-27.

Dalrymple D G (1986) Development and spread of high-yielding rice varieties in developing countries, Bureau for Science and Technology, U.S. Agency for International Development, Washington, D.C.

David C C, Otsuka K (1994) Modern rice technology and income distribution in Asia. Lynne Rienner Publishers, Inc., Colorado and London.

Desai G M (1989) Policy for rapid growth in use of modern agricultural inputs. Pages 204-218 in Mellow J M, Ahmed R (Eds.) Agricultural price policy for developing countries. The Johns Hopkins University Press, Baltimore and London.

Feder G R, Just E, Zilberman D (1985) Adoption of agricultural innovations in developing countries: a survey. Econ. Dev. Cult. Change 33:255-298.

Garrity D P, Flinn J C (1987) Farm-level management systems for green manure crops in Asian rice environments. In Sustainable agriculture: green manure in rice farming. International Rice Research Institute, P.O. Box 933, Manila, Philippines.

Goodell G (1984) Challenges to international pest management research and extension in the Third World: do we really want IPM to work? Bull. Entomol. Soc. Am. 30:18-26.

Griffin K (1974) The political economy of agrarian change: an essay on the Green Revolution. Harvard University Press, Cambridge.

Hossain M, Fischer K S (1995) Rice research for food security and sustainable agricultural development in Asia: achievements and future challenges. GeoJournal 3:286-298.

Hosssain M, Laborte A (1994) Differential growth in rice production in Eastern India: agroecological and socioeconomic constraints. International Rice Research Institute, P.O. Box 933, Manila, Philippines.

Mellor J W, Ahmed R (1989) Agricultural price policy for developing countries. The Johns Hopkins University Press, Baltimore and London.

Pandey S (1996) Rainfed lowland rice research: challenges and priorities for the 21st century. International Rice Research Institute, P.O. Box 933, Manila, Philippines.

Pandey S, Masicat P, Velasco L, Villano R (1997) Risk analysis of rainfed rice production systems in Tarlac, Central Luzon. International Rice Research Institute, P.O. Box 933, Manila, Philippines.

Pingali P L, Hossain M, Pandey S, Price L (1995) Determinants of farmer adoption of knowledge-intensive nutrient management technologies. Paper presented at the Mini Workshop/Think Tank on Nutrient Use Efficiency of the Cropping System(s), 13-15 Dec, International Rice Research Institute, Los Banos, Laguna, Philippines.

Rosegrant M W, Roumasset J A (1987) Economic feasibility of green manure in rice-based cropping systems. In Sustainable agriculture: green manure in rice farming. International Rice Research Institute, P.O. Box 933, Manila, Philippines.

Rosegrant M W, Svendsen M (1992) Irrigation investment and management in Asia. Paper presented at the IRRI-IFPRI Planning Workshop on Rice Supply and Demand Project. International Rice Research Institute, Los Baños, Laguna, Philippines.

Roumasset J A, Rosegrant M W, Chakravorty U N, Anderson J R (1989) Fertilizer and crop yield variability: a review In Anderson J R, Hazell P B (Eds.) Variability in grain yields: implications for agricultural research and policy in developing countries. The Johns Hopkins University Press, Baltimore.

Ryan J, and Subramanian KV (1975) Package of practice's approach in adoption of high-yielding varieties: an Appraisal. Econ. Polit. Weekly 10(4):A101-110.

Singh Y, Khind C S, Singh B (1991). Efficient management of leguminous green manures in wetland rice. Adv. Agron. 45:135-189.

Simmonds N (1985) Farming systems research: a review. World Bank Technical Paper No. 43. World Bank, Washington, D.C.

Tomich T P, Kilby P, Johnston B F (1995) Transforming agrarian economies: opportunities seized, opportunities Missed. Cornell University Press, Ithaca and London.

Waibel H (1986) The economics of integrated pest control in irrigated rice. Springer-Verlag, Berlin.

Government policy support for technology promotion and adoption: a case study of urea tablet technology in Indonesia

E. PASANDARAN[1], B. GULTOM[2], J. SRI ADININGSIH[3], HAPSARI[1], and SRI ROCHAYATI[3]

[1]*Center for Agricultural Research Programming*

[2]*Center for Agricultural Machinery Development*

[3]*Center for Soil and Agroclimatic Research, Agency for Agricultural Research and Development, Jakarta, Indonesia*

Key words: food security, intensive rice farming, deep-placement machine, farmer perception, technology constraints, adoption constraints, benefit-cost ratio, dissemination strategy

Abstract

Broadcasting of prilled urea not only is technically inefficient (recovery rate is only 30% of applied N); it also causes a substantial monetary loss to farmers and a high environmental cost to society. Urea tablet deep-placement technology has shown a 25% saving in N fertilizer rates, an average increase of 400 kg ha^{-1} in rice yield, and a benefit-cost ratio of 4-15 in various on-station and on-farm trials in Indonesia. The technology was introduced and promoted in Java island in 1992. Hand applicators and deep-placement machines were introduced to ease the application of urea tablets. Both consumption and area coverage of urea tablets increased substantially from 1992 to 1995. Farmers recognize the economic advantage of deep placement of urea tablets, but they think that it is less practical and more labor-intensive than broadcasting of prilled urea. There are also problems of storage, distribution, and handling of urea tablets, though these can be technically overcome. Appropriate dissemination strategies and government policy support on pricing, coordination of various fertilizer industries and distributors, and credit are vital for promoting the urea tablet technology further.

Introduction

One of the consequences of the Green Revolution is the increased use of modern inputs such as chemical fertilizers, pesticides, and massive irrigation systems. The development of large-scale rice intensification programs in various versions--Bimas of 1968, modified Bimas of 1971, Special Intensification (Insus) of 1979, and Super Intensification (Supra Insus) of 1987--is the prime mover to achieving and maintaining rice self-sufficiency. The main component of the intensification program is the use of high-yielding rice varieties which are very responsive to fertilizers, especially N. Thus, fertilizer use plays an important role in maintaining self-sufficiency in rice production.

Rice self-sufficiency and food security are considered strategic for Indonesia. Any shortages of this staple food can easily become a nationwide sociopolitical issue and a major security problem. In this regard, fertilizer subsidy is used as an important policy instrument in addition to a fertilizer distribution program which aims to make chemical fertilizers available at the farm level to support the intensification programs.

The use of urea and other inputs has significantly increased rice production. However, when urea is broadcast on flooded ricefields (the more commonly used method), utilization efficiency is low because of losses resulting from ammonia volatilization, nitrification-denitrification, runoff, and leaching. It is not uncommon for farmers to lose 70% of the applied fertilizer N (Wetselaar et al 1984). Therefore, efforts are being exerted to achieve more efficient use of N fertilizers and thereby reduce N losses, save on energy,

V. Balasubramanian, J. K. Ladha and G. L. Denning (eds.), Resource Management in Rice Systems: Nutrients, 181–190.

increase rice production, increase farmer income, lessen environmental pollution, and improve the national economy.

Present status of fertilizer production and consumption

The role of urea

Over the last two-and-a-half decades, the aggregate consumption of fertilizers has increased substantially. The use of urea increased from 372,260 t in 1969 to 3,696,260 t in 1995 (Table 1), mainly accelerated by the government's fertilizer subsidy. However, there remains a strong evidence of excessive fertilizer use at the farm level. By the mid-1980s, fertilizer application rates in Indonesia were much higher than those in other Asian countries (World Bank 1993). Moreover, research has shown that in many places in Java, the high N fertilization gives declining and, possibly, negative rates of return. In some areas of Java, farmers use double the recommended dose. For example, in some areas of intensification, farmers apply urea at 300-500 kg ha^{-1}, although the recommended dose is 200-250 kg ha^{-1} (CSAR 1996). Urea consumption was 12% higher than recommended rates in West Java, 28% higher in Lampung, and 89% higher in South Sulawesi (Table 2).

Research conducted by the Center for Soil and Agroclimatic Research has shown that broadcasting prilled urea resulted in very low plant uptake--only 30%. The rest, about

Table 1. Urea production and consumption in Indonesia ('000 t).

Year	Production	Import	Export	Consumption
1969	84.17	233.91	0	372.26
1970	98.41	225.26	0	399.67
1971	104.76	285.87	0	460.24
1972	130.15	561.11	0	555.04
1973	177.11	490.72	0	715.15
1974	262.67	1003.91	0	687.85
1975	451.17	1470.13	0	736.84
1976	400.39	15.93	0	765.89
1977	861.20	17.65	400.20	1027.30
1978	1505.78	39.09	230.46	1193.59
1979	1888.20	31.30	299.30	1348.85
1980	2083.54	266.98	162.43	1849.85
1981	2100.96	261.13	38.89	2167.65
1982	na	na	na	na
1983	na	4.57	320.78	2300.46
1984	3048.78	304.57	157.74	2739.52
1985	3795.33	0	627.15	2819.89
1986	4285.70	0	1513.63	2913.07
1987	4302.11	0	1023.28	3174.74
1988	4419.4	0	943.59	3145.28
1989	5149.43	0	1683.57	3204.63
1990	5242.50	0	1562.28	3207.33
1991	5127.39	0	1628.85	3207.33
1992	4950.28	0	1257.80	3692.98
1993	5132.72	0	1526.06	3606.65
1994	5289.11	0	1733.00	3556.11
1995	5891.41	0	2195.15	3696.26

Sources : DG. Food Crop, BP Bimas, APPI, press release: Fertilizer Procurement and Distribution (1995).

Table 2. Ratio of actual to recommended rates of urea consumption in irrigated rice farms in Indonesia. (*Source*: CASER, unpubl.).

Location	Actual/recommended rates (%)
West Java	112
Lampung	128
South Sulawesi	189

70% of N, is lost from the system by volatilization, denitrification, and leaching. Deep placement of urea in the form of urea tablets can reduce N need by 25% of the recommended dose for prilled urea and increase rice production by about 300-450 kg ha^{-1} $season^{-1}$. These research findings indicate that the introduction of urea tablet deep-placement technology could reduce urea consumption, increase rice production, and lessen N-related environmental pollution. It is estimated that with the use of urea tablets, urea consumption will be reduced by about 60 kg ha^{-1} and rice production will increase by an average of 400 kg ha^{-1}. In Java, which has an area of about 5.3 million ha, the savings in urea will be about 318,000 t.

Since deep placement (about 7-10 cm below soil surface) of prilled urea was not possible, the use of urea tablets was proposed. Even manual deep-placement of urea tablets is tedious, labor-intensive, and time-consuming. The factors that determine the success of the urea tablet technology include the effectiveness of the methods of placement (either by hand or with some kind of machine); the development, testing, and modification of deep-placement applicators suited to a range of farmers' field conditions; the labor requirement and cost of each application method; the suitability of other application methods to the current cultural practices as well as to the social and economic environments; the farmers' capacity to adopt the technology; and appropriate government support.

The government has disseminated the urea tablet technology to farmers in some provinces such as West Java, Central Java, Yogyakarta, and East Java (AARD 1993).

Fertilizer production and consumption

Indonesia has six fertilizer manufacturers, five of these produce urea and one produces both triple superphosphate (TSP) and ZA. The aggregate production capacity of these plants is 6.8 million t. All of these manufacturers, except Asean Aceh Fertilizer (AAF) which is a joint venture with the governments of other ASEAN countries, are government-owned (Delima 1996). In 1990, total urea production was 5.2 million t, or about 102% of the installed production capacity. Pupuk Kaltim and PUSRI are the major producers, accounting for 34 and 30% of the total urea production, respectively. Urea is produced in various formulations--urea granule, prilled urea, urea briquette, and urea tablet. They differ in physical characteristics but have the same chemical character.

The agricultural sector accounted for about 97% of the country's fertilizer consumption in 1991. Urea consumption in the food crop subsector increased steadily from 1983 to 1990 and stabilized around 2.5-2.7 million t during 1991-95 (Table 3). In 1989-90, total urea consumption was about 2.7 million t, 72% of which was consumed in Java (a major

Table 3. Urea consumption in the food crop subsector, 1983-95.

Year	Consumption (t)
1983	2,004,154
1984	2,366,167
1985	2,300,007
1986	2,366,199
1987	2,529,968
1988	2,529,293
1989	2,622,934
1990	2,690,896
1991	2,483,608
1992	2,483,608
1993	2,703,164
1994	2,393,855
1995	2,675,575

Sources: Vademekum Sumberdaya 1992, DGFC 1995, press release: Fertilizer Procurement and Distribution (1995).

fertilizer-consuming region), 18% in Sumatra, and 4% in Bali and Nusa Tenggara. Potassium chloride (KCl) is used mostly in Sumatra. Triple superphosphate is used more on estate crops than on food crops.

Indonesia is a net exporter of urea. Total domestic demand for urea reached 3.2 million t, compared with 5.2 million t production in 1990. About 1.86 million t from the surplus were exported to other countries. Among the urea producers, AAF and Pupuk Kaltim are major exporters, accounting for 36 and 32% of the total urea exports in 1990. Indonesian urea is competitive in the world market and the fertilizer industry is now regarded as one major source of foreign exchange earnings.

Since urea tablets were introduced in several pilot projects and development areas in the 1992 planting season, urea tablet application has made a significant progress. In West Java, urea tablet application jumped from only 0.4% in 1992 to 69% of the total urea consumption in 1995. The pattern was repeated in Central Java and East Java where urea tablet consumption increased from 0.6 and 0.5% in 1992 to 49 and 54% of the total urea use in 1995, respectively. In terms of rice area covered by urea tablet application, the same pattern was observed. West Java, Central Java, and East Java used urea tablets on 0.3, 0.2, and 0.2% of the rice area in the 1992 planting season. These figures increased to 77, 49, and 50% in the 1995 planting season (Table 4).

Table 4. Planted area and urea tablet consumption in Java, 1992 -95 (*Source:* Bimas 1995).[a]

Planting season	West Java		Central Java		East Java	
	PA (%)	UT (%)	PA (%)	UT (%)	PA (%)	UT (%)
1992	0.3	0.4	0.2	0.6	0.2	0.5
1992/93	11	9	7	7	10	15
1993	27	24	24	24	14	16
1993/94	58	47	50	50	62	67
1995	77	69	49	49	50	54

[a]PA=planted area covered by urea tablet (% of total planted area), UT=urea tablet (% of total urea consumed).

Perspective of urea tablet deep-placement technology

Economic perspective

Research has shown the following advantages of deep-placed urea tablet over broadcast prilled urea: (a) it reduced the rate of urea application by 25%; (b) it increased rice yields at research stations (range was from 300 to 1900 kg ha^{-1}) (Table 5); and (c) it increased rice yields in on-farm trials (range,7-677 kg ha^{-1}; mean, 400 kg ha^{-1}) (CASER 1996). The benefit-cost ratio ranged from 4.1 to 14.6 for on-farm trials, a good indication of the profitability of the technology (Table 6).

Table 5. Estimation of yield increase and net profit with recommended dose of 84 kg N ha^{-1} for prilled urea and other large-size urea fertilizer in Karawang, Klaten, and Madium, Java, 1990-91 (*Source:* CSAR 1996).

Urea form	Yield increase (kg ha^{-1})		Net profit (Rp 000 ha^{-1})	
	Broadcast	Deep placement	Broadcast	Deep placement
Karawang				
Prilled urea	19.9	11.9	477	253
Urea granule	14.1	19.7	321	462
Urea briquette	13.8	22.5	310	522
Urea tablet	12.2	23.3	268	542
Klaten				
Prilled urea	12.3	15.2	277	348
Urea granule	13.4	13.5	302	298
Urea briquette	16.8	16.4	390	363
Urea tablet	12.5	23.3	276	542
Madium				
Prilled urea	13.5	23.4	310	564
Urea granule	17.9	28.5	421	694
Urea briquette	20.7	33.9	491	823
Urea tablet	21.5	32.5	513	784

Table 6. Benefit-cost ratio (BCR) analysis of urea tablet application in East Java and West Java, Indonesia, 1995 (*Source*: CASER 1996).[a]

Village/ Kabupaten	Additional labor cost (Rp)	Fertilizer saving (Rp)	Additional production value (Rp)	Marginal BCR
East Java				
Tambeleng, Jombang	(14,400)	19,960	184,850	14.6
Karang Tengah, Nganjuk	(24,250)	45,465	251,640	12.3
West Java				
Jatimulya, Subang	(6,740)	(2,410)	37,470	4.1
Sukra, Indramayu	(4,200)	4,359	29,610	8.1

[a]Figures in parentheses are additional costs.

In spite of these advantages, placing urea tablets manually at the 5-10-cm soil depth is tedious, time-consuming, and labor-intensive. An inexpensive and effective urea applicator is needed.

Development of urea tablet applicators

Urea tablets have to be deep-placed in the rice root zone (about 10 cm deep in the soil). Tools for applying urea tablets have been progressively improved. Initially, a simple form of applicator was made from bamboo by farmers. Some prototypes of deep-placement machines have been developed by the Center for Agricultural Machinery Development (CAMD) of AARD, taking into account the following criteria:

- It should have the precision needed to enable urea tablet application at the recommended rates and desired depths (rice root zone).
- It should be simple and easy to operate.
- It should be light.
- It should be made from locally available materials.
- It should be made by local artisan/workshop.
- It should be cheap.

AMINI-I, the first prototype developed, was similar to the jab planter. Its field capacity was about 48 h ha^{-1}. The nonuniform shape and size of urea tablets initially manufactured caused uneven transfer of urea tablets to the soil. It was thus not popular among farmers. The second prototype, AMINI-II, also had many problems and was likewise not promoted to farmers.

AMINI-III is a two-row, manually drawn urea tablet applicator with a field capacity of 11 h ha^{-1}. Two models of this type were developed, one for transplanted rice (AMINI-IIIT) and the other for direct-row,wet-seeded rice (AMINI-IIID). AMINI-IIIT places three urea tablets between four hills $clump^{-1}$ of rice; AMINI-IIID immerses one urea tablet every 13.3 cm between two rows of rice plants. These applicators are equipped with a furrow opener, a furrow closer, and a metering device.

AMINI-III was designed in three sizes to meet different labor capacities (strengths). The single-row AMINI-III is suitable for women; the two- and three-row AMINI-III are for men, depending on their capacity to operate the machine. AMINI III can immerse the urea tablets accurately in the rice root zone and reduce drudgery. It is simple and easy to operate and transport. The materials used are available in the local market and the equipment can be made in small workshops at a cost affordable to average farmers.

In support of the urea tablet technology, the government initiated the Presidential Aid Program in 1993 to promote the wider use of urea tablet applicators. Through this program, a total of 1500 AMINI-I applicators were distributed to various provinces in Java. AMINI-I, which faced several problems in the deep-placement of urea tablets, was no longer recommended. At the same time, a simple hand applicator was developed and used by farmers to place urea tablets into the soil.

Through the Indonesian rice-based farming systems, AARD has introduced improved technologies on 44,000 ha over the last 2 yr to improve the productivity and efficiency of rice production. To support this program, the CAMD in 1996 produced 105 units of AMINI-IIID and distributed them to 14 provinces in Indonesia. Although not too many, these units are expected to be adopted by local manufacturers. AMINI-III is also appropriate for small farms and potentially can be owned, managed, and used by a group of farmers, a cooperative, or a commercial enterprise. A detailed technical description of AMINI-III is presented in Appendix 1.

Constraints to adoption of urea tablet deep placement

Procurement and distribution constraints. The introduction of urea tablets undoubtedly has improved farmers' income and welfare. It also has changed the urea application pattern from 2-3 split applications of prilled urea to a single application of urea tablets during the early stage of crop growth. Consequently, farmers have to obtain all their urea tablets at one time every planting season. Additional warehouses and appropriate urea tablet processing machines have to be installed to meet the one-time farmers' demand every season (AARD 1993, 1996). In fact, in some instances, the demand for urea tablets cannot be met. In Kabupaten Subang, for example, demand for urea tablets was about 400 t d^{--1}, while production is only 100 t d^{-1}.

Innovation and technology constraints. It took some time for farmers to adopt the urea tablet technology. Experience has shown that a new technology will be well adopted if it has more perceptible advantages and is easier to apply than the present technology. For small farmers in Java, broadcasting prilled urea is really more practical and not as labor-intensive. Here are some farmers' perceptions on the advantages and disadvantages of urea tablet application compared with prilled urea broadcasting:

Advantages		Limitations	
1.	Reduced rate of fertilizer	1.	More labor requirement
2.	Higher productivity per hectare	2.	Less practical to use
3.	More profit per hectare	3.	Higher cost
4.	Less pollution		

Another problem is the lack of farmers' interest to use deep-placement machines because of inconsistency in placing the tablets below the soil surface (a maximum of 56% for AMINI-II). Moreover, changing the farmers' attitude and behavior is not easy. Farmers tend to apply urea tablets at more than the recommended rates, similar to what they do with prilled urea (Table 7).

The hygroscopic character of urea has made urea tablets more perishable, and thus more difficult to store before application. In addition, unfavorable conditions such as humid weather or long storage create a bridging effect--i.e., the tablets cling together. It

Table 7. Urea tablet application rate compared with recommend dose in East Java and West Java, 1996 (*Sources:* Bimas 1995, CASER 1996).

Location	Dose (kg ha^{-1})		% of recommended
	Recommended	Actually applied	
East Java			
Tambelang, Jombang	187.5	180	96
Karang Tengah, Nganjuk	187.5	209	111
Seloredjo, Nganjuk	187.5	246	132
West Java			
Jatimulya Subang	187.5	199.6	106
Sukra, Indramayu	187.5	228.3	122

implies that urea tablet production should be decentralized to several centers of consumption and that distribution should be carried out in the shortest time possible during the planting season.

From the distributors' perspective, urea tablets should be handled with care. Unlike prilled urea, urea tablet bags cannot be stored in large (or tall) piles because the tablets would cling together. Considering this situation, more and better storage facilities should be provided by distributors; this in turn will increase the price of urea tablets (AARD 1996).

To reduce the hygroscopic character of urea, covering urea with sulfur or silica is suggested.

Diffusion/dissemination constraints. Diffusion of the current urea tablet technology at the farm level, especially outside the project area, is still low (Table 8). The most likely reasons are the abovementioned technical disadvantages of urea tablets. Farmers hesitate to adopt urea tablets when they see no improvements in the storage and application aspects.

Table 8. Expected and actual urea tablet application, 1995 planting season to 31 Mar 1996 (*Source*: Bimas 1995).

Province	Expected/targeted		Actual	
	Area (ha)	Distribution (t)	Area (%)	Distribution (%)
West Java				
Pilot project	773,813	134,457	90.3	81.0
Extended area	107,327	20,127	62.7	51.7
Central Java				
Pilot project	399,117	74,838	107.9	100.8
Extended area	148,756	27,889	106.3	101.0
East Java				
Pilot project	612,539	111,982	79.8	87.9
Extended area	71,500	13,048	110.1	117.0
Yogyakarta				
Pilot project	39,092	7,330	101.1	97.5

Government policy support

Procurement, production, storage, and distribution policies

Because prilled urea is used as raw material for urea tablets, a strong linkage is needed between distributor (PT PUSRI) and producer (private company). The objective is to secure the timely delivery of urea tablets to farmers.

To accelerate urea tablet development in the near future, the production capacity of urea tablet machines (UTMs) has to be optimized to meet the increasing demand in various regions. The quality of UTMs has to be improved to enable production of uniform-size tablets, which in turn will improve the performance and consistency of urea tablet applicators. In 1995, urea tablet production was 30% of the 112 units of UTM. Again, better coordination among PT PUSRI, urea tablet producers, and village cooperative) is vital. The government has appointed a private company to produce and install UTMs close to PT PUSRI Warehouse Line III. In 1995, this company distributed 122 units of UTM in Java.

Diffusion and dissemination policies

Labor-intensive innovations such as urea tablet deep-placement technology are not quite popular yet among farmers. Provision of a deep-placement machine that is effective, efficient, practical, and inexpensive is necessary for widespread farmer adoption of urea tablets in Indonesia.

To accelerate farmer adoption of urea tablets, mass media, especially radio and television, can be utilized. Large-scale field demonstration is another way to convince farmers about the effectiveness of the urea tablet technology. In addition, information dissemination and organized campaigns on the technology should be carried out continuously to make farmers understand its benefits and to encourage them to use it. Meanwhile, farmers' practical experiences on the use of urea tablet deep-placement technology can improve their knowledge, skill, and attitude about the application of this technology.

Adoption and diffusion can also be enhanced through provision of credit and a favorable price policy to enable farmers to purchase urea tablets at one time. In the short run, a price policy in the form of "lowest retail price" (HET=*Harga Eceran Terendah*) can be implemented to make urea tablets and prilled urea available to farmers at a competitive price. Through a policy of selling prilled urea at a higher price than urea tablet, prilled urea consumption, particularly in regions using high rates of urea, will decrease. In contrast, increasing urea tablet consumption will increase social benefits. Therefore, the HET policy will accelerate urea tablet adoption by farmers. If this policy is properly applied, the gap between private and social cost will be minimized.

Extension materials and special programs should be directed to change the farmers' habit of applying excessive urea to the rice crop.

Because urea tablet application needs adequate irrigation, water supply should be improved in irrigated areas.

References

AARD--Agency for Agricultural Research and Development - Badan Litbang Pertanian (1993) Penggunaan Urea Tablet. Analisis Masalah dan Saran tindak Lanjut (Urea tablet application. problem analysis and recommendation). (unpubl.)

AARD--Agency for Agricultural Research and Development - Badan Litbang Pertanian (1996) Perspektif penggunaan dan pengembangan pupuk urea dalam peningkatan produksi pertanian (Perspective of urea tablet application in increasing agriculture production). (unpubl.)

Bimas (1995) :Pelaksanaan program urea tablet TA 1995/1996 (Urea tablet implementation program 1995-96). Seasonal report.

CASER--Center for Agro Socio-Economic Research (1996) Annual report on technology assessment. CASER, Bogor, Indonesia.

Center for Agricultural Machinery Development (1996) Government policy support for technology adaptation and promotion: a case study of urea tablet deep placement technology in Indonesia. (unpubl.) 10 p.

CSAR--Center for Soil and Agroclimatic Research (1996) Penggunaan pupuk N, P, dan K secara efisien pada lahan sawah. (unpubl.)

Delima A H (1996) Fertilizer policy in Indonesia."Paper presented at the Regional Workshop on Fertilizer Policies and Subsidies.

Wetselaar R, Sri Mulyani N, Hadi Wahyono, Prawirasumantri J, Damdam A M (1984) Deep point-placed urea in flooded soils: research result in West Java. Proceedings of a workshop on urea deep-placed technology. AARD - IFDC.

World Bank (1993) Fertilizer pricing in Indonesia: some possible options in 1993. (mimeogr.).

PART III

Approaches and methodologies for evaluation and promotion of crop and resource management technologies

High-impact strategy: useful principles and practices for the effective delivery of research products

G. L. DENNING and M. A. BELL
International Rice Research Institute (IRRI), P.O. Box 933, Manila, Philippines.

Key words: agricultural research, impact, institutions, proactivity, purpose, priorities, partnership, people.

Abstract
Publicly funded agricultural research institutions must increasingly generate impact and demonstrate accountability to a diversity of stakeholders. Drawing on experience at IRRI, Centro Internacional de Mejoramiento de Maiz y Trigo (CIMMYT), and national agricultural research programs, the authors propose a high-impact strategy framework, which is simplified in terms of the five Ps--proactivity, purpose, priorities, partnerships, and people. With personnel and institutional commitment to the five Ps, agricultural research can generate high impact, which is crucial to attracting and maintaining financial and political support to research.

Introduction

Public sector institutions are increasingly being scrutinized for their ability to deliver impact. In an environment of broad fiscal austerity in government, reduced public expenditures in the agriculture sector, and increasing accountability required of publicly funded institutions, there is growing interest in understanding, adapting, and applying strategies and approaches that make public institutions more effective. Agricultural research institutions should be especially concerned, because they potentially suffer from both an urban bias against the agriculture sector and a perception bias against research. In contrast with plant breeders and engineers, those trying to develop better approaches to sustainable natural resource management are particularly vulnerable to criticism.

This paper introduces a collection of useful principles and practices that can accelerate the impact of research. We call this the high-impact strategy (Fig. 1). The approach was derived from an analysis of Stephen Covey's popular management guide: *The Seven Habits of Highly Effective People* (Covey 1990). Drawing on experience at IRRI, CIMMYT, and national agricultural research programs, we have attempted to adapt and modify Covey's approach to the needs of agricultural research institutions.

The high-impact strategy framework can be simplified in terms of five Ps: proactivity, purpose, priorities, partnerships, and people.

Proactivity--freedom to choose

There are two broad categories of people--those who react and those who are proactive. What does this mean? Think of the people you work with and know. Do they feel like they are in control of what they do, or do they feel helpless to set their own agenda? Perhaps they feel they are victims of their job and their circumstances. They feel they

V. Balasubramanian, J. K. Ladha and G. L. Denning (eds.), Resource Management in Rice Systems: Nutrients, 193–198.

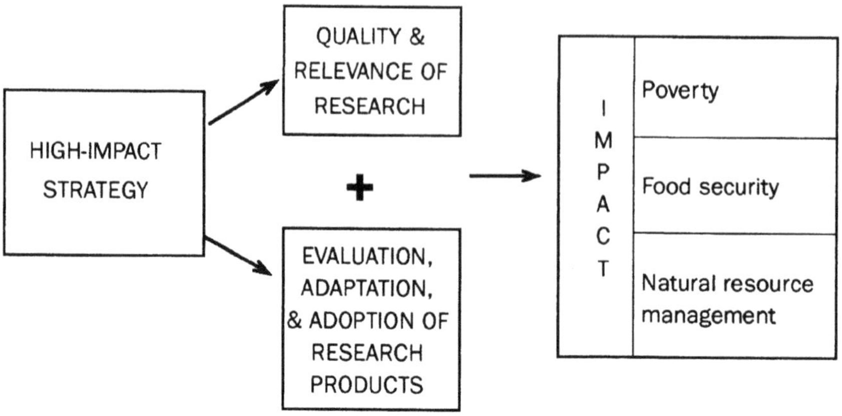

Figure 1. High-impact strategy.

have no say in what they can do; they simply react to what they are told to do. They often try to work hard but, in the end, they feel they are simply not in control. This situation often results in demoralized and unmotivated staff.

On the other hand, there are those who work and seem to succeed despite the circumstances. They view problems as opportunities, not stumbling blocks. They can be considered as the "doers," the ones that say they **will** make a difference, no matter what the circumstances are. Well-known examples include Norman Borlaug, of the Wheat and Green Revolution fame, and Mechai Viravaidya, whose "death control" activism has drastically reduced the spread of HIV/AIDS in Thailand. These people model circumstances to them, work with the circumstances, and act to make a difference, no matter what the circumstances are. They often end up molding rather than being molded by the circumstances.

Activity: Will a proactive or reactive person have greater impact? Where do you feel you and your staff fit? What could you do to change your attitude?

Purpose--begin with the end in mind

When you or your colleagues are asked "What do you do?", what will your answer be? Can you articulate clearly and succinctly how and why your program will have impact, why and where it will make a difference? Can you visualize the impact, or are you tossing up ideas and hoping that one will land and take root? Do you begin a project with the end in mind?

A clarity of purpose promotes teamwork, focuses efforts, and increases the likelihood of impact. Preparation of a mission statement can help articulate the purpose of your institution or group within the institution. The purpose should be framed as an end result or impact, not as a process or activity. The purpose should also indicate the anticipated beneficiaries of the work.

Activity: Take time and articulate briefly what you and your staff would hope to achieve through your work. This should be a group exercise.

Priorities--first things first

Can you think of a project that is well focused and one that is poorly focused? What is the difference?

Much of the difference is often simply related to whether the time has been taken to analyze the situation. Such analysis ensures that we are working on real problems with possible solutions that have potential for impact. If the problem is clearly identified and the potential for impact clarified, then priorities can be set. The steps are common sense, but often overlooked. Simply consider the following:

- What are the problems?
- What are the "true" causes of the problems?
- Can viable solutions be identified (in what time frame)?
- If solutions are found, what will be the expected extent of adoption and impact?

When these questions are asked and answered, the priorities can be properly set. A useful tool in helping set priorities and allocate resources is the logical framework matrix (McLean 1988, Schubert et al 1991). This approach, developed to improve project planning, has been adapted and used for agricultural research. The approach requires teamwork among participants as they develop research plans through the following steps:

- *Analysis of interest groups:* analyzing the program target group (e.g., farmers) and other organizations or individuals that affect or are affected by the research plans;
- *Identification of core problem and research output:* identifying the core problem to be addressed by the research plans and translating it into a core objective, identifying constraints to achieving the core objective, recognizing related concerns (e.g., environmental impact, impact on women and children), and identifying potential research outputs;
- *Development of research planning matrix:* establishing a hierarchy of goal, objectives, outputs, and activities for each program, along with corresponding sets of progress indicators, means of verification, and assumptions; and
- *Development of work plan and resource allocation:* developing work schedules and corresponding resource allocations (human and financial) for each activity.

An important element of the method is visualization of ideas through the use of color cards and pinboards. The adapted logframe approach described here was used by IRRI scientists in the development of the Institute's work plans for 1990-94 and for 1994-98.

Activity: Write down your primary activities. Write a companion list showing the key problems faced by farmers. Compare the two lists. Indicate on the first list whether the problem should be a high, medium, or low priority.

Partnership--exploiting differences

Reflect on the mission of your organization. Consider the immediate outputs of your research program--new varieties, better approaches to fertilizer management, a new equipment design for increased labor efficiency. Now visualize the endusers, the rice farmers, using new varieties, applying fertilizers more efficiently, and reducing labor costs by using new equipment. How were these results achieved? What partnerships are needed to achieve impact?

The profile of IRRI's collaborating partners has changed since the early 1960s. Although the Rockefeller Foundation, the United States Agency for International Development, and other bilateral agencies remain important donors and national agricultural research systems continue to be IRRI's principal partners in research and training, new opportunities for value-added partnership have emerged: other international agricultural research centers (IARCs), advanced research institutions, universities, nongovernment organizations (NGOs), and the private sector (Denning and Bernardo 1995).

The growth of complex, multiparty partnerships in rice research also has implications on the organizational structure and systems of concerned institutions, such as

- creation of a unit or units responsible for liaison with external parties;
- personnel policies (e.g., recruitment procedures, evaluation and reward systems) that support an externally oriented institution;
- governance arrangements (e.g., Board composition) that reflect the importance of key partners; and
- modern communication systems that facilitate regular, low-cost dialogue among partners.

Research must be viewed as a continuum--from the researcher to the farmer. Often, farmers are viewed as coming at the end. They must, however, come at the beginning, the middle, and the end. This continuum is taking on a number of exciting new variants in various countries. In Latin America, the public sector has turned to a more private industry approach with increased resources being levied, but increased pressure to produce result (Echeverria et al 1996). In Vietnam, research institutions have long sought partnerships with local government authorities to develop impact-oriented technologies.

Activity: Make a list of your active current R & D partners: in setting the overall strategy and work priorities of your institution; in implementing the work of your institution; and in delivering the outputs of your institution to the endusers. Draw a flow diagram showing how each one works. Identify the potentially weak links.

People--a force for change

What is your institution's most valuable asset? The buildings? The vehicle fleet? The land? The budget? The reputation? Most of you (we hope) will instinctively reply "people." But assets can be productive or can be underutilized or can even depreciate in value. Effective institutions value and nurture their people. They realize that motivated and skilled people are crucial to impact.

Activity: Consider: Through what policies and recent actions does your institution demonstrate that people are its most valuable asset?

Conclusion

The achievement of high impact is crucial to attracting and maintaining support for agricultural research. Impact can no longer be assumed. In today's economic and political environment, researchers must seek new ways of ensuring that research products reach the endusers. We argue that staff and institutional commitment to proactivity, purpose, priorities, partnerships, and people will deliver the results being demanded by stakeholders. These principles and practices must be learned through exposure, experiences, and reflection. We suggest the use of an "impact audit," which can be simply framed around the following questions:

1. Seek out proactive people and identify proactive institutions. What circumstances/conditions made them proactive?
2. Identify the purpose of the research and/or the delivery process being described. Are the expected end results clearly articulated?
3. Consider how the priorities were set and how they might be different. Can you identify the strategic impact areas in each case?
4. Identify the partnerships involved. Which partnerships stand out as highly effective in terms of impact and why?
5. What lessons have you learned of personnel management that might be useful in increasing the impact of your work and that of your institution?

References

Covey S R (1990) The seven habits of highly effective people: powerful lessons for personal change. Fireside/Simon & Schuster, New York.

Echeverria R G, Trigo E J, Byerlee D (1996) Institutional change and effective financing of agricultural research in Latin America. World Bank Technical Paper No. 330. The World Bank, Washington, D.C.
Denning G L, Bernardo F A (1995) Partnership in rice research: rationale, experience and implications. GeoJournal 35.3: 374-383.
McLean D (1988) The logical framework in research planning and evaluation. ISNAR Working Paper No. 12. International Service for National Agricultural Research, The Hague, Netherlands.
Schubert B, Nagel U J, Denning G L, Pingali P L (1991) A logical framework for planning agricultural research programs. International Rice Research Institute, P.O. Box 933, Manila, Philippines.

From-research-to-farmer practice: a case study in rice pest management

K.L. HEONG
International Rice Research Institute, (IRRI), P.O. Box 933, Manila, Philippines

Key words: communication, decisions, IPM, leaffolders, heuristics, cognitive dissonance

Abstract
In Asian rice farmers' pest management, knowledge gaps between what farmers need to know and what they actually know are the main constraints, and pesticide misuse is due primarily to these gaps. Over the last two decades, changes in pest management decisions and practices of about 300 million rice farmers in Asia are hardly evident. Knowledge gaps may be due to lack of research or lack of synthesis of research results. Even when research is well-targeted, the results may not be adopted because of delivery problems. To facilitate adoption, research must distill research results further, developing them into forms that can facilitate communication. In addition, research will need to investigate factors constraining adoption and explore the utility of communication methods and systems. This paper discusses a case study illustrating the continuum, from research to practice, in leaffolder management. Large proportions of farmers' insecticide sprays are directed at rice leaffolder. Research has shown, however, that these sprays are unnecessary and are, in fact detrimental to ecological balance causing secondary brown planthopper problems. Farmers spray for leaffolder control primarily due to misperception. A simple rule-of-thumb, or "heuristic," was distilled from years of research by many scientists--*In the first 40 days after sowing, leaffolder control is not necessary."* When presented to farmers, the heuristic can be easily transformed into a research question, which can be tested. When farmers were invited to participate in experiments to test this heuristic, most found no yield difference between pesticide-treated and untreated plots, changed their attitudes toward leaffolders, and stopped spraying to kill them. In a pilot project in Vietnam, various media materials were used to encourage farmers to test the heuristic. The target audience was 20,000 farm families in Long An Province. An evaluation conducted 16 mo after the project was initiated showed that farmers' insecticide application dropped from 3-4 to 1-2. Perceptions of leaffolder damage and the need to spray early for control also changed.

Introduction

Numerous research and extension programs directed at improving pest management in tropical rice have been carried out over the last 20 yr. Research institutes have concentrated on developing pest-resistant cultivars and accumulating knowledge on pest ecology, pest-yield relationships, computer models of pest populations, and various management options. Implementation agencies, on the other hand, have conducted training courses, published books, and disseminated technical materials, leaflets, and posters to improve farmers' knowledge and decision making. In the meantime, the chemical industry has developed new pesticides and application technologies and conducted marketing campaigns to promote its products. Through these two decades, substantial changes in pesticide products and cultivars have occurred, but the rice farmers' pest management decision making has stagnated.

Insecticide misuse has continued to be rampant (Heong et al 1994, 1995). A large percentage of farmers in Asia still use "highly" hazardous pesticides, such as methyl parathion, monocrotophos, and methamidophos (Heong et al 1997), and apply insecticides unnecessarily to control leaf-feeding insects and other highly visible pests (Heong et al 1995; Vo Mai et al 1993). Although farmers may have adopted pest-resistant cultivars, most continue to spray these crops with insecticides (Escalada and Heong

V. Balasubramanian, J. K. Ladha and G. L. Denning (eds.), Resource Management in Rice Systems: Nutrients, 199–211.

1993). In the Philippines, for instance, despite the high adoption of pest-resistant cultivars, insecticide use increased markedly (IRRI 1994). Similarly, most farmers in the Mekong River Delta of Vietnam grew brown planthopper (BPH)-resistant cultivars, but about 34% of the insecticide sprays were for BPH control (Heong et al 1994).

Most rice-growing countries have accepted the philosophy of integrated pest management (IPM) and have implemented IPM programs in various forms (Ooi et al 1992). However, changes in the pest management decision making and practices of the 300 million-odd rice farmers in Asia are hardly evident. In reality, implementation of IPM that can lead to significant practical improvements in farmers' practice has proved far more difficult.

Norton (1993) suggested that the problem may be due to design, where research and development (R&D) is directed at the wrong questions or at developing inappropriate practices. If R&D is well-targeted, the results may not be reaching farmers: a problem of delivery. Various concepts and techniques to address these two problems are now available (e.g., Norton and Mumford 1993). In this paper, I illustrate the utility of these concepts and techniques in addressing an insect management problem of Asian rice farmers. More emphasis is placed on the delivery problem, although this should strictly be seen as an integral part of the design as well.

Analysis of the pest problem

The vegetative stages of the rice crop are often subjected to infestations of insects, which include species of the orders Orthoptera, Lepidoptera, Coleoptera, Thysanoptera, and Diptera (Fig. 1). Most of them are herbivores feeding on the leaves and stems of the rice plant. In extreme cases, these pests can cause plant death. Perhaps the most frequently encountered group of insects at early stages of the rice crop are the leaf feeders--species of Lepidoptera, Coleoptera, Diptera, and Thysanoptera. They damage the leaves and reduce the photosynthetic area, either by cutting or scraping. Such damages may sometimes reduce vigor, increase water loss, or cause secondary infections by fungus and bacteria.

More than 340 species of leaf feeders are found in rice ecosystems. However, only a few species are commonly found (Table 1). The important defoliators are the rice leaffolders (*Cnaphalocrocis medinalis*), the rice caseworm (*Nymphula depunctalis*), thrips (*Stenchaetothrips biformis*), armyworms (*Spodoptera* spp.), and the whorl maggot (*Hydrellia* spp.). Under extreme conditions, which are naturally rare, each of these species can destroy the rice crop. The whorl maggot, for instance, usually attacks the rice crop in the first 2 wk after transplanting. As much as 100% of the hills could be damaged, making the crop look miserable, yet no yield loss could be detected (Viajante and Heinrichs 1986).

Perhaps the most common and most visible pests to farmers at the early crop stages are the rice leaffolders. Under favorable conditions, a larva can consume about 25 cm^2 of the leaf area or about 25% of the total leaf (Heong 1990). Consumption varies with rice variety and crop age, and is reduced to about 6 cm^2 or 6% in mature leaves (Table 2). Average leaffolder densities in rice crops are usually less than 2 larvae $hill^{-1}$ (Wada and

Table 1. Common leaf-feeding insects found in the early crop stages of rice in Asia.

Diptera		
Whorl maggot	*Hydrellia* spp.	Larvae tunnel within tillers and deformed leaves emerge
Leaf miner	*Agromyza oryzae*	Larvae tunnel within leaf blades, consuming leaf tissues
Lepidoptera		
Caseworm	*Nymphula depunctalis*	Larvae cut leaves and form tubular cases.
Hairy caterpillar	*Rivula atimeta*	Larvae cut out sections of leaf blades.
Semilooper	*Naranga aenescens*	Larvae cut out sections of leaf blades.
Armyworms and cutworms	*Spodoptera mauritia, Spodoptera exempta, Mythimna separata, Spodoptera litura*	Larvae feed by cutting off leaf sections or panicles. In extreme cases, large portions of the crop are defoliated.
Leaffolders	*Cnaphalocrocis medinalis Marasmia patnalis, Marasmia exigua*	Larvae fold leaves and feed by scraping, leaving white patches.
Coleoptera		
Rice hispa	*Dicladispa armigera*	Adults and larvae feed on leaves by scraping, making elongated damage marks.
Leaf beetle	*Oulema oryzae*	Injury caused by larvae and adults by scraping.
Thysanoptera		
Thrips	*Stenchaetothrips biformis*	Feed on leaves by rasping tissues, which dry up and turn brown.
Orthoptera		
Grasshoppers	*Oxya* spp.	Nymphs and adults feed on leaf tissues, consuming large sections from the edges.
Locusts	*Locusta migratoria*	Infest crops in swarms.
Crickets	*Euscyrtus concinnus*	Cut out center portions of leaf blades.
Katydids	*Conocephalus longipennis*	Feed on grains at milky stage.

Table 2. Total leaf area (cm^2) consumed by one larva of the rice leaffolder *Cnaphalocrocis medinalis* in its life span on different cultivars and crop ages.

Crop age[a]	MR84	IR36	IR70
40 DAS	24.5	11.9	14.9
60 DAS	19.9	7.9	12.9
80 DAS	18.9	9.9	10.2
100 DAS	16.7	5.8	-

[a]DAS = days after sowing.

Shimazu 1978, Guo 1990). However, in highly fertilized crops, the density can be as high as 5 larvae $hill^{-1}$ (De Kraker 1996). At such high densities, the proportion of damaged leaves may reach 50%. However, japonica rice (which generally has smaller leaves) with 67% damaged leaves at tillering did not suffer any yield loss (Miyashita 1985). Using computer simulations, Fabellar et al (1994) showed that larval densities needed to reach 15 $hill^{-1}$ before any yield loss could be detected. This is due to the rice crop's high compensation ability to injuries caused by insects (Rubia et al 1996).

A large percentage of farmers in Asia, however, spray insecticide at the early crop stages. In Vietnam, about 80% of farmers administered their first insecticide sprays in the first 40 d after sowing (Vo Mai et al 1993). Similarly, in Leyte, Philippines, 77% of farmers applied their first sprays within the first 30 d after transplanting (Escalada and Heong 1993). Their main targets were rice leaffolders and other lepidopterous leaf feeders, which were collectively termed as "worms." Here, farmers' actions clearly reflected only their perceptions of the insect problem and not the actual situations. Farmers had overestimated losses caused by the leaf feeders. Indigenous attitudes, such as the belief that all insects are harmful, seem to have made them victims of insecticide abuse (Bentley 1989).

Exploring for solutions

Using Norton's (1982) pest management decision model (Fig. 1), we can further explore the apparent pest problem from farmers' perspectives. A farmer's decision may be affected by four factors: options; his/her perceptions of the pest attack; damage and control efficacy; and objectives. The option that can best satisfy farmers' objectives is chosen based on farmers' perceptions of the situation. After implementing the control option, they will evaluate the outcome to see if the results are satisfactory.

As discussed in the preceding paragraph, rice farmers in Asia seem to strongly believe that leaf-feeding insects, particularly leaffolders, are damaging and yield reducing. Because of this perception, they have chosen to kill the pests with insecticides to protect their yields. The options available seem limited to insecticides that would satisfy farmers' objectives in "killing" the pest. Over the years, farmers have learned to use insecticides with high toxicity to larvae but with low costs, such as methyl parathion, methamidophos, and monocrotophos. Because the farmers' objectives seem focused on "killing the pests," insecticides have remained their main option. This, and the perception that leaf-feeding insects are detrimental to yields, have seemingly trapped rice farmers in Asia in a vicious cycle.

Throughout the growing season and from season to season, farmers are faced with similar problems. Each decision the farmer makes is affected by experience and the previous decisions made. Mumford and Norton (1984) introduced the dynamic decision model (Fig. 2) to describe this. According to this model, a decision made during the period (t +1) can be influenced by decisions of the previous period (t). Because pest problems and control options often remain much the same from year to year, farmers' perceptions are likely to remain similar. This could explain the persistence of farmers' decisions to spray for leaffolder control.

Because the outcome of period (t) can influence perceptions of the problem in period (t +1), introducing an action that may have a different outcome may be used to influence farmers' perceptions. At the same time, farmers may be persuaded to observe other criteria, like yields or profits, rather than insect deaths, when comparing options.

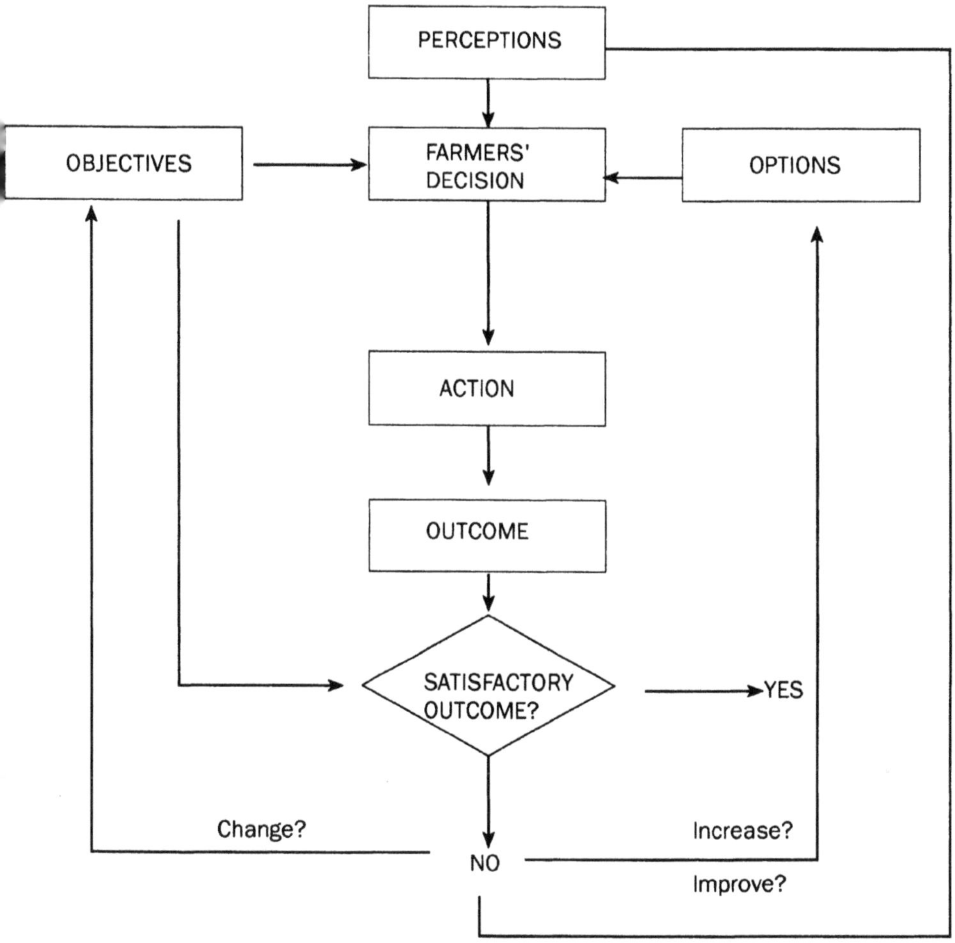

Figure 1. A model of pest management decision-making (Norton 1982).

Information and knowledge gaps

A major constraint to improving pest management is knowledge gap, the difference between what farmers need to know and what they actually know. As clearly pointed out by Bentley (1989), "What farmers don't know can't help them." There are strengths and weaknesses in farmers' knowledge of pest management, and what seems to be important is to identify what is needed at the farm level and that which farmers are aware of (Norton and Mumford 1982). Often, information may be successfully disseminated and received, but not internalized. This is especially the case for information concerned with new opinions, attitudes, and behavior. The new information needs to be integrated into the

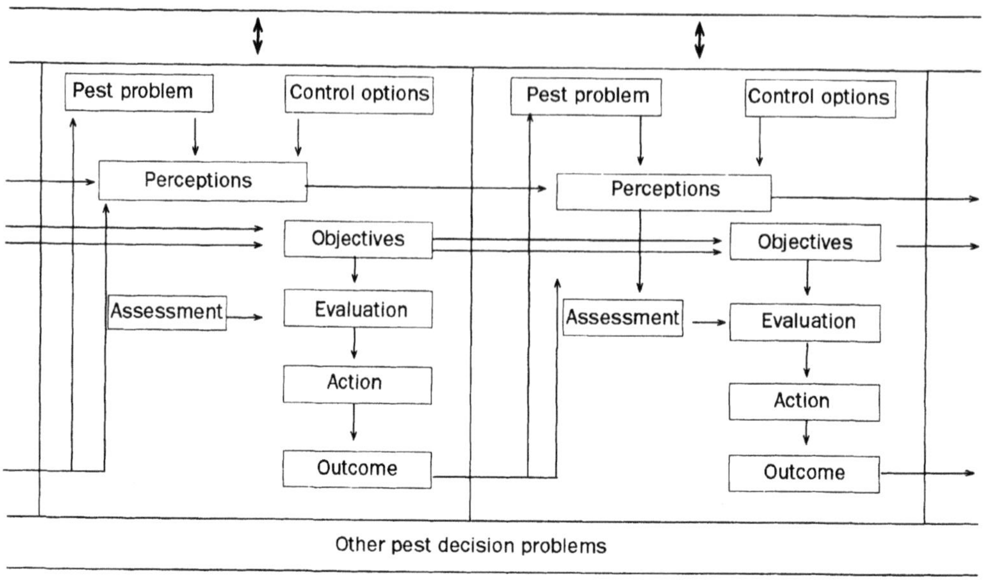

Figure 2. The dynamic decision model (Mumford and Norton 1984).

existing knowledge systems for it to be utilized in making decisions. According to Anderson's (1980) information integration theory, a new opinion is adapted through formation of a general impression which is based on many information pieces. Some of this information may support a change in opinion, attitude, or behavior, while other information may argue against a change. Also, information sources may vary in credibility and objectivity. How the new information is processed will also depend on how it fits into the individual's existing cognitive structures or patterns.

One way to consider knowledge gaps is to explore the factors that might contribute to them. Norton and Mumford (1982) suggest four possible reasons for information gaps. I would like to expand this concept further by examining various processes that might help in bridging knowledge gaps. These processes are research, synthesis, distillation, communication, and extension (Fig. 3).

Research

A radical change in research approach should be adopted, with emphasis on understanding farmers' pest management knowledge and practices and the root causes of such problems. Details of such approaches are discussed in Norton and Mumford (1993), Heong (1996), and Bentley and Andrews (1996). Research had focused on technology generation and evaluation, with limited emphasis on the adaptation, integration, dissemination, and adoption, in what should be a continuous and dynamic process

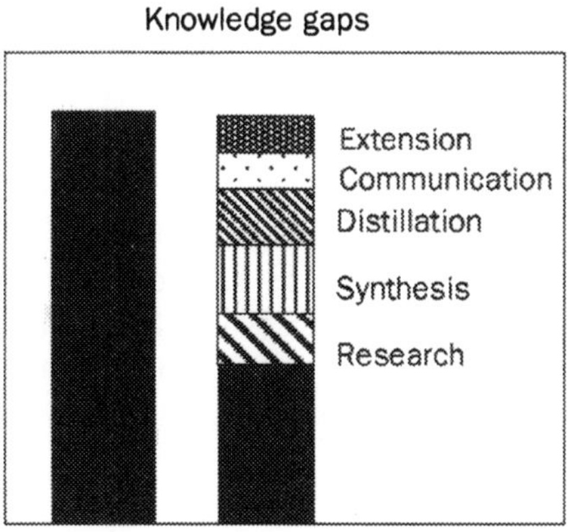

Figure 3. Various processes that may help bridge knowledge gaps.

between science and practice (Roling 1990). Often, research stops midway through, leaving a "fatal gap" while researchers focus on publications as their main products. The important role of research to transform farmers' needs into researchable problems and communicate the results back to farmers (Pray and Echeverria 1990) seems to have been neglected.

Synthesis

Information gaps due to this reason have been discussed by Norton and Mumford (1982). Often, research focus is narrow, concentrating on single factors, variables, or species in controlled environments. Information may be available in the literature but is usually not sufficiently synthesized in the context of the problem. Application of systems analysis and decision analysis may be useful.

Distillation

The information after synthesis is often still rather abstract, directed mainly at the scientific community. A process is needed to "distill" or simplify the information further, into a form that can be best communicated to farmers. The form in which the information is presented to farmers is also important. A well-distilled piece of information may also be able to bridge the reception gap discussed by Norton and Mumford (1982).

Communication

Information not reaching the farmers is due to a dissemination gap (Norton and Mumford 1982). Often, information gets through, but farmers are not able to utilize it--a reception gap. For information to be disseminated, received, understood, internalized, and utilized in decisionmaking by the recipients, communication must be improved. The communication process is made up of various elements (source, information, channel, receiver) and behaviors (encoding, decoding, responding to new information). These demands are interesting and interdependent. An important element in the process is noise, which leads to information distortion. Sources of noise may be internal to the communicator, such as in the use of language and selective processes. The use of technical terms and jargon is an example of language noise. Another source is in the medium, such as static or interference in audio channels and physical distortion in information presentation. A third source is receiver noise due to selective attention, selective perception, and selective retention (e.g., Festinger 1957, Krech and Crutchfield 1971, Sears and Freedman 1971, Schramm 1973). In pest management, research into the contributing factors and methods to overcome these constraints would be useful.

Extension

In most Asian countries, agricultural research and extension are separate institutions. Although they may be located in the same ministry, they operate in different ways. Researchers are expected to develop new technology, which is then turned over to extension for demonstration and diffusion to farmers. Such systems have functioned reasonably well in transferring technologies such as new varieties, fertilizers, pesticides, and equipment, and they have benefited mainly the resource-rich farmers who have the capacity to adapt these technologies. Tichenor et al (1970) have shown that the impact of a technology transfer program often leads to a widening of the knowledge effect gap among the target audience. However, in technologies related to resource management skills, which are knowledge-intensive, the conventional extension systems have not been as successful. Many resource-poor farmers have continued to use inputs such as fertilizers and pesticides without adequate knowledge. To be cost-effective, appropriate extension methods that will have increased opportunities for farmers to participate in and adapt need to be designed.

Decision rules and information distillation

A common strategy and method that humans use in solving problems is to use simple rules. Kahneman and Tversky (1972, 1973) introduced the term "***heuristic***" to refer to an informal rule-of-thumb used by people to simplify information processing and decision-making. Heuristics are usually acquired through experience, educated guesswork about possible answers to a problem, and integration of various ways in knowledge-manipulation capacities of humans. These simple rules are at the heart of our thinking framework when we are able to interpret observations without bringing up all the historical data stored in our memories. For example, "If it has been raining, then traffic conditions in Route 1 will be bad and Route 2 is the wiser option." Thus, a well-chosen rule represents a diligent attempt to integrate, compare, and analyze a large amount of data to produce a conclusion. The application of heuristics stored in computer knowledge bases is the central programming method used in building computer expert systems (Harmon and King 1985, Heong 1990).

Volumes of scientific information related to the rice leaffolder problem are found in journals, books, research reports, and various literature. However, these pieces of information are discussed in separate publications. For instance, the feeding rates of the leaffolder are described in one paper, while the insect's biology and biological control are discussed in other papers. Plant compensation from injuries may be found in another journal and works on farmers' perceptions of the leaffolder reported elsewhere. Such pieces of information may be thought of as different reagents which a chemist can use in a distillation process to derive a useful compound. This product generated from the process of information distillation may be presented as a simple decision rule. An example in rice pest management is "*Spraying for leaffolder control in the first 40 days after sowing (or 30 days after transplanting) is not necessary.*"

This rule is in direct conflict with farmers' current beliefs. Festinger's (1957) cognitive dissonance theory states that when a person is presented with information that is in conflict with existing attitudes, it can lead to psychological dissonance. This state of dissonance usually does not immediately dissipate even when the person has embarked on the chosen option (Eiser 1986). For psychological dissonance to dissipate, the person must engage in some cognitive work which will lead to re-evaluation of the relevant cognitive elements. An information presented as a conflict rule can be easily transformed into an evaluation question, i.e., "Is spraying for leaffolder in the first 40 days necessary?" Answering this question through re-evaluation will help resolve the state of dissonance. The rewards that come when the question is resolved serve as important incentives. In this case, farmers can reduce costs and labor if insecticide applications in the first 40 days are not needed.

Farmer participatory research

We presented 101 farmers in Leyte, Philippines, with the simple rule, and invited them to evaluate it by participating in an experiment. Each participant marked out a portion of his or her field that would not be treated with insecticides for leaffolder control in the first 30 days after transplanting. The rest of the field received normal treatments. At the end of the season, farmers compared yields of their two plots. Statistically, the yields were not significantly different (t = 1.01 p<0.05), although the experimental plots gave higher mean yield (5.1 t ha^{-1}) than the main plot (4.8 t ha^{-1}). Most farmers (77%) had either the same or higher yields in their experimental plots.

After the experiment, the number of farmers applying insecticides in the first 30 days after transplanting was reduced from 68 to 20%; a year later, it was further reduced to 11%. Farmers' attitudes toward leaf-feeding insects changed. Before the experiment, farmers believed that leaf-feeding insects would cause severe damages (77%) and yield losses (75%), and that spraying had to be done early in the season (62%). After the experiment, these figures were reduced to 28, 9, and 10%, respectively.

Farmer participatory research can thus help resolve conflicts in existing knowledge. As pointed out by De Bono (1977), the conflict method for changing ideas works well where new information can be evaluated objectively.

Communication--reaching out to millions of farmers

For most farmers, pest management decisions are influenced by biased perceptions of pests, their damages, and potential losses. Indigenous beliefs that pests, especially highly visible ones, are harmful, had made farmers victims of pesticide abuse (Bentley 1989). Farmers also tend to associate pesticide use with modernism (Escalada and Heong 1993, Bentley 1989, Kenmore et al 1985) and high yields (Lazaro and Heong 1995). Using logic analyses, Tjornhorm (1995) showed that increase in pesticide misuse in the Philippines was associated with the high value farmers place on advice from chemical company representatives, membership in a cooperative, and visits by local technicians. Encouragement through chemical company advertisements, extension programs, and the simplicity of chemical applications will continue to reinforce farmers' selective perceptions that pesticide use is beneficial.

In the case of the leaffolder problem, it is evident that farmers are responding to insect damages based on misperceptions. The sprays they use are often broad-spectrum insecticides, which can cause secondary pest problems like the brown planthopper (Schoenly et al 1994), and are hazardous to farmers' health (Rola and Pingali 1993). To a great extent, the problem is thus social. Solving social problems requires changing psychological processes dealing with perceptions and attitudes. Because the unnecessary use of insecticides for leaffolder control among farmers is widespread in Asia (Heong et al 1997), an immediate intervention strategy seems appropriate. Because farmers' beliefs and attitudes are deeply entrenched, the use of communication systems to initiate changes in perceptions and attitudes needs to be evaluated.

In Vietnam, a pilot project in two districts of Long An Province evaluated the use of media materials to encourage farmers to test the simple rule. Printed materials (such as a leaflet and a poster) and a radio drama were designed, pretested, reproduced, and distributed to 20,000 farm families. Two months later, a management monitoring survey of 4,640 farmers showed that 92% of the target farmers were aware of the simple rule. More than half (59%) tested the simple rule. An evaluation of 450 farmers conducted 16 months after the pilot project was initiated showed that farmers' insecticide applications dropped from 3-4 to 1-2. The number of farmers applying their first insecticide sprays in the first 6 wk after sowing was reduced from 96% before the project to 62%. Similarly, farmers' perceptions of leaffolder damage and the need for early control changed (Fig. 4). It is evident that the use of communication media to present a simple rule as a conflicting idea, together with a method for evaluating it, stimulated thousands of farmers to respond toward change.

Discussion

This case study illustrates the important role of research "to transform farmers' needs into researchable problems and communicate the results back to farmers" (Pray and Echeverria 1990). If research is indeed charged with improving farmers' pest management practices, efforts must be exerted to adapt and integrate research findings and communicate them. Because pest management requires farmers to make knowledge-

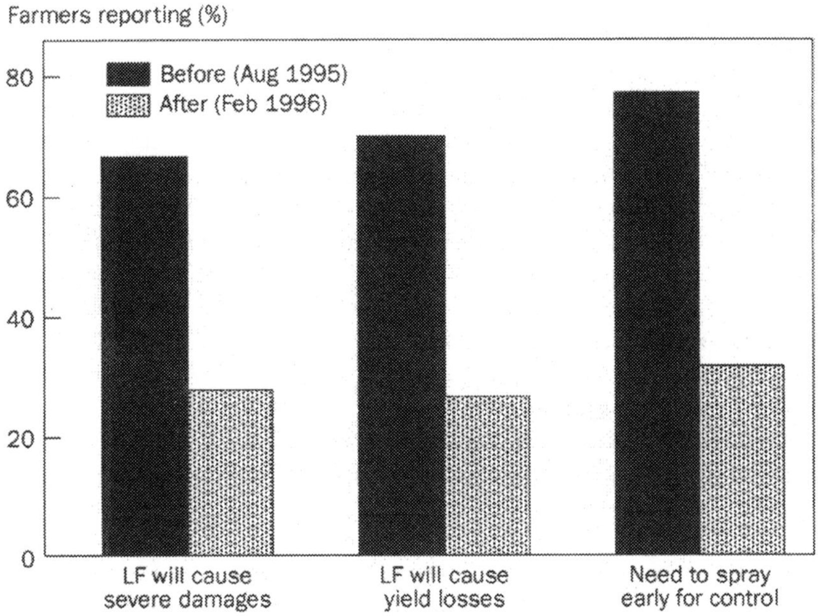

Figure 4. Farmers' attitudes toward the leaffolder and early-season insecticide spraying before and after the farmer-participatory research.

intensive decisions, heuristics can be useful. Volumes of research information may be distilled into heuristics, which are easier to communicate. To encourage farmers to integrate new information into existing cognitive patterns that will influence their decisionmaking, we need to discover and evaluate different ways of presenting information. In this study, the heuristic was presented as a conflict information that can stimulate response. An evaluation method, in the form of a simple experiment, was presented with the conflict. The approach stimulated changes in farmers' perceptions and practices. Perhaps not all information and concepts in pest management are easily transformed into heuristics and have a convenient evaluation method that farmers can participate in. Bentley (1992) provided a convenient matrix to classify farmers' indigenous knowledge and suggested ways for intervention. For instance, for pests that are important but difficult to observe, like diseases, he suggested three ways of intervention. First, farmers' taxonomies of such problems may be expanded by discovering existing knowledge and expanding on them. Second, farmers' observations may be enhanced through the use of tools of observation, such as the microscope. Third, existing beliefs may be challenged. This approach may be difficult, hence farmer experiments may be useful.

For research to meet the challenges of effecting improvements in farmers' practices, a paradigm shift is needed in research and extension. Heong (1996) discussed the use of a diagnostic approach to facilitate this shift and proposed alternatives to four prevailing paradigms. Bentley and Andrews (1996) used a radical approach, placing emphasis on learning the gaps in farmers' knowledge before research. Approaches, techniques, and tools described by Norton (1982) and Norton and Mumford (1993) also placed emphasis on understanding the pest problem from the farmers' perspective before research is contemplated. Besides addressing the technical aspects of the pest problem, research must place more emphasis on communication of the results. Activities to transform research results into appropriate forms that will encourage adaptation and adoption by farmers are within the scope of research. This transformation process involves a truly holistic understanding of the ecological and socioeconomic aspects of the pest problem. Such research activities seem vital in the transformation of research into practice.

Acknowledgements

I would like to express thanks to the Swiss Agency for Development and Cooperation (SDC) which funded the Rice IPM Network from which most of these ideas developed, and to Dr. M. M. Escalada, who shared with me her insights in communication science.

References

Anderson N H (1980) Integration theory applied to cognitive responses and attitudes. In Petty R E, Ostrom T M, Brock T C (Eds.) Cognitive responses in persuasion. Erlbaum, New York.

Bentley J W (1989) What farmers don't know can't help them: the strengths and weakness of indigenous technical knowledge in Honduras. Agric. Hum. Values 6:25-31.

Bentley J W (1992) The epistemology of plant protection: Honduran campesino knowledge of pest and natural enemies. In Crop protection for resource-poor farmers. NRI, Chatham, UK.

Bentley J, Andrews K (1996) Through the road blocks: IPM and Central American smallholders. Gatekeeper Series No. 56. International Institute for Environment and Development, London, UK.

De Bono E (1977) Lateral thinking. Penguin Books, Hermondsworth, England.

De Kraker J (1996) The potential of natural enemies to suppress rice leaffolder populations. Ph D thesis, Wageningen Agricultural University, The Netherlands.

Eiser J R (1986) Social psychology: attitudes, cognition and social behaviour. Cambridge University, Cambridge, UK.

Escalada M M, Heong K L (1993) Communication and implementation of change in crop protection. Pages 191-207 in Crop protection and sustainable agriculture. Ciba Symposium 177. Wiley, Chichester, UK.

Fabellar L T; Fabellar N, Heong K L (1994) Simulating rice leaffolder feeding effects on yield using MACROS. Int. Rice Res. Newsl. 19:7-8.

Festinger L (1957) A theory of cognitive dissonance. Row, Peterson, Evanston Ill.

Guo Y (1990) Larval parasitization of rice leaffolders (Lepidoptera: Pyralidae) under field and laboratory conditions. Ph D thesis, University of the Philippines Los Baños, Laguna, Philippines.

Harmon P,King D (1985) Artificial intelligence in business expert systems. John Wiley & Sons, Inc., New York.

Heong K L (1990) Computer expert systems for improving insect pest management. Rev. Agric. Entomol. 78:1-11.

Heong K L (1996) Pest management in tropical rice ecosystems: new paradigms for research. Pages 139-154 in Hokyo N, Norton G. (Eds.) Proceedings of the International Workshop on Pest Management Strategies in Asian Monsoon Agroecosystems. Kyushu National Agricultural Experiment Station, Ministry of Agriculture, Forestry and Fisheries, Japan.

Heong K L, Escalada M M, Dedolph C (Eds.) (1997) Pest management practices of rice farmers in Asia. International Rice Research Institute, P.O. Box 933, Manila, Philippines.

Heong K L, Escalada M M, Lazaro A A (1995) Misuse of pesticides among rice farmers in Leyte, Philippines. Pages 97-108 in Pingali P L, Roger P A (Eds.) Impact of pesticides on farmers' health and the rice environment. Kluwer Press, California.

Heong K L, Escalada M M, Vo Mai (1994) An analysis of insecticide use in rice: case studies in the Philippines and Vietnam. Int. J. Pest Manage. 40:173-178.

IRRI--International Rice Research Institute (1994). Integrated pest management: the IRRI perspective. IRRI Information Series No. 3. International Rice Research Institute, P.O. Box 933, Manila, Philippines.

Kahneman D, Tversky A (1973) On the psychology of prediction. Psychol. Rev. 80:237-251.

Kahneman D, Tversky A (1972) Subjective probability: a judgement of representativeness. Cognitive Psychol. 3:450-454.

Kenmore P E, Heong K L, Putter C A (1985) Political, social and perceptual aspects in integrated pest management programmes. Pages 47-66 in Lee B S, Loke W H, Heong, K L (Eds.) Integrated Pest Management in Malaysia. Malaysian Plant Protection Society, Kuala Lumpur.

Krech D, Crutchfield R S (1971) Perceiving the world. Pages 235-264 in Schramm W, Roberts D F (Eds.) The process and effects of mass communication. Revised edition.. University of Illinois Press, USA.

Lazaro A A, Heong K L (1995) Analysis of factors influencing farmers' insecticide sprays. Int. Rice Res. Notes 20: 2.

Miyashita T (1985) Estimation of the economic injury level in the rice leaf roller, *Cnaphalocrocis medinalis* Guenee (Lepidoptera:Pyralidae). I. Relation between yield loss and injury of rice leaves at heading or in the grain filling period. Jpn. J. Appl. Entomol. Zool. 29:73-76.

Mumford J D, Norton G A (1984) Economics of decisionmaking in pest management. Annu. Rev. Entomol. 29:157-174.

Norton G A (1982) A decision analysis approach to integrated pest control. Crop Prot. 1: 147-164.

Norton G A, Mumford J D (Eds.) (1993) Decision tools for pest management. CAB International, Oxford, UK.

Norton G A, Mumford J D (1982) Information gaps in pest management. Pages 589-597 in Proceedings of the International Conference on Plant Protection in the Tropics. Malaysian Plant Protection Society, Kuala Lumpur, Malaysia.

Norton G A (1993) Philosophy, concepts, and techniques. Pages 1-21 in Norton G A, Mumford J D (Eds.). Decision tools for pest management. CAB International, Oxford, UK.

Ooi P A C, Lim G S, Ho T H, Manalo P L, Waage J. (Eds.) (1992) Integrated pest management in the Asia-Pacific region. CAB International, Oxford, UK.

Pray C, Echeverria R (1990) Private sector agricultural research and technology transfer links in developing countries. Pages 197-216 in Kaimowitz, D. (Ed.) Making the link: agricultural research and technology transfer in. developing countries. Westview Press, Boulder, Co.

Rola A C, Pingali P L (1993) Pesticides, rice productivity, and farmers' health: an economic assessment. International Rice Research Institute, P.O. Box 933, Manila, Philippines.

Roling N (1990) The agricultural research-technology transfer interface: a knowledge systems perspective. Pages 1-42 in Kaimowitze, D (Ed.) Making the link: agricultural research and technology transfer in developing countries. Westview Press, Boulder, Co.

Rubia E G, Heong K L, Zalucki M, Gonzales B, Norton GA (1996) Mechanisms of compensation of rice plants to yellow stem borer *Scirpophaga incertulas* (Walker) injury. Crop Prot. 15:335-340.

Schoenly K G, Cohen J E, Heong K L, Arida G S, Barrion A T, Litsinger J A (1994) Quantifying the impact of insecticides on food web structure of rice-arthropod populations in a Philippine farmer's irrigated field: A case study. Pages 343-351 in Polis G A, Wisemiller K (Eds.) Food webs: integration of patterns and dynamics. Chapman and Hall, London.

Schramm W (1973) Men, messages and media: a look at human communication. Harper & Row, New York.

Sears D O, Freedman J L (1971) Selective exposure to information: a critical review. Pages 209-234 in Schramm W, Roberts, D (Eds.) The process and effects of mass communication. Revised edition. University of Illinois Press, USA.

Tichenor P J, Donahue G A, Olien C N (1970) Mass media flow and differential growth in knowledge. Public Opinion Q. 38(2):159-170.

Tjornhorm J D (1995) Assessment of policies and socio-economic factors affecting pesticide use in the Philippines._MS thesis, Virginia Polytechnic Institute, Blacksburg.

Viajante V, Heinrichs E A (1986) Rice growth and yield as affected by the whorl maggot, *Hydrellia philippina* Ferino (Diptera: Ephydridae). Crop Prot. 5:176-181.

Vo Mai, Thu Cuc N, Hung N Q, Chau N M, Chien H V, Escalada M M, Heong KL (1993) Farmers' perceptions of rice problems and management tactics used in Vietnam. Int. Rice Res. Newsl. 18:2.

Wada T, Shimazu M (1978) Seasonal population trends of the rice leaf roller, *Cnaphalocrocis medinalis* Guenee (Lepidoptera:Pyralidae). Appl. Entomol. Zool. 17:278-281.

Ecosystem analysis-based methodology for technology extrapolation

V.P. SINGH[1], V. Q. MINH[2], A.N. SINGH,[3] and S.P. KAM[1]

[1]*International Rice Research Institute (IRRI), P.O. Box 933, Manila, Philippines*

[2]*Soil Science Department, University of Cantho, Cantho, Vietnam*

[3]*Uttar Pradesh Remote Sensing Application Center, Lucknow, India*

Key words: temporal and spatial variability, agroecosystem analysis, geographic information systems, technology extrapolation domain

Abstract

Agroecosystem analysis and geographic information systems (GIS) are used to delineate target domains for extrapolation of a given technology. Biophysical, socioeconomic, and policy requirements of a chosen technology should match similar conditions of potential target areas for the technology to be extrapolated successfully in such areas. Technology extrapolation requires knowledge about the technology and the characteristics of experimental sites and potential target areas. In agroecosystem analysis, biophysical conditions of test sites can be characterized in greater detail and in quantitative terms, while those of larger target areas can be assessed only in broader, qualitative terms. Both spatial and temporal variabilities become significant as we move up the scale in the characterization. GIS is used to combine data sets from different sources and to overlay maps to develop suitability classes of target areas for the chosen technology and to identify technology application domains. Ground truthing and detailed on-farm studies are necessary to validate the final results. A case study of double rice cropping technology is presented to illustrate the procedure in characterizing the technology utilization requirements (TURs), analysis of target area qualities (TAQs), and matching TURs with TAQs to delineate potential target areas.

Introduction

New and promising agricultural technologies are usually tested or demonstrated at specific sites in research stations or farmers' fields. It is desirable to introduce these technologies to farmers over a broader geographical area, and it is expedient to determine ex-ante where these technologies might be suitable and adopted. Garrity et al (1990) used land form analysis as a basis for extrapolating the cropping systems technologies. Minh (1995) developed a geographic information system (GIS)-based framework and structure for technology extrapolation using soil and agrohydrological parameters in the rainfed environments and tested the methodology on a limited scale. Singh and Singh (1996) and Bhuiyan and Ahmed (1993) found the tools for ecosystem analysis useful in extrapolating the technologies at certain scales. This paper reviews some of the available methods and proposes a generic methodology for technology extrapolation, at different scales, particularly in rainfed environments.

Conceptual framework

All technologies are generated under a certain set of conditions and circumstances, and they should, therefore, be successful under similar situations. These situations consist of biophysical and socioeconomic/cultural factors. The extent of matching the experimental

V. Balasubramanian, J. K. Ladha and G. L. Denning (eds.), Resource Management in Rice Systems: Nutrients, 213–229.

site and the target area determines the level of extrapolation of a technology. In a broader sense, therefore, the knowledge about the technology, the experimental site where it is generated, and the area where it is to be applied are essential components of the extrapolation. The conceptual framework used in developing the proposed methodology is based on these principles and on the concepts of ecosystem analysis (Singh et al 1992, IRRI 1992), rice ecosystem classification (IRRI 1984), and land evaluation criteria for rainfed agriculture (FAO 1983). The extrapolation process essentially includes the analysis of the technology including the experimental site(s) for determining the technology utilization requirement (TUR) and the target area for determining the target area quality (TAQ) and the matching of TURs with TAQs. The basic steps are shown in Figure 1. In attempting to characterize the experimental sites and the target area, the following issues should be considered.

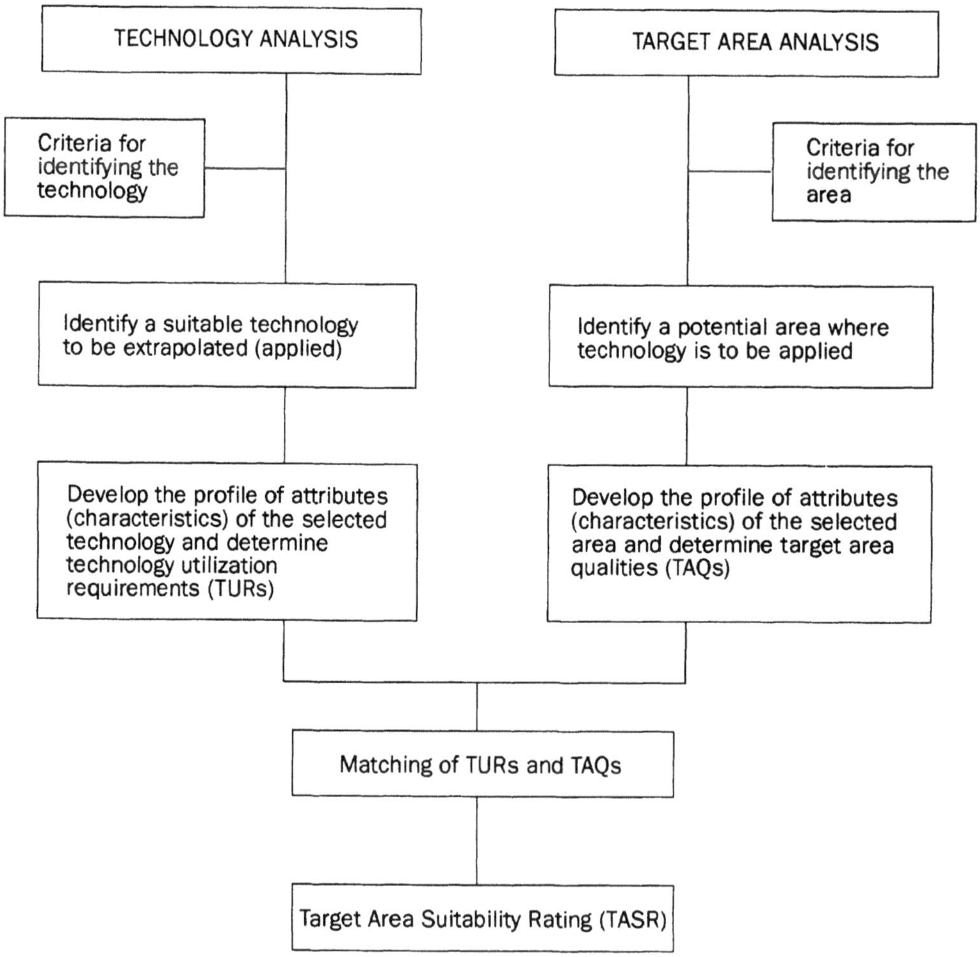

Figure 1. Schematic diagram of the methodology of technology extrapolation.

Characterization of test sites vs characterization of target areas

Technologies intended for extrapolation should be successful in areas whose characteristics are similar to those of the test sites. This requires an agroecological characterization at different levels (Singh et al 1992) and preferably in a continuum. The biophysical conditions, especially at the sites, can be characterized in reasonable detail, because these sites are practically discrete, point-based locations relative to the geographical scale of the target areas. However, the characterization of target areas practically cannot achieve the same data density as that of the test sites.

Scale, resolution, and variability

For wetland rice cultivation, a slight change in microtopography affects considerably the hydrology, which in turn not only influences the rice varieties grown but also has implications on water and nutrient management. Determining target domains for such practices needs to take cognizance not only of the spatial heterogeneity and the spatial resolution at which it matters but also of the temporal variability at the appropriate temporal resolution.

Figure 2 depicts the data, scale, and resolution dilemma in environmental characterization and determination of domains for technology extrapolation. The scale hierarchy for technology extrapolation domains also corresponds with the agroecological hierarchy, represented here by the soil system but could also include vegetation/land cover and climatic and hydrological systems. Scaling down the agroecological hierarchy to a lower level of characterization implies greater detail of dynamic parameters, e.g., while it might be sufficient to use rainfall parameters to characterize length of growing period for crops at the continental level, additional data on lithology and landform would be needed to interpret interactions with rainfall to determine site-specific soil moisture constraints to crop growth duration at the farm or field level. The scales relevant to technology extrapolation would span between the field/farm level to the subdistrict/district level (corresponding to the micro- and meso-macro levels indicated in Singh et al [1992]).

Test site characterization is usually carried out at the field/farm level of the scale hierarchy, where it is relevant and feasible to measure and record data in quantitative terms for a greater variety of static as well as dynamic parameters. For example, the drought tolerance of specific cultivars at the test sites is expressed not in terms of cumulative water availability over the entire growing season but in terms of within-season drying-up periods which are influenced by microlevel, site-specific morpho-hydrologic characteristics.

On the other hand, technology extrapolation in practice is carried out at a higher level in the scale hierarchy, say, at the subdistrict/district level (depicted by the upper plane in the scale diagram of Figure 2, although there may be a progression of planes up the scale continuum). At a higher level of the scale hierarchy, which also corresponds to the agroecological hierarchy, certain dynamic parameters at the lower levels become irre-

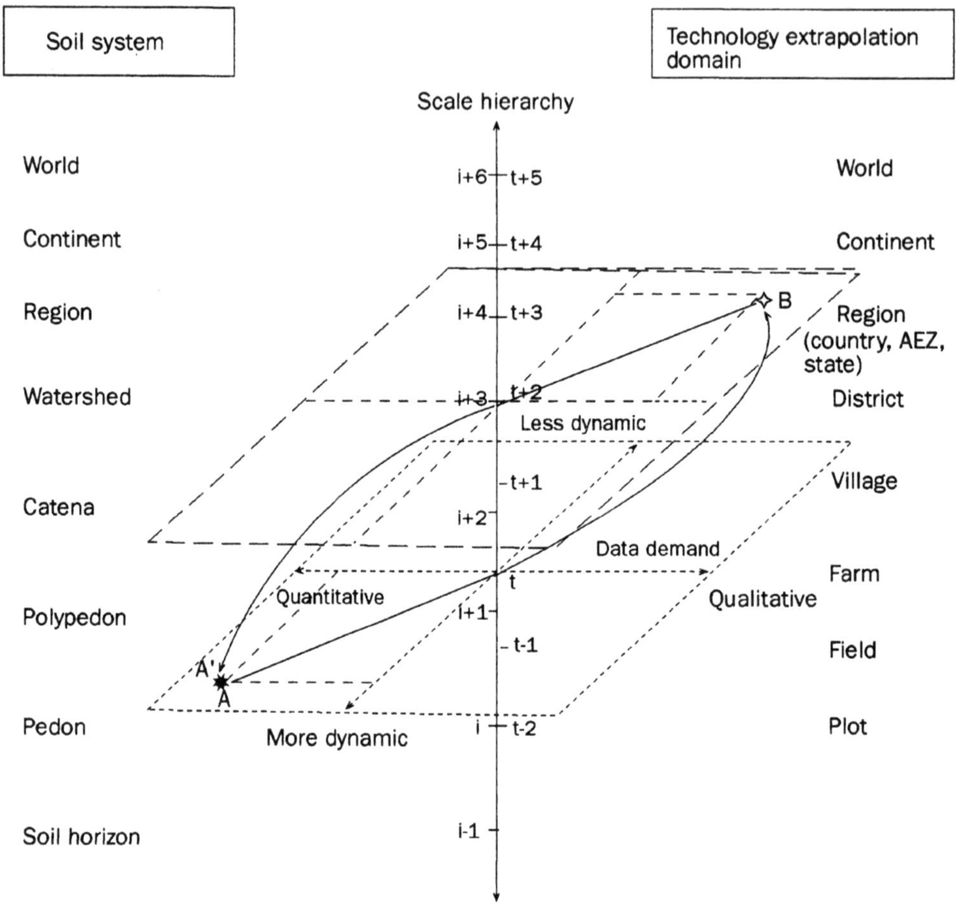

Figure 2. Scale diagram for three-dimensional classification of technology extrapolation approaches with regard to their knowledge demand (modified from the model classification diagram proposed by Hoosbeek and Bryant 1992).

levant. Also in reality, it is not feasible to obtain detailed quantitative data for a large geographical coverage of the target areas for technology extrapolation.

Hence, point A on the lower plane and point B on the upper plane depict the relative positions of test sites and target area in the data and characterization complexity dimensions, relative to the scale hierarchy. In carrying out technology extrapolation, one needs to distill the location-specific information of the test sites into some form that is more amenable to broader area characterization of the target areas (i.e., move from point

A on the lower plane to point B on the upper plane), yet not lose the relevant specificity of conditions which are necessary and sufficient to determine if the target area is suited for the technology being considered.

This movement from point A to point B is not linear along the diagonal, but rather in a curvilinear manner, whereby there is progressive distillation and spatial aggregation of the measured variables at the lower level to arrive at key variables at the higher level. These key variables used for characterizing the target area are then used for ex-ante identification of possible locations where the technology might be appropriately adopted. It is then necessary to carry out more detailed characterization of these candidate sites and to validate their suitability with farm-level studies; this process is depicted by the validation trajectory from point B to point A' in Figure 2.

This multiscale approach to technology extrapolation may be summarized by the following main steps:

(1) Select the particular technology of interest, e.g., certain ecosystem-specific rice cultivar or ecosystem-specific agronomic practice.
(2) Identify the main biophysical conditions and socioeconomic circumstances under which the technology is known or expected to be appropriate. These would be identified as being the driving factors (opportunities and constraints) and the constants (the limitations) at the level of the scale hierarchy at which the test site characterization is done.
(3) Construct indices/criteria relevant to extrapolation of the technology for the next higher level of the scale hierarchy, based on the more detailed characterization data obtained at the lower level. Upscaling might mean that constants or exogenous factors considered at the lower level might become endogenous at the higher level.
(4) Capture variability at the higher level where relevant, either through the indices constructed or statistical measures.
(5) Seek analogous, although not necessarily homogeneous, locations from within the target area thus characterized, which are then the candidate sites for testing the chosen technology, on the basis of the criteria established. These are the technology extrapolation domains.
(6) Validate the identified target locations with more detailed characterization and farm-level studies and experiments.

GIS as a tool for technology extrapolation

The spatial dimension is inherently important in determining technology extrapolation domains; hence, GIS is seen as a useful tool to provide this spatial dimension as well as to provide the means for integrating map data of different subject matters and from disparate sources. Two issues are pertinent in the mapping and use of GIS in the multiscale approach:

(1) To implement the multiscale approach, mapping should be done to reflect the nested hierarchy of the agroecosystems and the extrapolation domains. Mapping allows for

data collected at a lower level to be used for constructing indices, criteria, summary statistics, etc., at the next higher level.

(2) The GIS data structure used should be amenable to the multiscaling required and should capture, at a particular level of the scale hierarchy, the relevant variability inherent in the next lower level. The relative merits of using vector and raster data structures for data capture and analysis will need to be investigated.

One must remember that GIS places certain demands on data, such as explicit geographical referencing, reasonable geographical accuracy at the appropriate mapping scale, and adequate geographical coverage, and that the benefits accruing from spatial integration of data can be realized only if the input data sets are properly designed for the purpose.

Technology analysis

Identifying and selecting the technology to be extrapolated

Technology identification is an issue linked with research prioritization and technology generation process. In the extrapolation process, the technology is selected according to the specific requirements: e.g., as a solution to a particular problem, for a specific group of users or for a specific agroecological situation. In a broader sense, the technology to be extrapolated should be the one that could improve farming systems performance and farmers' income.

Based on the present farming situation in the rainfed areas, the double rice crops (dry-direct seeded as first crop and transplanted as second crop) technology was selected as a case study for developing this methodology. Single rice crops with low yields dominate the rainfed lowland rice areas; two crops of rice are cultivated only in places that have favorable soils and agrohydrology. Soil toxicities and adverse agrohydrological conditions, such as drought, less rainfall, and shorter rainy season, preclude the adoption of this technology in other places. However, such soil and hydrological variations are not uniform and adjustments in the crop management practices, such as the time of planting and use of appropriate varieties, provide ample opportunities for increasing the cropping intensity.

The farmers favor two crops of rice because of their familiarity and experience with the rice crop, and importance of rice for local consumption and for export (supported by government policies). Storage, processing, and marketing of rice are also within the reach of average farmers and all the household members can contribute to and benefit from the crop.

Determining technology utilization requirements

For a technology to be extrapolated, it is essential to know under which conditions it was generated and under which situations it should be applied to be successful. This requires characterizing test sites and developing profiles of attributes of the technology using biophysical and socioeconomic criteria and analyzing the advantages and disadvantages of the selected technology. The basic criteria may include climatic factors (rainfall, temperature, solar radiation), physical factors (land, soil, hydrology, duration of growing season), biological factors (crop and crop varieties, insect pests and diseases), socioeconomic factors (farmer preference, local demand and market potential, cost of inputs and farmers' access to the needed resources), and policy (government policy and support). This profile provides the technology utilization requirements (TURs), which may be arranged into three major groups--biophysical, management (socioeconomic/cultural), and environmental (conservation).

Biophysical requirements

- Climatic: rainfall amount and pattern (including beginning, end, and length of rainy season), temperature, humidity, daylength, and solar radiation
- Hydrological: water depth, soil moisture, flood and drought occurrence
- Soil: pH, nutrient/fertility status, soil depth, soil workability and load-bearing capacity, and other soil conditions affecting germination or crop establishment
- Biological: crop duration; variety; rooting behavior/conditions; conditions for crop ripening; weed, insect pest, and disease pressure; and levels of crop/varietal tolerance for these stresses

Management/socioeconomic requirements

- Farmers' capacity and access to provide the level of needed inputs and agronomic practices
- Landholding size
- Timely availability of sufficient labor (household/hired) or potential for mechanization for critical operations and low level of competition with other sectors
- Availability of storage, processing, and marketing infrastructure
- Lower competing demands for household labor of all ages and gender groups

Environmental/conservation requirements

- Minimal soil erosion, flooding, and soil degradation hazards
- Minimal contamination of surface and groundwater resources and competition with other sectors
- Low level of competition with and degradation of the common property in the community

The main features of TURs for the chosen technology are the following:

- It can be practiced in the rainfed areas, which do not have access to irrigation water for dry season crop.
- Rainwater is the main source for supplying the crop water requirement.

- The rainfall amount in the area should be greater than 1,400 mm yr^{-1}, and the rainy season longer than 5 mo.
- The moisture availability period is longer than 180 d.
- Severe drought does not occur or drought does not exceed 10 d during the growing period of the rice crop. Otherwise, supplementary life-saving irrigation may be needed for the second crop.
- The soil and groundwater are nonsaline. To some extent, slightly acid soils can be used, but yields would be low because of soil toxicity, especially at the beginning of the rainy season. Such areas, therefore, would require rice genotypes that are tolerant of acidity, or some level of soil reclamation would be necessary.
- Short-duration rice varieties (<110 d), such as IR42, are available for both seasons, thereby maximizing the use of available moisture. Where the maximum inundation depth during the first rice crop is greater than 60 cm, or the rainy period in the area is longer than 6 mo, traditional tall rice varieties can be used for the first rice crop.
- The technology would require dry seeding for the first crop in fields prepared before the start of the rainy season, implying that labor and equipment, including energy sources if mechanical land preparation is done, are available and accessible to farmers. Dry seeds will be broadcast on unpuddled dry (or moist) soil and will remain in the soil until adequate moisture from the rain is accumulated for germination. Therefore, the seeds have to be covered with soil to protect them from birds and other damages. After emergence, the rice plants have to grow in relatively low moisture (upland) conditions until enough rainfall has been received to flood the field. This situation may cause some degree of moisture stress to the crop and high weed competition. The rice variety used should have some degree of drought tolerance, and weed management must be an essential component for this technology to succeed. If chemical weed control is not employed, labor will be needed for manual weeding. After weeding, some seedling redistribution may also be necessary to obtain a uniform crop stand in the field. Thereafter, with subsequent rains, fields will often have adequate moisture and standing water, unless there is a long dry spell.
- Before harvesting the dry-seeded crop, the seedlings of the second rice crop of improved variety are to be raised in another place to have them ready at the right age for transplanting within 1 wk after harvest of the first crop. Adequate provisions including a piece of land, seeds, and labor for growing seedlings must be made.
- After the first rice crop is harvested, the field will be plowed and puddled, and the seedlings transplanted immediately (usually within 1 wk). This implies that there must be adequate labor for simultaneous harvesting of the first crop and establishing of the second crop.
- The second rice crop may have to grow in flooded condition (water depth of about 30 cm and occasional short-duration submergence) until the end of rainy season, i.e., late November or December. Therefore, the rice variety used for this crop must have some kind of submergence tolerance.
- Each of the rice crops would require 100-150 kg urea ha^{-1}, 50 kg diammonium phosphate, and 25 kg of other fertilizers per hectare. The amount of fertilizer for the second rice crop may be less than that for the first rice crop, because in cloudy-rainy weather, rice often does not respond to high levels of fertilizer application and there

is a higher risk due to climatic adversities, e.g., crop lodging, submergence due to high rainfall accompanied by strong winds.

- In most cases, the first rice crop would need an application of herbicides, and both crops would require adequate pest control measures. Sufficient resources must be available in time for the purchase of needed inputs which must be available on time in the local market.

Target area analysis

A prerequisite to determining the domains for technology extrapolation is the characterization of the target areas. Relevant characteristics, whether biophysical or socioeconomic, should be known, so that we can judge if these characteristics constitute or provide the enabling conditions for a particular technology to be successfully adopted.

Target area identification

After the technology and the TURs have been defined, it is now necessary to identify the target area where the technology is to be applied. Since the selected technology in this paper is concerned with the increased cropping intensity of rainfed rice, potential rainfed lowland rice areas can be identified as the target areas for extrapolation, by using the same criteria used for selecting the technology and for developing the TURs. By using these criteria, we can select potential areas where

- only one rice crop is grown in a year;
- rainfall amount is greater than 1,400 mm yr^{-1};
- the rainy season is longer than 5 mo;
- moisture availability period is longer than 180 d;
- soils are neutral or only slightly acidic;
- possibilities of life-saving supplementary irrigation exist;
- drought does not occur during the first rice crop, and if it occurs, it does not exceed 10 d during the cropping season;
- increasing the cropping intensity is vital to grow more rice for local consumption and for market;
- farmers are willing to and can apply needed inputs; and
- support and infrastructure are available for transport, storage, processing, and marketing.

Procedures for groecosystem analysis (IRRI 1992, Garrity et al 1996) can be applied to identify and characterize the potential area. Macrolevel analysis can be used to identify and delineate rice ecosystems, and more detailed, mesolevel analysis can be used to identify and delineate rainfed lowland rice subecosystems in the rainfed ecosystem.

The macrolevel analysis can identify and delineate the rice area into irrigated, rainfed upland, rainfed lowland, and deepwater ecosystems using irrigation sources, soil, agrohydrology, climate, and major rice farming systems. The homologous zones are delineated on the basis of the following information collected from secondary sources:

- rice farming techniques and cropping systems, including methods of crop establishment and cropping calendars;
- main soil constraints and agrohydrological factors, including flooding pattern and surface water depth;
- soil type distribution and chemical characteristics of different soil types;
- maximum inundation depth, rainfall distribution, drought occurrence, length of moisture availability periods and rainy season; and
- rainfall distribution data (daily or monthly basis) from major stations in the rainfed areas.

A base map of the study area is created and stored in a digital format and used in ground truthing (verification) and field data collection and for validating secondary data. Maps collected from secondary sources are also digitized by using a suitable GIS software package. Maps at different scales are transformed to the same scale and projection so that the data processing can be easily done by personal computers. Mapping operations can also be done manually but these are tedious and cumbersome.

After digitizing, converting, and integrating with other data, the map information is used for identifying and delineating rice ecosystems and subecosystems, and for making a suitability assessment by using the criteria for specific uses. The basic approach for creating various desired maps is illustrated in Figure 3.

Identifying and delineating the rice area

Based on the existing land use map, a consolidated map of the rice area is developed by using a computer-based software program or other means that could group the land use map into specific locations of the ricefields. The rice area from the detailed land use map is delineated by mapping all areas that had rice as a component of the farming systems. This may include two rice crops, one rice crop only, one or two rice crops + one upland crop, or any other combination with rice.

Identifying and delineating rice ecosystems and subecosystems

The rice ecosystem map of the selected areas can be created on the basis of the definition and terminology of rice environments developed by IRRI (1984, 1992). All the areas where even partial irrigation is possible are classified into irrigated zone, and by subtracting it from the total rice area, the remaining is classified as rainfed rice. The process of identifying and delineating rice ecosystems is shown in Figure 4.

Each of the irrigated and rainfed lowland rice ecosystems is analyzed in more detail at the mesolevel to differentiate the rice area of a given ecosystem into its subecosystems using specific subecosystem criteria as specified in IRRI (1984, 1992). The rainfed low-

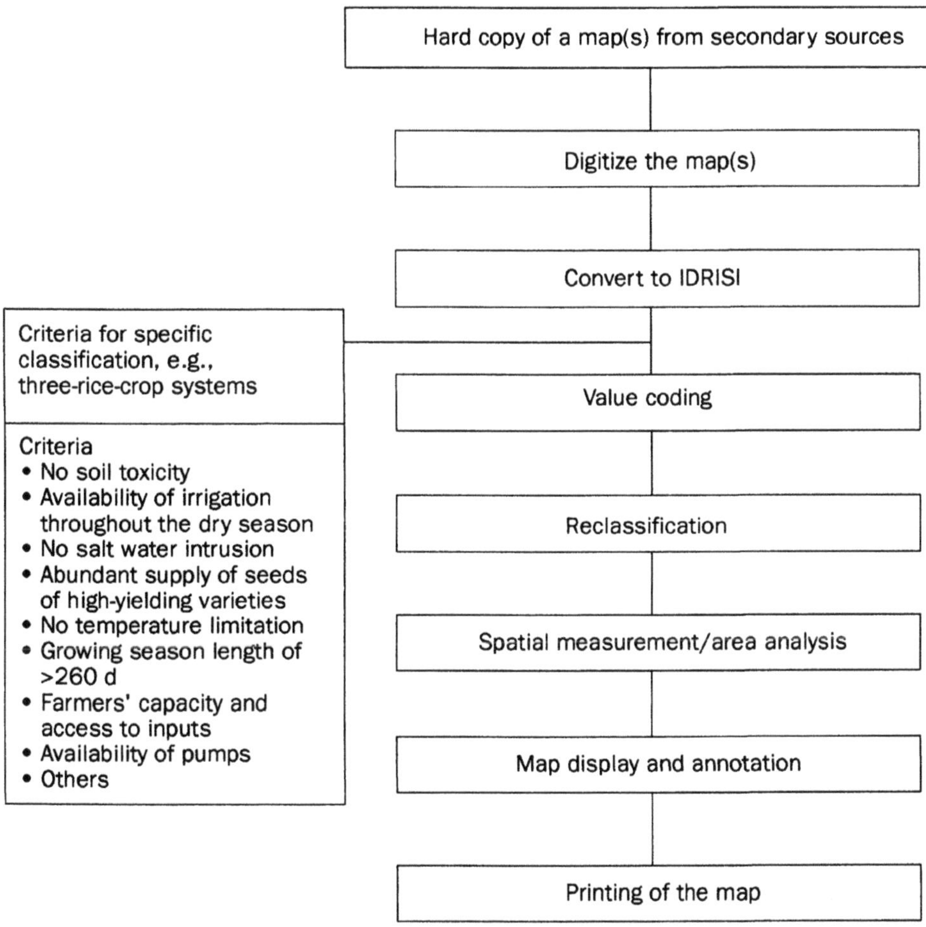

Figure 3. Basic flow diagram for creating various desired maps.

land rice subecosystems, for example, are generally identified and delineated using the maximum inundation depth, moisture availability period, rainy season duration, and drought occurrence.

The results of ecosystem and subecosystem zonation and their characterization are checked and verified through ground truth survey, particularly during the wet season. In addition to the ocular inspection, an interview of a large number of farmers may be needed to get their expert opinion and experiences. The data gathered from the interview may include major rice cropping systems, cultivation techniques, soil and agrohydrolo-

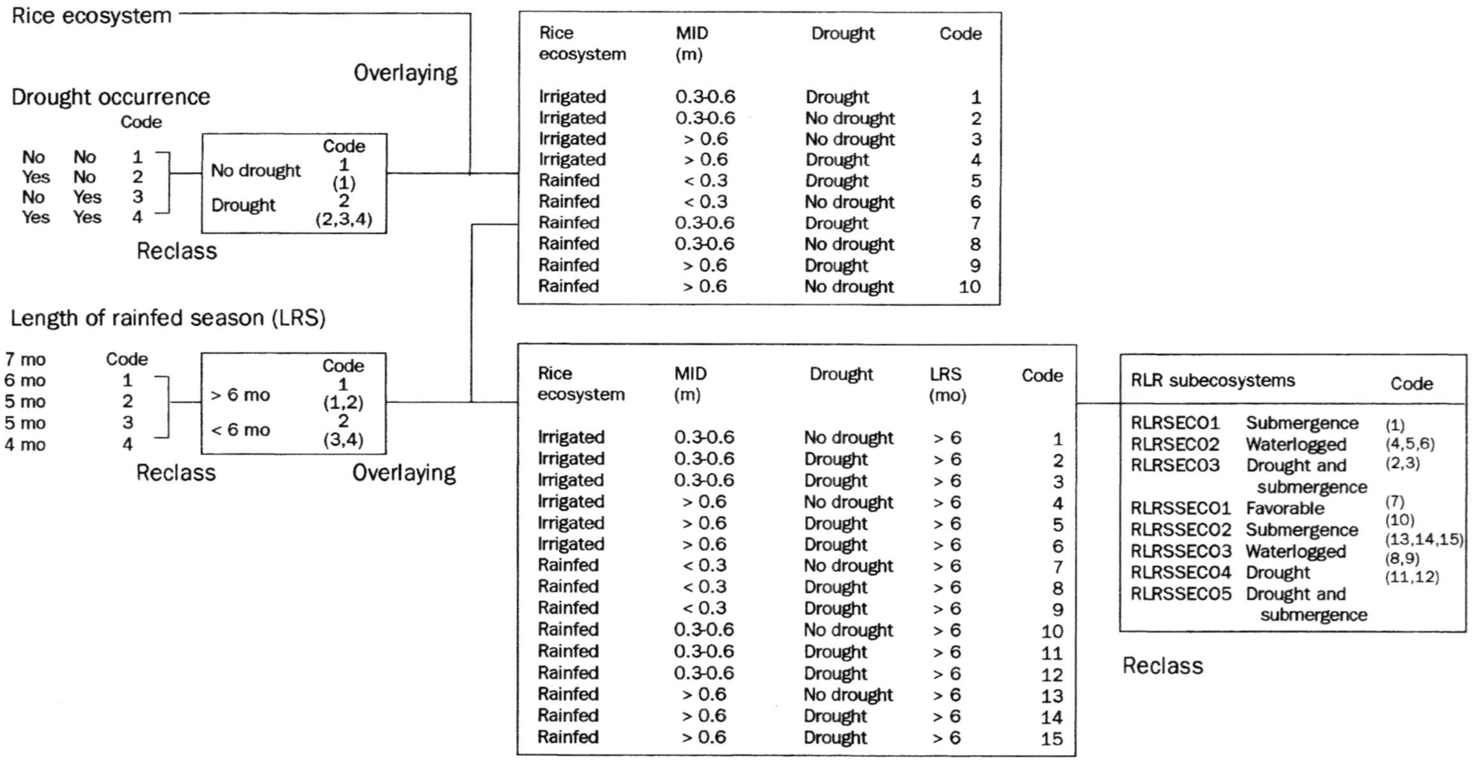

Figure 4. The process of identifying the rainfed lowland rice subecosystems.

gical characteristics, main constraints to crop production, level of capital investment, recurrent costs, produce consumption and sale patterns, average income, labor requirement for producing one crop, area for each crop, and crop production cost.

During the ground truthing process, some secondary data are also collected from government institutions to get a broader picture of the land characteristics, land uses, socioeconomic situations, main constraints to crop production, and government agricultural policy in the study areas. The data on soil, climate, and cropping patterns are additionally collected, wherever possible. Soil analysis results from previous reports are used to identify and relate the soil characteristics with the existing cropping systems in the respective rainfed lowland subecosystems and to evaluate the land of the rainfed lowland areas.

The climatic data collected include rainfall, rainy season length, flooding depth, moisture availability period, and drought occurrence during the crop seasons, at the field, village, and/or district levels in a given subecosystem.

The socioeconomic data collected are used for calculating capital investment, recurrent costs, labor needs, and gross and net benefits for each rice cropping system, in the potential target area. The relationship of rice yield with the classes of land characteristics (soil groups), rainfall distribution pattern, drought occurrence, and maximum inundation depth can be determined by using a statistical software to calculate the rating value of land qualities, which is used in evaluating area suitability and in calculating the area and its percentage in respective suitability ratings.

Determining target area qualities

After identifying and characterizing a potential target area, a profile of its characteristics is developed using biophysical, management (socioeconomic), and environmental (conservation) criteria. The factors included in each of these categories are similar to those used in the profile of the technology to be extrapolated. The profile of the target area broadly includes the following:

Biophysical conditions

- Climatic: amount and pattern of rainfall and rainy season duration, temperature, humidity, solar radiation, and daylength
- Hydrological: seasonal soil moisture, surface water depth, water table depth, and availability of ground and surface water
- Soil: pH, fertility status, specific deficiency or toxicity, soil depth, soil moisture, soil workability, and conditions that affect germination and/or crop establishment
- Biological: incidence of weeds, pests, and diseases affecting growth; crop duration

Management/socioeconomic conditions

- Landholding size
- Farmers' access and capability to apply needed inputs and agronomic practices
- Availability of storage, processing, and marketing infrastructure
- Availability of labor and potential for mechanization

- Common property uses and management systems
- Household and community interaction

Environmental/conservation conditions
- Level of soil erosion, flooding, and soil degradation
- Water quality
- Forest management

Matching TURs with TAQs

To evaluate and classify the suitability of a target area to a technology, it is important to compare the characteristics of the target area with the requirements of the selected technology. Based on the factor ratings of target area qualities (TAQs), the technology extrapolation suitability classification is done by matching the TAQs and TURs of the selected technology. The suitability rating of the target area depends on the degree to which each of its characteristics/qualities satisfies each TUR (FAO 1976, 1983). The suitability of the target area for the selected technology is classified as highly, moderately, or marginally suitable, or unsuitable.

The suitability classes of all the qualities of the target area pertaining to the characteristics of the technology are converted into factor rating values, as suggested for soils by the FAO (1984). The factor rating values are combined by the method of multiplication in arithmetic procedure (FAO 1983) and are used in matching the suitability. The suitability orders and classes of the TAQs used in technology extrapolation can be the same as those provided in the FAO framework and guidelines of 1976 and 1983.

By using GIS tools, maps of qualities of the target area are processed and matched with the criteria of TURs to produce a suitability map of the target area, and the application domains of the selected technology are identified and delineated in the target area.

The steps in identifying and delineating the suitability classes of the target area are illustrated in Figure 5. The process of developing the suitability classes is shown in Figure 6.

Target area and technology suitability rating

Depending on the purpose of the extrapolation, a technology can be extrapolated to one or several target areas, or several technologies can be extrapolated to one target area. Both of these, however, require target area and technology suitability ratings. In the target area suitability rating, several target areas are evaluated and all are rated in different classes for a particular technology. In the technology suitability rating, several technologies are evaluated and rated in different classes for a particular area.

The relationship between the yield of rice and selected biophysical and socioeconomic characteristics from secondary data sources can be determined by using a statistical software, to pinpoint the important factors affecting the yield of rice. From this analysis, factors found to significantly affect rice yields are used as land qualities and diagnostic factors in evaluating the suitability of the technology in the study area.

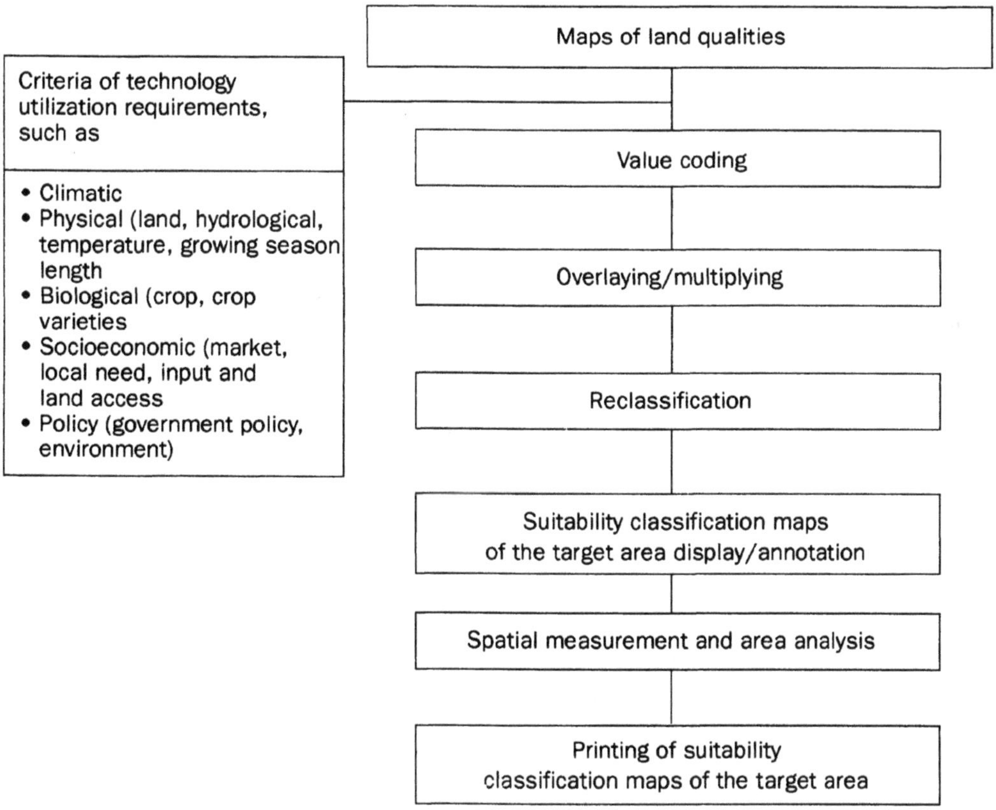

Figure 5. Steps in identifying and delineating the suitability classes of the target area for selected technology.

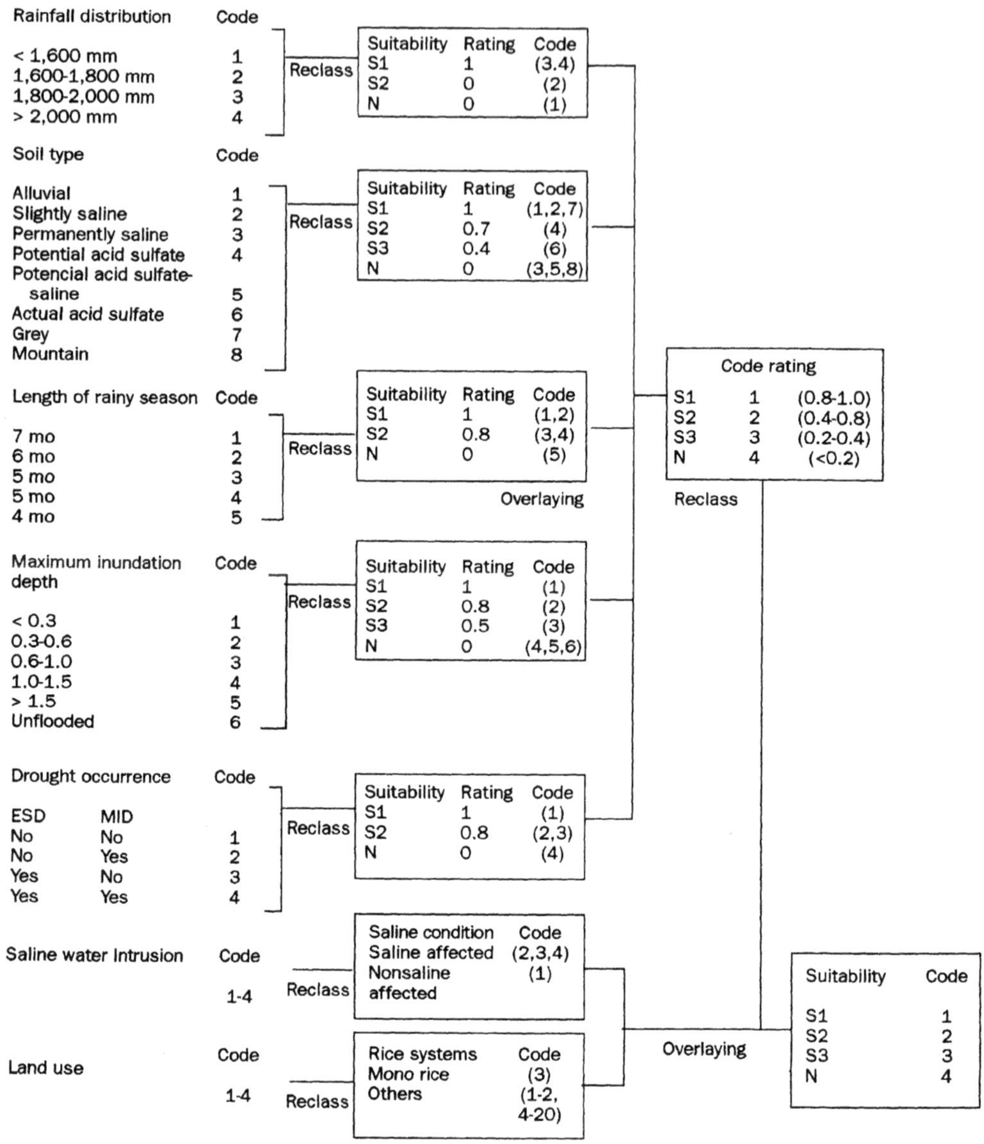

Figure 6. Process of developing the suitability classes of the target area and technology application

Ground truth collection for checking the results

The results of technology extrapolation assessment should be validated through a field survey to check if the extrapolated technology is fit for and acceptable to the users in the target area.

The application of this methodology has been validated in India and Vietnam, and is found to be useful for various other technologies and in heterogeneous environments (Minh 1995, IRRI-IFAD 1995, Singh and Singh 1996).

References

Bhuiyan A M, Ahmed N U (1993) Agroecological zonation and extrapolation of field level research in Bangladesh: present status and future prospects.

FAO--Food and Agriculture Organization of the United Nations (1976) A framework for land evaluation. FAO Soils Bulletin No. 32. Rome. 72 p.

FAO--Food and Agriculture Organization of the United Nations (1983) Guidelines: land evaluation for rainfed agriculture. FAO Soils Bulletin No. 52. Rome. 233 p.

FAO--Food and Agriculture Organization of the United Nations (1984) Guidelines: land evaluation for rainfed agriculture. Forestry Paper No. 48. Rome. 123 p.

Garrity D P, Garcia A G, Agustin P C, Dacumos R Q (1990) Methods for the extrapolation of cropping systems technologies. Paper presented at the planning workshop on ecosystem analysis for extrapolation of agricultural technologies, 22-25 May 1990, Tuguegarao, Cagayan, Philippines.

Garrity D P, Singh V P, Hossain M (1996) Rice ecosystems analysis for research prioritization. Rice research in Asia: progress and priorities. CAB International and International Rice Research Institute, P.O. Box 933, Manila, Philippines. p 35-58.

Hoosbeek M R, Bryant R (1992) Towards the quantitative modeling of pedogenesis--a review. Geoderma 55:183-210.

IRRI--International Rice Research Institute (1984) Terminology for rice-growing environments. International Rice Research Institute, P.O. Box 933, Philippines. 35 p.

IRRI--International Rice Research Institute (1992) Program report: rainfed lowland rice ecosystem. P.O. Box 933, Manila, Philippines. p 62-66.

IRRI-IFAD--International Rice Research Institute-International Fund for Agricultural Development (1995) Collaborative research and development of sustainable rice farming systems in South Asia (Phase II). Progress Report. 96 p.

Minh V Q (1995) Use of soil and agrohydrological characteristics in developing technology extrapolation methodology. A case study of the Mekong Delta, Vietnam. MS thesis, University of the Philippines Los Baños, Philippines.

Singh VP, Hunt E D, and Singh AN (1992) Rice environmental characterization, research prioritization and technology extrapolation. Paper presented at a Seminar on Satellite Remote Sensing and Geographic Information Systems in Agricultural Research, 20-22 Jul 1992, International Rice Research Institute, Los Baños, Philippines.

Singh V P, Singh A N (1996) Agricultural technology extrapolation through environmental characterization. Paper presented at the Second International Crop Science Congress. 17-23 Nov 1996, New Delhi, India.

Assessment Institute for Agricultural Technology (AIAT) for evaluation and dissemination of crop production technologies in Indonesia

A. DIMYATI[1] and A. SURYANA[2]
[1]*Assessment Institute for Agricultural Technology (AIAT), Lembang, Indonesia*
[2]*Center for Agricultural and Socio-economic Research (CASER), Bogor, Indonesia*

Key words: location-specific research, technology adaptation with farmers, research-extension-farmer linkage, agribusiness orientation

Abstract
There have been widespread complaints that agricultural research in Indonesia does not properly address farmers' problems and that technology transfer is slow because of weak linkage among research, extension, and users. To rectify this situation, farm trials and demonstrations involving researchers, extension workers, officials from agriculture-related offices, and farmers have been conducted. These efforts resulted in the creation of 17 Assessment Institutes for Agricultural Technology (AIAT) covering the 27 provinces in the country in April 1995. These institutes are tasked with doing research on location-specific agricultural technologies, gathering and channeling feedback from clients to researchers to improve agricultural research programs, and disseminating research results as extension material. AIAT's main strength is the integration of researchers and extension personnel under one umbrella at the provincial level. New mechanisms and institutional linkages for generating, evaluating, adapting, and applying location-specific agricultural technologies have been installed. A new paradigm adopting a participatory bottom-up planning approach is being implemented. A number of operational instruments such as implementing guidelines, innovative research methods, and protocols for enhancing users' participation in research activities should be continuously developed. The implications of the new approaches on various research management issues are discussed. A case of technology adaptation and promotion by AIAT through an assessment of a rice-based farming system with agribusiness orientation is described.

Introduction

Indonesia's success in agricultural development during the First Long-Term Development Period (LTDP I) has been widely acclaimed. Its strongest point was the self-sufficiency in rice achieved in 1984 and its maintenance in the following years. This monumental achievement was supported by several government policies and investments in the form of intensification programs, research, and extension. At present, dynamic changes in the strategic environment, both domestic and international, have rendered some of these policies ineffective and these same policies have, in fact, created obstacles which need to be addressed:

1. Adoption rate of new or improved technology is low.
2. Agricultural research agenda do not appropriately address actual problems faced by farmers and other constituents.
3. Mechanisms for technology transfer are neither effective nor efficient.
4. Participation of farmers and other technology users in the selection of technologies and in the general decision-making process is low.

V. Balasubramanian, J. K. Ladha and G. L. Denning (eds.), Resource Management in Rice Systems: Nutrients, 231–242.

Efforts have been made to rectify the situation. The Agency for Agricultural Research and Development (AARD) has implemented several farming systems research projects to develop location-specific technology packages for different agroecosystems such as the fragile uplands, watershed areas, irrigated lowlands, and swampy areas. The commodities involved here are numerous and are from different subsectors, as are the institutions involved in implementing the projects.

In the beginning of the second LTDP, several new approaches were introduced to enhance technology transfer to farmers. On-farm trials and demonstrations were conducted for various commodities. AARD researchers conducted almost all the trials and demonstrations, together with extension agents from the former provincial Agricultural Information Center, officials from other agriculture-related offices, and farmers, taking into account the interests of all agencies involved. The most successful and widely acclaimed activities using this approach were facilitated through the Research-Extension Linkage Pilot Project (reviewed comprehensively by Sail [1995]).

Implementing new concepts

The application of "blanket technology" through the top-down intensification programs in Indonesia and elsewhere has proved successful in spurring the Green Revolution (Ramasamy 1995). However, crop productivity has plateaued in many areas, and additional sources of productivity growth are needed to nourish the growing population. Declining productivity growth has been observed in irrigated rice systems where return to investment on production inputs, especially fertilizers, declined steadily (Hermanto et al 1993, Roling and van de Fliert 1994). In this situation, every potential possibility to increase productivity should be explored and utilized (CRIFC 1992). One of the feasible sources of increasing productivity relates to the spatial variation in soil fertility and climatic characteristics. These variations have not been effectively utilized due to wide application of blanket technology. This is reflected in the statement of one leading farmer in Indramayu, West Java: "Higher yield and income are obtained through the use of locally modified technologies deviating from the widely recommended ones" (Ruman S, 1996, pers. commun.). In short, the application of nationwide blanket technologies has induced inefficient use of agricultural resources; there have been both overuse and underuse of inputs. This is counterproductive to the efforts of producing agricultural commodities with high competitive advantage, which is a must in an era of globalized economy. A case of inefficient use of fertilizers was revealed by Zaini et al (1993). They suggested that soils be categorized into three classes based on phosphorus (P) availability as high, medium, and low; and that different rates of P fertilization be used: 50, 75, and 125 kg of triple superphosphate (TSP) ha^{-1}. By so doing, TSP consumption in Java could be reduced by up to 200,000 t $season^{-1}$, and this can be used on approximately 1 million ha of marginal lands outside of Java.

To fully utilize spatial variation in land productivity, location-specific technologies should be applied. This calls for the local modification, adjustment, and correction of nationally recommended technologies in each location. In other words, the technology to be recommended should be developed in the specific conditions representing the target area or resource management domain. The highest level of specificity in the agricultural system is reached by the lowest hierarchy, namely household farming. Therefore, highest technical efficiency will be reached by applying the technology developed for each farm. Because it will not be financially feasible for the government to do all research activities intended to develop household-specific technologies, the farmers themselves should be capable of conducting their own adaptations and innovations.

Application of location-specific technologies should be accompanied by greater capability of farmers in diagnosing their problems, identifying their technology needs, and selecting the appropriate technologies. In addition, their capability for generating, adapting, and applying new technologies should be improved. Farmers have to be involved in the process of generating, adapting, and disseminating new technologies (Lacy 1995, Uphoff 1995). Their involvement is particularly important in the decisionmaking process; they should be involved in identifying problems to be addressed and evaluating the outputs (Eponou 1993). In Indonesia, a good example of successful farmers' involvement in diagnosing problems, identifying technology needs, selecting appropriate technologies from a number of available alternatives, and applying them in their circumstances has been demonstrated by the Farmer Field School (FFS) program for integrated pest management started in 1989 (Roling and van de Fliert 1994). The FFS has been considered an attempt to correct the failure of the Training and Visit system, which was widely applied in the rice intensification program. A conservative estimate indicated that by the end of 1995, about 650,000 farmers had been trained at the FFS. The success has induced agriculturists from various countries to adopt the same approach (Oka 1996).

Three new principles are being established to rectify the drawbacks of the previous policies and practices : decentralized decisionmaking, bottom-up planning, and greater people's participation. Indonesian legislators and government authorities have realized that there should be fundamental transformation of research and extension institutions to accommodate these new principles into agricultural development. Five legal pronouncements have been issued to achieve this transformation (Table 1).

Thus, AIAT's mission is to support regional agricultural development and optimal use of agricultural resources through the application of location-specific technologies. There are some consequential changes in the basic philosophy for applying location-specific technology. AIAT has adopted the following approaches to accomplish this mission (AARD 1995):

- incorporation of farmers' and other users' interests in the generation, evaluation, dissemination, and application of technologies by involving farmers and other users in all stages of the process,
- active participation of researchers and extension workers from within AIAT as well as from provincial and district agricultural offices,
- strong backstopping from commodity-based national research institutes (NRI) within the AARD system, and
- collaboration with provincial universities and other sources of expertise and information.

Table 1. Legal pronouncements pertaining to transformation of research and extension systems and their summarized content relevant to the subject of this paper.

Legal pronouncement	Relevant summarized content
Legislation No. 12/1992 on Crop Cultivation	• Greater freedom of farmers • Consideration of people's interest in government programs • Assurance of income for farmers joining government programs • Encouraging greater people's participation in government's programs
Decree of the Minister of Agriculture No. 798/1994	• Establishment of 17 AIATs covering 27 provinces • Assignment of AIATs tasks-- research on location-specific commodities and assessment of location-specific technologies • Supervision of AIATs by the Director of the Center for Agricultural and Socio-economic Research (CASER) and administrative coordination by the Representative of the Ministry of Agriculture (MOA) in each province • Working together of researchers and extension agents under the single roof of AIAT
Decree of the Minister of Agriculture No. 804/1994	• Delegation of authority of agricultural affairs to the district government • Affairs delegated include evaluation and application of technology as well as extension.
Decree of the Minister of Agriculture No. 804/1995	• Definition of various technical terms • Adoption of the concepts "location-specific technology" and "from farmers back to farmers" • Job delineation among agricultural institutions • Functional relationships among these institutions • Mechanism for generating, evaluating, and applying agricultural technologies and channeling feedback from users to researchers • Role and constituents of the provincial Advisory Committee on Agricultural Technology Assessment
Joint Decree of the Minister of Home Affairs and the Minister of Agriculture No. 54 and No. 301/1996	• Establishing guidelines for conducting extension at the national, provincial, district and subdistrict levels • Establishment of Agricultural Information and Extension Center (AIEC) in each district • Assigning all extension agents in a district to AIEC • Assigning the functions of AIEC to include technology evaluation and application

These philosophical approaches should be effectively incorporated in the following implementation strategies of AIAT:

a) research on location-specific commodities,
b) evaluation and adaptation of location-specific technologies,
c) dissemination of successful research and evaluation results through simple and easy-to understand extension materials,
d) gathering and channeling of feedback from users to researchers to improve research programs, and
e) technical services pertaining to technology assessment.

Ministerial Decree No. 804/1995 established a new institutional arrangement and the linkages and mechanisms for technology generation, adaptation, and application. The decree replaced the old system set forth by Ministerial Decree No. 439/1989 which empowers a strongly centralized authority to generate, evaluate, and recommend agricultural technologies. In the new system, the authority to evaluate and adapt the technologies (which was used to be centralized at the hand of the subsectoral directorates general) was distributed to the 17 AIATs. Technology recommendation, which was a function of the directorates general, was delegated to the provincial representative of the Ministry of Agriculture (MOA) with the support of the provincial Advisory Committee for Agricultural Technology Assessment. The Committee is chaired by the director of the provincial representative of the MOA and the members are the directors or representatives from the subsectoral provincial offices for agricultural affairs, provincial Board for Development Planning, AIAT, and other agriculture-related institutions. The Committee gives suggestions and opinions to the director of the provincial representative of the MOA regarding the recommended technology, based on results of research and evaluation done by AIAT. Legalized recommendations are transformed by the respective subsectoral offices at the provincial and district levels into technical guidelines which are further translated into simple extension materials by the Rural Extension Center (REC) at the subdistrict level. The extension materials are the main guidelines used by farmers in applying the technology.

While applying the technology in their farms, farmers assess the emerging problems and discuss constraints with fellow farmers from the same group. The problems are brought to the attention of extension workers and the extension forum of the REC. The REC compiles the identified problems, synthesizes them into agricultural technology needs at the subdistrict level, and brings them to the district subsectoral offices of agriculture. These offices compile them as subsectoral problems and synthesize them into subsectoral technology needs at the district level. At the provincial level, the Advisory Committee on Agricultural Technology Assessment will discuss the identified technology needs, formulate the necessary assessment program to be conducted by AIAT, and propose commodity research programs to be conducted by the national research institutes (NRIs). Research proposals from various provinces are compiled by these commodity-based NRIs and synthesized into commodity research programs. These commodity research programs are discussed and approved by the National Agricultural Research Committee, which is composed of the directors general of AARD and of all subsectors and is led by the Minister of Agriculture. The procedure for generating, evaluating, and applying location-specific agricultural technologies is described in Table 2.

The role of the newly established Agricultural Information and Extension Center (AIEC) in each district was not prescribed by Decree No. 804/1995 which was published before the establishment of the AIEC. However, because this institute is charged with coordinating all RECs in the district, it serves as a bridge between the provincial agricultural offices including AIAT and the RECs and a channel through which extension materials and feedback are sent. The AIEC is also assigned to do verification trials, a significant segment in the flow-of-innovation. According to the newly released guidelines from the Minister of Home Affairs and the Minister of Agriculture, AIEC has a "consultative relationship" with AIAT regarding this function. This statement may not be sufficient as a straightforward instruction for AIAT to coordinate AIEC's activities, but it allows the two institutes to take initiatives to interact closely on many aspects of their

Table 2. Procedures for generating, evaluating, and applying location-specific agricultural technologies according to the decree of the Minister of Agriculture No. 804/1995.

Major step	Output	Responsible institution/person
Basic research	New knowledge or information	National research institutes (NRI)
Applied research	Applicable knowledge or information	NRI
Strategic research	Mature technology component	NRI and AIAT
Adaptive research	Adapted technology component	AIAT
Technology packaging	Technology package	AIAT
Technology evaluation	Mature technology package	AIAT
Technology recommendation meeting	Recommendation material	Advisory Committee for Agricultural Technology Assessment
Formulation of technical guidelines	Technical guidelines	Provincial and district offices of subsectoral agricultural affairs
Development of extension material	Extension material	Rural Extension Center
Extension	Skillful farmers	Rural extension workers
Technology application	Successful farming	Farmers
Feedback gathering and problem identification	Identified feedback and problems	Rural extension workers
Identification of technology needs	List of technology needs at subdistrict level	Rural Extension Center
Consolidation of district level technology needs	Consolidated list of technology needs at district level	District offices of subsectoral agricultural affairs
Consolidation of technology needs at provincial level	Consolidated list of technology needs at provincial level	Provincial offices of subsectoral agricultural affairs
Formulation of provincial technology assessment program	Provincial technology assessment program	Provincial Advisory Committee for Agricultural Technology Assessment

functions, especially in planning, executing, and evaluating research and technology assessment activities as well as in disseminating findings from those activities. The AIEC is also assigned to be the permanent secretariat of the district leading farmers and fishermen's group. Many AIATs are also the permanent secretariat of provincial leading farmers and fishermen's group. Through these assignments, the functional relationship between AIEC and AITs is enhanced.

Improvements required

Implementation of AIAT's tasks and functions requires a number of adjustments, modifications, and improvements in various components of the agricultural systems. Institutional transformation is being undertaken but further steps are needed for its completion. Decentralization of extension and the establishment of AIEC at the district level are conducive to the development and application of location-specific technologies. Although a functional relationship between AIAT and AIEC has been prescribed, the actual operation needs a lot of adjustment. In addition, AIAT as a new institution with different kinds of tasks and functions, has to work hard to establish itself in the existing system. Linkages between AIAT and various agricultural institutions within the national, provincial, and district government have been legally prescribed in Ministerial Decree

No. 804/1995. However, actual implementation requires further operational guidelines set forth by legal pronouncements. Linkages with agricultural colleges and universities as well as with nondepartmental research institutions should also be legally and operationally established.

Above all, the key to success in generating and applying location-specific technologies is the human resources involved in carrying out the various tasks and functions in the process. Because AIAT is the major regional institution for technology generation and dissemination, the quality of its human resources determines the fate of the whole process. The knowledge, attitude, and skills of AIAT staff will help steer and motivate people in other institutions. The readiness of people in other institutions is also crucial. After a long experience in applying top-down blanket technology approaches through various intensification programs, people have to adjust to new endeavors and new environments. Observations at the initial stage of the application of the new approaches indicated that most of the people are eager to participate in the new endeavors but are not sufficiently prepared because operational guidelines which can be used as references are not sufficient (Suriapermana et al 1996). Most of the available operational guidelines are products of the old system.

A major part of the work in generating, evaluating, and adapting location-specific technologies requires efficient use of human and physical resources. To accomplish this, a large number of protocols must be developed for various steps of research and technology assessment. Components of the protocols are similar to those of experimental design in basic and applied research, but their characteristics and sizes are different. In general, there should be a steady reduction of complexity and number of treatments, along with increased number and size of experimental units (plots). Table 3 summarizes proposed protocols for evaluating and adapting location-specific rice production technologies. Similar exercises should be carried out with other commodities. Obviously, the criteria for component characteristics and sizes will be unique to each commodity.

There will also be a tremendous need to construct and improve buildings and to procure equipment. A large number of meetings of various group sizes will require meeting rooms. Increased interaction with various AIAT's stakeholders will require numerous communication facilities including transportation. Existing facilities are being reassessed and procurement needs are being identified.

Organization of the construction and procurement needs as well as implementation of AIAT's functions will require additional funds. The funds may come from the national budget, provincial budget, private businesses, farmers' contributions, and external sources. Currently, the largest portion of AIAT's budget comes from the national budget and from external assistance. Contributions from other sources should be steadily increased.

Technology adaptation and promotion through SUTPA

Since the establishment of AIAT, a new technology assessment method has been introduced, tested, and improved. Generically the new method is called Assessment of

Table 3. Summarized protocols for the evaluation and adaptation of location-specific rice production technologies.

General term	Major step	Component	Nature	Field size	Responsible agents
Strategic research	Technology-component maturation	Area treatment Plot replication Parameter analysis	Farmers' field Production input Constructed plot Within same field Yield, components Technical	0.2 ha 4-6 50 m^2 5-10 - -	NRI and AIAT
Adaptive research	Technology-component adaptation	Area treatment Plot replication Parameter analysis	Farmers' field Production input Farmers' plot Farmers Yield Technical	1-2 ha 3-4 500 m^2 10-15 - -	AIAT, agricultural officers, farmers
Farming system research	Technology packaging	Area treatment Plot replication Parameter analysis	Farmers' field Cropping patterns Farmers' plot Farmers Yield Technical, financial	5-10 ha 3-4 1000 m^2 10-15 - -	AIAT, agricultural officers, farmers
Technology demonstration	Demonstration of technical performance of new technologies	Area treatment Plot replication Parameter analysis	Farmers' field Technology package Farmers' plot Single Yield Technical, financial	5-10 ha 1 - 1 - -	AIAT, agricultural officers, farmers
Technology application trial	Adaptation of newly recommended technologies at certain localities	Area treatment Plot replication Parameter analysis	Farmers' field Technology package Farmers' plot Single Yield Technical, financial	5-10 ha 1 - 1 - -	AIEC, agricultural officers, farmers
Commercial farming system assessment	Institutional and economic assessment of technology application	Area treatment Plot replication Parameter analysis	Farmers' field Technology package Commercial scale Single Yield, adoption rate Technical, financial, institutional, social	500 ha - 500 ha 1 - -	AIAT, agricultural officers, farmers

Commodity-based Farming Systems with Agribusiness Orientation (referred to as SUTPA, an acronym of Sistem Usahatani berbasis Padi berorientasi Agribisnis). The essential features of this method are the following:

- Performance of new technology is assessed on a commercial scale, depending on the nature of the commodity involved.
- Assessment is done under actual farmers' circumstances--biophysical conditions; cooperation within a group and among different groups of farmers; accessibility to input supply, capital, and market; and all other socioeconomic characteristics.
- Assessment is done jointly by researchers, extension workers, agriculture officials, and farmers.

- Researchers and extension workers are deployed to all assessment units and they live in farmers' communities.

The method was introduced as an additional tool to the existing methods of research and development. In addition to research components, the method also bears some aspects of technology dissemination. Another dissemination method which has been used by AIAT on a smaller scale is technology demonstration, in which technology performance is assessed on a farmers' group scale. Conceptually, technology demonstration can be considered the preceding phase of commercial-scale technology assessment. Assessment of technology under technology demonstration emphasizes technology performance under farmers' biophysical conditions but does not deal considerably with the institutional and economic facets.

Types of commodity-based farming systems are determined on the basis of priority commodities in various regions. The largest of the commodity-based farming systems is the SUTPA.

Several new rice cultivation techniques were introduced in SUTPA. The first phase of SUTPA, carried out for three planting seasons from the end of 1995 to the second half of 1996, involved 12 AIATs in 14 rice-producing provinces. The total area of assessment was 46,000 ha covering 92 assessment units. The major focus of the technologies introduced was on cropping systems or farming systems that can increase cropping intensity up to 300%. The farming systems introduced in the program were

a) Rice - rice - *palawija* (nonrice food crops)
b) Rice - rice - vegetables
c) Rice/fish - rice/fish - palawija
d) Rice/fish - rice/fish - vegetables

The introduced rice cultivation technologies include newly released cultivar Memberamo, direct seeding method, use of mechanical seeder, use of wide spacing with or without fish intercrop, use of azolla, and rational use of fertilizers based on soil analysis. In addition, advanced cultural practices for other food crops and horticultural crops were introduced (AARD 1995).

Direct seeding was found promising in many on-station and on-farm trials (Malian et al 1995). The following advantages of direct seeding have been mentioned by Dimyati (1995b) and Suriapermana et al (1996):

a) The growing period can be shortened by 15-20 d $season^{-1}$, thus allowing a more flexible third crop after dry-season rice and increasing cropping intensity.
b) Uniform maturity results in better quality grain.
c) Labor for crop establishment and weeding is reduced.
d) Proper scheduling of planting may enhance water use efficiency.
e) It is compatible with available mechanized cultural practices.

The same authors cited some constraints which may hamper the use of direct seeding (Dimyati 1995a,b; Suriapermana et al 1996):

a) Seed drift may occur if seeds are planted during or just before a heavy rain.
b) Some cultivars are more susceptible to lodging when direct seeded.
c) There may be heavy weed infestation if herbicide is not properly applied.
d) The crop is more vulnerable to rat and golden snail attacks.

Another rice cultivation technique found promising is wide spacing, called *legowo* planting system. This technique can be used to enhance rice/fish and rice/azolla cultures which improve total land productivity while maintaining or even improving rice yields. This technique is compatible with direct seeding and transplanting (Suriapermana and Syamsiah 1995).

Farmers' responses to and officials' enthusiasm for various technologies introduced through SUTPA were satisfactory, despite hectic and complicated administrative arrangement during the preparatory stage of the program. In most provinces, the realized area of direct seeding and legowo planting systems in the first season was higher than targeted. This was partly due to overenthusiasm of officials and farmers in responding to the new technologies, often without sufficient care for the inherent risks (Dimyati 1995a,b). This overenthusiasm proved to be counterproductive to the long-term adoption of the technologies. In the second season, the area of direct seeding and Memberamo declined considerably because many problems encountered by the farmers were not sufficiently anticipated by the researchers, extension workers, and officials in the areas. Farmers complained that (Suriapermana et al 1996)

- rats attacked germinating seeds and young seedlings which caused drastic losses;
- Memberamo was susceptible to lodging and had fewer filled grains due to high proportion of empty hulls, leading to lower price of the cultivar compared with those of IR64 and Cisadane;
- labor for land preparation increased;
- unexpected floods occurred in certain areas due to heavy rainfall during the early planting season; and
- labor for weeding increased due to inappropriate herbicide use.

In addition to the agronomic problems, farmers complained of social and institutional problems:

- personnel of agricultural offices and extension had not prepared for the implementation of the new participatory and bottom-up systems;
- farmers were not prepared to accept drastic changes from well-arranged rows of transplanted rice to possibly messy rows of direct seeded rice plants; and
- many earlier legal pronouncements and policy instruments, which do not favor the new practices, were still in place.

Adoption rate of legowo has been better than that of direct seeding and Memberamo. This was indicated by the fact that the system has been used with transplanting method using other cultivars such as IR64 and Cisadane (Suriapermana et al 1996).

Some corrections have to be made in the second phase of SUTPA to improve its technical performance and to make the procedures consistent with the new participatory concept. These include

- proper use of participatory planning protocols,
- accurate selection of assessment sites,
- greater flexibility in selection of technology components, and
- proper positioning of SUTPA in the technology generation and application system.

Conclusions and recommendations

1) The Government of Indonesia has responded to the pressures of globalized economy which requires efficient use of resources and more competitive agricultural production through a number of policy instruments including decentralized regional agricultural development. Regionalization of research and extension is part of that policy.
2) Seventeen AIATs were established to enhance the generation and application of location-specific agricultural technologies through participatory research and technology assessment activities. AIAT's main strength is the integration of researchers and extension workers under one roof at the provincial level.
3) Institutional rearrangement of research and extension has been under way, but further improvements are still required through the development of operational guidelines, innovative research and extension protocols, human resource development, construction of facilities, procurement of equipment, and provision of sufficient operational budget.
4) The key to the successful accomplishment of AIAT's mission is personnel--theirs and those of the partner institutions. Therefore, familiarization of personnel with the concepts and training them in such practices are vital.
5) Protocols of evaluation, adaptation, and verification trials as well as extension functions to be conducted by AIAT and AIEC should be prescribed by taking into account linkage and division of labor between the two institutes. This functional linkage will be established naturally through interdependence between the two but will be enhanced through an appropriate formal arrangement.
6) SUTPA has been used by AIAT as one of the vehicles for commercial rice-based farming system assessment. The method has been effective in disseminating several technology components of rice cultivation such as wide spacing, legowo system, direct seeding, and rice - fish culture. However, for better performance of the program, some corrections need to be made in site selection, selection of technology components, proper use of participatory planning protocols, and proper positioning of the program within the technology generation and application system.

References

AARD--Agency for Agricultural Research and Development (1995) Strategic plan, Agency for Agricultural Research and Development, 1995-2005. AARD, Jakarta, Indonesia. 64 p.

CRIFC--Central Research Institute for Food Crops (1992) Hasil utama penelitian tanaman pangan 1987-1991. Pusat Penelitian dan Pengembangan Tanaman Pangan. Bogor, Indonesia. 95 p.

Dimyati A (1995a) Introduction of direct-seeding and *legowo* spacing systems to improve rice-based farming system. Paper presented at the Study Meeting on Prospective Agricultural Technologies, 25 Jul-5 Aug 1995, Tokyo, Japan. 18 p.

Dimyati A (1995b) Teknik semai langsung dalam perspektif pemeliharaan swasembada pangan dan peningkatan pendapatan petani. Prakarsa, Nov 1995. p 20-27.

Eponou T (1993) Integrating agricultural research and technology transfer. Public Admin. Dev. 13:307-318.

Hermanto R S, Rifai, Hendiarto, Indraningsih K S (1993) Perkembangan produktivitas sektor pertanian. Series No. 12. Pusat Penelitian Sosial Ekonomi Pertanian, Badan Penelitian dan Pengembangan Pertanian. Bogor, Indonesia.

Lacy W B (1995) Strategies for change: research, extension and users' partnership. Paper presented at a Workshop on Extension Dynamics and Future Role in Indonesia, 4-5 Jul 1995, Bogor, Indonesia.

Malian A H, Supriatna A, Supriadi H (1995) Kelayakan ekonomi budi daya padi sebar langsung di Pantai Utara Jawa Barat. Pages 248-256 in Zaini Z, Hermanto, Djauhari A, Adnyana M O, Pasaribu D (Eds.) Sistem Usahatani Berbasis Pangan. Puslitbang Tanaman Pangan. Bogor.

Oka I N (1996) Integrated crop pest management: one way to empower farmers to develop efficient and environmentally sound agricultural practices. Indon. Agric. Res. Dev. J. 18 (1): 1-12.

Ramasamy C (1995) Impact and problems in the adoption of green revolution technology. Paper presented at the Study Meeting on Prospective Agricultural Technologies, 25 Jul-5 Aug 1995, Agricultural Productivity Organization, Tokyo.

Roling N G, van de Fliert E (1994) Transforming extension for sustainable agriculture: the case of integrated pest management in Indonesia. Agric. Hum. Values, Spring-Summer 1994:96-108.

Sail R M (1995) Impact assessment on research-extension linkage pilot projects. Final report. Agency for Agricultural Research and Development, Jakarta, Indonesia. 39 p.

Suriapermana S, Syamsiah I (1995) Tanam jajar legowo pada sistem usahatani minapadi-azola di lahan sawah irigasi. Pages 74-83 in Zaini Z, Hermanto, Djauhari A, Adnyana M O, Pasaribu D (Eds.) Sistem Usahatani Berbasis Pangan. Central Research Institute for Food Crops, Bogor, Indonesia.

Suriapermana S, Nurbaeti B, Nurhati I, Dewi O (1996) Mengungkap dinamika pengkajian SUTPA I 1995/1996 dan Persiapan pengkajian SUTPA II 1996-1997 di Provinsi Jawa Barat. Makalah pada Pertemuan Koordinasi dan Sosialisasi SUTPA II, 22 Agustus 1996. Balai Pengkajian Teknologi Pertanian, Lembang.

Uphoff N (1995) Institutionalizing user participation in a system of linkage among research, extension, and farmers. Paper presented at a Workshop on Extension Dynamics and Future Role in Indonesia, 4-5 Jul 1995, Bogor.

Zaini Z, Adnyana M O, Adiningsih S, Darmawan D H, Widjaja-Adhi I P G, Suhartatik E, Erwidodo (1993) Kajian peningkatan efisiensi sistem produksi pangan dalam upaya peningkatan pendapatan petani dan penghematan subsidi pupuk. Agency for Agricultural Research and Development, Jakarta, Indonesia. 43 p.

Front-line demonstrations for evaluating and promoting new technologies in rice systems in India

K. KRISHNAIAH
Directorate of Rice Research (DRR), Rajendranagar, Hyderabad 500030, India

Key words: block demonstration, high-yielding varieties, technology packages, farmer-to-farmer seed exchange, rainfed lowland ecology, eastern India

Abstract
Rainfed rice systems in eastern India are characterized by low yields, low adoption of high-yielding varieties (HYVs), and low fertilizer use. This paper illustrates the role of front-line demonstrations in identifying and popularizing viable and suitable modern rice technologies among farmers in rainfed lowland ecology, under the direct supervision of scientists. Front-line demonstrations, initiated in eastern India in 1990, were also intended to supplement ongoing special rice development programs in identifying location-specific HYVs and related production technologies based on simultaneous on-farm evaluation and popularization in the targeted area. A total of 337 compact block front-line demonstrations covering 2,346 ha were organized across eastern India during the past 5 yr (1990-95). As many as 80 rice cultivars, along with improved crop management practices, have been demonstrated/verified in farmers' fields. About 40 HYVs were found suitable for various rainfed systems. The demonstrations proved that the availability of suitable HYVs is not a limiting factor for the enhancement of productivity in this region. Rough-rice yields of these HYVs under improved management ranged from 4.5 to 5.0 t ha^{-1} in shallow water (0-30 cm inundation), 3.0-3.5 t ha^{-1} in semideep water (30-50 cm inundation), and 2.0 - 3.0 t ha^{-1} in upland rainfed rice systems, as against 1.0-1.5, 1.0-1.3 and 0.9-1.2 t ha^{-1} for traditional varieties and management in the 3 ecosystems. Thus, front-line demonstrations identified appropriate HYVs plus management packages that could outyield local varieties + traditional practices by nearly 1.5-2.0 t ha^{-1} (60-150%). These yield gains on an average produced an additional household income of US$114 ha^{-1}.

Introduction

Modern rice technologies, which evolved and became popular in the mid-1960s, have primarily benefited irrigated environments, contributing significantly to overall increases in rice output during the post-1966 era (Barker and Herdt 1985, David and Otsuka 1994). A favorable policy environment and suitable infrastructure have also been created in irrigated regions to increase farmers' adoption of new rice varieties and production technologies. However, rice production in irrigated environments suffers from declining productivity growth, diversion of land and labor away from rice, and increased incidence of biotic and abiotic stresses (Krishnaiah et al 1996). Such emerging concerns threaten the sustainability of intensive rice production in India. Further, the rainfed regions in most of eastern India suffer from low productivity. Identification and popularization of suitable high-yielding varieties (HYVs) and production technologies in rainfed areas appear to be the most potential approach for sustaining self-sufficiency in national rice production.

Eastern India, comprising the states of Assam, Bihar, eastern Uttar Pradesh, (UP), eastern Madhya Pradesh, (MP) Orissa, and West Bengal, has a gross cropped rice area of about 27.8 million ha, which accounts for about 58% of the total rice area and contributes about 48% to the total rice production. Leaving 6 million ha of irrigated rice area, the rest is occupied by 4.4 million ha of upland, 15.6 million ha of shallow and semideep

V. Balasubramanian, J. K. Ladha and G. L. Denning (eds.), Resource Management in Rice Systems: Nutrients, 243–252.

lowland, and 2 million ha of deepwater rice. Technical constraints that limit productivity in eastern India include lack of wider choice of HYVs to fit into the different land types and growing conditions; nonavailability of good-quality seeds of promising varieties; lack of farmers' and extension agents' exposure to better varieties and crop management technologies; the looming agro-energy crisis and the consequent delay in planting; inadequate use and improper management of agricultural inputs such as fertilizers and pesticides; climate-related constraints such as flooding or droughts; soil-related constraints such as salinity and multinutrient deficiencies; and physiological disorders.

This highly diverse agroecology requires more location-specific variety-technology packages. The special rice/food production programs, integrated cereal/rice development programs, and the IRRI-IFAD Collaborative Program implemented in this region have given encouraging results. To complement these ongoing programs, front-line demonstrations were initiated in 1990 in the states of Assam, Bihar, Orissa, West Bengal, eastern MP, and eastern UP.

Of the various technology transfer methods, such as national demonstrations, operational research projects, lab-to-land programs, farm science centers, institute-village linked programs introduced at various times in the country, the front-line demonstration is considered the most suited for the All India Coordinated Crop Improvement Projects (AICRIP) implemented by the Indian Council of Agricultural Research (ICAR) in collaboration with state agricultural universities. The main reason for selecting front-line demonstrations is that they facilitate the large-scale evaluation of varieties and management technologies developed through the national multilocation testing program organized across the country to verify/select technologies for different locations. Because the demonstrations are conducted by the scientists generating the technologies, they provide direct feedback to the scientists on the merits and demerits of introduced technologies and offer them an opportunity to refine their technologies. The large size of the demonstration gives better visibility of the performance of the chosen technology to farmers and extension workers and forms the best source for multiplying and supplying the seeds of new varieties.

Goal and objectives

The primary goal of the front-line demonstration program is to evaluate and promote promising rice varieties and production technologies in farmers' fields in the rainfed regions of eastern India. The specific objectives are to

- provide better varietal choice for the diverse agroecosystems in eastern India;
- acquaint extension personnel and local farmers with improved varieties and management technologies suitable for different agroecosystems;
- complement the ongoing efforts of the Ministry of Agriculture and other agencies for achieving improved rice productivity in eastern India; and
- facilitate the rapid multiplication of quality seeds and to organize the spread of newly introduced high yielding, pest-resistant cultivars.

Technologies demonstrated during 1990-95

Rainfed upland rice

- Short-duration, drought-tolerant varieties suited for direct seeding
- Line sowing to ensure optimum plant stand and economic weed management
- Deep summer plowing
- Stale seedbed
- Moderate fertilizers: $N-P_2O_5-K_2O$ 'at 50-25-25 kg ha^{-1}
- Seed treatment with carbendazim and soil application of BHC dust in termite endemic areas
- Need-based chemical control against blast, stem borer, and rice (gundhi) bug

Rainfed lowland rice (shallow [0-30 cm] and semideep [30-50 cm] ecology)

- Approved photoperiod-sensitive HYVs
- Transplanting wherever water depth does not exceed 30 cm
- Direct dry seeding in medium and deep water conditions
- High stand establishment through line sowing with high seed rate and/or closer planting
- Application of $N-P_2O_5-K_2O$ at 60:20:20 kg ha^{-1}
- Rectification of Zn and S deficiency
- Weed control by herbicides wherever economical
- Seed treatment with carbendazim
- Need-based chemical control of insect pests and diseases

Program implementation

Implementing states

Assam, West Bengal, Bihar, Orissa, eastern MP, and eastern UP. Ecology-wise/technology-wise allocation of demonstrations is decided in the national annual workshops. The cooperating scientists of state agricultural universities facilitate the supply of recommended seeds and organize the technical supervision in collaboration with extension institutions. Actual implementation is by local extension functionaries of the State Department of Agriculture.

Size and number of demonstrations

Compact areas of 8 ha each of contiguous plots, involving multiple farmholdings preferably owned by a number of small and marginal farmers, were selected. The large size facilitated high visibility of the technologies being demonstrated. Demonstration sites were fixed by the cooperating scientists, in consultation with local staff of the Department of Agriculture and the farming community. A total of 337 demonstrations co-

Table 1. Front-line demonstrations from 1990 to 1995, by state.

State	Area (ha)				
	1990	1991	1994	1995	Total
Assam	130	72	123	162	487
West Bengal	194	226	178	–	598
Bihar	130	154	68	87	439
Orissa	–	73	74	–	147
Eastern M.P.	81	154	35	97	367
Eastern U.P.	114	32	81	81	308
Total	649	711	559	427	2346

vering 2,346 ha (Table 1) were organized during 1990, 1991, 1994, and 1995 in Assam, West Bengal, Bihar, Orissa, eastern MP, and eastern UP.

Critical inputs/funding of the demonstrations

Depending on the nature of demonstrations, necessary critical inputs such as seed, fertilizer, and pesticide are identified for different locations and supplied free to the participating farmers through the implementing agencies. Maximum allowed cost of input is pegged at US\$86 ha^{-1} (Rs 3,000 ha^{-1}). The funds provided by the Ministry of Agriculture of the Government. of India are routed through the Directorate of Rice Research of ICAR and the organizing agency of the state agricultural university or the State Department of Agriculture.

Training programs/field days

For effective transfer of technology, seasonal training of the local farmers and extension personnel was conducted by the cooperating scientists. The training was followed up with field days at appropriate stages of crop growth to expose the local farmers and extension workers to the main focus of the demonstrations.

Results

The ecology-wise breakdown of the demonstrations is shown in Table 2. The performance of the new rice varieties under improved crop management technologies is summarized as follows:

Rainfed shallow water system (0-30 cm inundation)

Assam, West Bengal, Bihar, Orissa, eastern MP, and eastern UP were selected for rainfed shallow water system with 0 to 30 cm inundation during a major part of the crop growth period. The statewise relative mean yields of improved vs traditional technology packages for different ecosystems are given in Table 3. The overall average performance of im-

Table 2. Front-line demonstrations from 1990 to 1995 in eastern India, by ecology.

Ecosystem	Area (ha)				
	1990	1991	1994	1995	Total
Upland	154	202	178	114	648
Shallow water	340	316	320	292	1268
Semideep water	155	194	61	20	430
Total	649	712	559	426	2346

Table 3. Relative average performance of improved practices vs local practices under different ecosystems.

State	Mean grain yield (t ha^{-1})		Increase in grain yield (%)
	Test varieties	Local varieties	
	Shallow water ecology		
Assam	4.85	2.97	64
W. Bengal	4.44	3.37	32
Bihar	4.12	2.24	84
Orissa	4.65	3.05	53
Eastern Madhya Pradesh	4.09	2.38	72
Eastern Uttar Pradesh	3.16	1.73	83
	Semi-deep water ecology		
W. Bengal	3.21	1.95	65
Bihar	2.90	1.45	100
Eastern Uttar Pradesh	2.15	1.19	81
	Upland ecology		
Bihar	1.93	0.84	130
Orissa	2.36	1.56	56
Eastern Madhya Pradesh	3.08	2.85	8
Eastern Uttar Pradesh	3.91	2.70	45

proved technology vs traditional practices for the three ecosystems of the entire eastern India during 1990, 1991, 1994 and 1995 is shown in Table 4.

Assam. Improved varieties such as Lakhmi, Biraj, Salivahana, IET8002, Kushal, Bahudar, and Ranjeet under improved crop management were compared with local varieties such as Pankaj, Manoharsali, Mahsuri, and Ahomsali under traditional practices. Improved varieties plus management packages recorded mean yields ranging from 4.6 to 5.4 t ha^{-1}, whereas local varieties plus traditional practices yielded from 3.0 to 3.4 t ha^{-1}. On an average, the improved package yielded 63% higher than local practices (Table 3).

West Bengal. Rice varieties IR42, Biraj, IET8002, Manasarovar, IET5914, Bipasa, Salivahana, IET12515, IET12690, and IET12130 were some of the test varieties demonstrated in this state. The local checks were Mahsuri, CR1006, and Radha. Average yields of the test varieties under improved management ranged from 3.9 to 5.0 t ha^{-1} while those of the check varieties under traditional practices varied from 2.6 to 4.5 t ha^{-1}. Pooled average yield for the test varieties was 4.4 t ha^{-1} compared with 3.0 t ha^{-1} for the

Table 4. Average rough rice yields using improved and local technologies in eastern India.

Ecosystem	Year	Rough-rice yield (t ha^{-1})		% increase over local
		Improved technology	Local practices	
Shallow water	1990	4.20	2.30	83
	1991	4.20	2.60	62
	1994	4.30	2.80	54
	1995	4.20	2.60	62
Semi-deep water	1990	2.62	1.25	110
	1991	3.25	1.80	81
	1994	2.68	1.72	56
	1995	2.15	1.41	53
Upland	1990	3.12	1.43	118
	1991	2.73	1.66	65
	1994	2.66	1.82	46
	1995	2.82	2.09	26

Sources: Directorate of Rice Research, various issues 1991, 1992, 1994, 1995.

local checks. Average yield increase of the test varieties under improved management was 32% (Table 3).

Bihar. Improved rice varieties such as Jayashree, Rajshree, and Kanak were compared with local checks Bakol, Patanai-3, Jaswa, T141, and BR34. Mean yields of improved varieties + management package ranged from 3.4 to 4.9 t ha^{-1} as against 1.9 to 2.5 t ha^{-1} for local checks. Mean yield increase of the test varieties over the checks was 84% (Table 3).

Orissa. Rice varieties OR621-6, Mahalaxmi, and Manika recorded mean yields ranging from 4.5 to 4.8 t ha^{-1} as against 3.0 to 3.1 t ha^{-1} of the local checks CR1014, Savitri, Dharitri, T1242, and Basuabhog. Mean yield increase of test varieties over checks was 53% (Table 3).

Eastern MP. Mean yields of improved cultivars Annada, Kranti Ruchi, MW10, IR36, Swarna, Surekha, Mahamaya, and RS74-11 ranged from 3.3 to 4.9 t ha^{-1}. Mean yield increase was 72% (Table 3) over local checks of Safri 17, Mahsuri, Chaptigurumatia, BD1200, and IR36.

Eastern UP. Test varieties such as Mahsuri, Jaya, Jayalakshmi outyielded local checks Sarju 52, Bishnuparaj, Lalmati, and Badshah. Mean grain yields of the test varieties ranged from 2.5 to 4.3 t ha^{-1}, while those of the local checks were 1.4 to 2.0 t ha^{-1}. Average yield increase of test varieties over the checks was 83% (Table 3).

On the whole, in the rainfed shallow water system of eastern India, improved varieties plus technologies outyielded local varieties under traditional practices by 83% in 1990, 62% in 1991, 54% in 1994, and 62% in 1995, (Table 4).

Rainfed semideep water system (30-50 cm inundation)

The states of West Bengal, Bihar, and eastern UP are included under rainfed semideep water system. The mean yields of demonstrations for the rainfed semi-deepwater system from 1990 to 1995 are given in Table 3. The statewise results are as follows:

West Bengal. Rice varieties Sabita, Jogen, Nalini, IET9065, and Dinesh were compared with local checks Mahsuri, Patnai-23, Mahamaya, Swarnadhan, Pankaj, Indrasail, Bansphool, and Kalma. Mean grain yields of test varieties ranged from 3.0 to 3.6 t ha^{-1}, while those of the checks ranged from 1.8 to 2.1 t ha^{-1}. Mean grain yield increase of improved technologies over that of checks was 65% (Table 3).

Bihar. Test variety Sudha was compared with local check Bakol. The mean yield of Sudha was 2.9 t ha^{-1} compared with 1.4 t ha^{-1} of Bakol, a 100% increase (Table 3).

Eastern UP. Improved varieties Chakia 59, Jalamagna, Madhukar, and Jalapriya were demonstrated alongside local checks Bhainslat, Rankajra, Srabhani, and Jaya. Mean yields of test varieties ranged from 1.9 to 2.4 t ha^{-1}, while check yields varied from 0.8 to 1.4 t ha^{-1}. Test varieties on an average recorded an 81% yield increased (Table 3) over the checks whose average yield was 1.2 t ha^{-1}.

On the whole, in the rainfed semi-deep water system of eastern India, improved varieties plus technologies outyielded local varieties under traditional practices by 110% in 1990, 81% in 1991, 56% in 1994, and 53% in 1995 (Table 4).

Rainfed upland system

The states of Bihar, Orissa, eastern MP, and eastern UP were included under rainfed upland system. The statewise mean yields of the demonstrations for rainfed upland are given in Table 3. The statewise performance of the varieties is as follows:

Bihar. Chakia, Jalamagna, Birsadahan Kalma-111, Birsa-101 and 201, Kalinga-3, Tulasi, and Vandana were some of the test varieties demonstrated alongside locally popular checks Gora and Chirikanta. Mean yields of test varieties ranged from 1.5 to 2.6 t ha^{-1}, while that of checks was 0.8 t ha^{-1}. The magnitude of increase in average grain yield of test varieties was maximum (130%) (Table 3).

Orissa. Varieties Kalinga, Parijat, Patnai, Nilgiri, Ganteswari, and Kandagiri were demonstrated alongside local checks Sankara, Deshmatia, Cheretka, and Saria. Average yields of improved varieties ranged from 2.0 to 2.7 t ha^{-1} while it was 1.3 to 1.7 t ha^{-1} for check varieties . The mean yield increase of the test varieties over checks was 56% (Table 3).

Eastern MP. Improved varieties Aditya, Tulasi, Annada, Kranti, IR36, and R281-31-1 were tested in farmers' fields; Chapti, Safri 12 and Chaptigurumatia were the local checks. Mean yields of test varieties ranged from 2.7 to 3.6 t ha^{-1} while the check varieties had a mean yield of 2.9 t ha^{-1} (Table 3).

Eastern UP. Saket 14, Narendra 18, Narendra 80, Narendra 97, and NDR118 were compared with the local checks Lalmati Bihnubaraj, Kalanana, Indrasain, Johsia, Nagina 22, Saket, and Jhora. Mean yields of improved varieties ranged from 3.7 to 4.5 t ha^{-1}, while those of local checks ranged from 1.9 to 2.0 t $ha^{-1.}$ The mean yield increase of the test varieties over the local checks was 45% (Table 3).

On the whole, in the rainfed upland system of eastern India, improved varieties plus technologies outyielded local varieties under traditional practices by 118% in 1990, 65% in 1991, 46% in 1994, and 26% in 1995 (Table 4).

Impact

Front-line demonstrations facilitated quick identification and rapid spread of improved rice cultivars adopted to the location-specific conditions of the diverse agroecosystems of eastern India. The important cultivars that have become popular are the following:

Rainfed shallow lands (0-30 cm inundation)

Assam: Lakhmi, Biraj, Salivahana, IET8002, Kushal, Bahadur, and Ranjeet
West Bengal: IR42,Biraj, IET8002, Manasarovar, IET12515, IET12690, and IET12130
Bihar: Jayashree, Rajasree, and Kanak
Orissa: OR621-6, Mahalaxmi, and Manika
Eastern MP: Annada,Kranti, Ruchi, MW10, IR36, Swarna, Surekha, and Mahamaya, RS74-11
Eastern UP: Mahsuri, Jaya , and Jayalakshmi

Rainfed semi-deepwater system (30-50 cm inundation)

West Bengal: Sabita, Jogen, Nalini, IET9065, and Dinesh
Bihar: Sudha
Eastern UP: Chakia 59, Jalamagna, Madhukar, and Jalapriya

Rainfed uplands

Bihar: Chakia, Jalmagna, Birsadhan, Kalma 111, Birsa 101Birsa 201, Kalinga 3, Tulasi, and Vandana
Orissa: Kalinga, Parijat, Patnai, Nilagiri, Gantteswari, and Kandagiri
Eastern MP: Aditya, Tulasi, Annada, Kranti, IR36, and R281-31-1
Eastern UP: Saket 14, Narendra 18, Narendra 80, Narendra 97, and NDR118

Introduction and popularization of the improved HYVs are reflected in the increased area under HYVs in several eastern states. In Assam, the area under HYVs increased from 1.0 million ha in 1990 to 1.3 million ha in 1994. Similarly, in Bihar, the area under modern rice varieties nearly doubled between 1990 and 1994. In Orissa, the extent of area under HYVs increased from 2.5 million ha in 1990 to 3.1 million in 1994. In Uttar Pradesh, the extent of increase under HYVs was modest (from 4.9 to 5.2 million ha). For Madhya Pradesh, the area under HYVs increased from 2.8 million ha in 1990 to 3.9 million ha in 1994.

With the introduction and popularization of HYVs and related production technologies, the overall productivity of rice in eastern India has also increased. For Assam, the overall productivity went up from 1,952 kg ha^{-1} to 2002 kg ha^{-1} in 4 yr, while for West Bengal, it rose from 2,371 kg ha^{-1} in 1990 to 2,925 kg ha^{-1} in 1994. In Bihar, productivity was stepped up from 1,823 kg ha^{-1} to 1,930 kg ha^{-1} in 4 yr. Likewise, for Orissa, rough rice productivity went up from 1,752 kg ha^{-1} in 1990 to 2,090 kg ha^{-1} in 1994. In Madhya Pradesh, the average rice productivity improved from 1,682 kg ha^{-1} in 1990 to 1,780 kg ha^{-1} in 1994. In Uttar Pradesh, it improved from 2,740 kg ha^{-1} in 1990 to 2,800 kg ha^{-1} in 1994.

In addition, seasonal training and field days havefacilitated informal exchange of quality seed among farmers, increased the awareness of farmers and extension workers about the state-of-the-art technologies available for obtaining high yields, and favored the fine-tuning of demonstrated technologies based on feedback from the endusers.

Summary and conclusions

Front-line demonstrations in eastern India during 1990-95 greatly helped in evaluating and promoting improved rice varieties and production technologies. Unlike in other methods of technology transfer, scientists in the various demonstration sites were directly involved in evaluating improved rice technologies. Farmers also actively participated in the process.

As a result of these demonstrations, area under HYVs increased significantly from 1990 to 1994. Fertilizer consumption also increased in eastern India during the same period. Consequently, rice productivity increased significantly by 20-25% in 4 yr. The demonstrations also facilitated an informal exchange of quality seeds among farmers. Farmers' awareness of modern rice technologies helped in the larger dissemination of useful technologies. Similar demonstrations can be organized in low-productivity regions or problem-prone areas to identify and popularize suitable technologies to mitigate the productivity constraints.

Acknowledgement

We gratefully acknowledge the financial support received from the Ministry of Agriculture and Cooperation of the Government. of India. Our thanks go to Dr. E.A. Siddiq, the former project director of DRR and the deputy director general (Crops), ICAR, and Dr. K.G. Pillai, the former director I/C of DRR, under whose overall guidance this program was organized successfully during the 1990, 1991, and 1994 kharif seasons.

We appreciate the cooperation and support received from scientists and officers of the cooperating state agricultural universities and state agricultural departments and the cooperating farming community. The assistance of Dr. M. Narayana Reddy and other scientists from DRR who helped in monitoring the demonstrations, compiling results, and preparing the script is gratefully acknowledged.

References

Barker R, Herdt R W (1985) The rice economy of Asia. Resources for the Future, Washington, D.C. p 47-48.

David C.C, Otsuka K, Eds. (1994) Modern rice technology and income distribution in Asia. Lynne Rienner Publishers, Boulder & London, and International Rice Research Institute, P.O. Box 933, Manila, Philippines. p 1-12.

Directorate of Rice Research (various years) Reports of frontline demonstrations on rice production technologies for eastern India. Directorate of Rice Research, Rajendranagar, Hyderabad, India.

Krishnaiah K A, Janaiah, Kundu D K, Rao A V (1996) Productivity trends of rice in intensive farming systems in India. Paper presented at the Expert Consultation on Technology Evaluation and Impact for Sustainable Rice Production in Asia and the Pacific, FAO Regional Office for Asia and the Pacific, 29-31 Oct 1996, Bangkok, Thailand.

A critical review of the outreach research program in technology evaluation, adaptation, and dissemination for rice farmers in Nepal

H. B. GURUNG, K. B. KADAYAT, R. P. SAH, and A. K. VAIDYA

Lumle Agricultural Research Center (LARC), Pokhara, Nepal

Key words: problem diagnosis, farmer participatory research, research-extension linkage, structural reorganization, policy guidelines

Abstract

Rice research in Nepal started in 1951, although a systematic coordinated national research program was initiated only in 1972. Outreach research (OR) started in the mid-1950s when japonica varieties were evaluated in farmers' fields. A formal outreach OR program started with the inception of coordinated commodity programs in 1972 and a more systematic approach developed by the Integrated Cereals Project in 1977 when the Central Agronomy Division conducted on-farm research in six field research sites representing different agroecological environments. In 1985, a Central Farming Systems (FS) Division was established to conduct FS research, which was reorganized into Central FS and OR Division in 1990. This was dissolved in 1992 to form the Coordination and Special Projects Division which was again dissolved in 1994 to form a new Central OR Division. Until 1990, OR was largely designed, managed, and implemented centrally by a central division or a commodity program. Now, however, the Nepal Agricultural Research Council (NARC) conducts OR through its network of research stations in OR sites. Two major problems with OR in Nepal have been the lack of stability--i.e., the OR structure underwent several changes--and the lack of adequate OR policy guidelines and their implementation. Although a better OR structure is in place now, it still suffers from several deficiencies: use of conventional approaches to research, inadequate farmers' participation, and an almost nonexistent research-extension linkage. To be effective, structural reorganization, both in research and extension, must accommodate OR in the organogram as an important function to facilitate research-extension linkage, and it must include training and information functions at the regional station level to improve information flow and dissemination. It is essential that the research system effectively implement OR policy and operational guidelines, conduct interdisciplinary research programs, and enhance farmer participatory approaches to diagnosis and research.

Introduction

Rice is the most prestigious food crop in Nepal. Rice straw is an important livestock feed. Rice is grown in about 60% of the total cultivated land, contributing about 52% of the total grain production in the country (NARC 1996a) and about 25% of the national gross domestic product (Sah 1996).

Although the Department of Agriculture was established in 1924 and several disciplinary divisions and agricultural farms and stations were established within the 1950-60 period (Kayastha et al 1989), rice improvement work in Nepal began only in 1951. A more systematic rice improvement research program started with the establishment of National Rice Research Improvement in 1972 at Parwanipur in the *Terai* region (a part of the Indo-Gangetic plain running east to west in southern Nepal).

The Nepal Agricultural Research Council (NARC) defines outreach research (OR) as on-farm adaptive research with a farming systems perspective, conducted in close cooperation with extension agents and farmers (NARC 1996b). However, OR in Nepal is commonly understood by agriculturists as on-farm trials in farmers' fields. Therefore, in the context of this paper, any research conducted in farmers' fields outside research stations is termed as OR.

V. Balasubramanian, J. K. Ladha and G. L. Denning (eds.), Resource Management in Rice Systems: Nutrients, 253–268.

The importance of OR in agricultural research and development of the country was emphasized when the Ministry of Agriculture in 1979 made it mandatory for scientists to spend 40% of their time in farmers' fields (Bhattarai 1990; NARSC 1987, 1988; ORD 1995). Efforts to strengthen OR and extension (NARSC and DOA 1988) were also made in the past two decades through a number of major projects. However, institutionalization of OR has been inadequate and ineffective. It is important to examine the developments made so far and identify impediments in order to increase the relevance and effectiveness of OR. This paper analyzes OR as applied to rice research and development in Nepal.

Outreach research development

Initial developments

Outreach research in Nepal began during 1952-54 (Kayastha et al 1989, Rai 1992, Joshi 1996). However, a more formal OR started with the inception of the national commodity research programs (commodity programs) in rice, wheat, and maize in 1972. The commodity teams developed and coordinated national-level farmers' field trials for multilocation testing and national-level evaluation for wide adaptability of new crop varieties.

Cropping Systems Program (1976-85)

The Integrated Cereals Project (1976-85) funded by the United States Agency for International Development (USAID) was primarily designed to improve the basic and adaptive research capabilities of commodity research programs and to implement a cropping systems program (CSP) for a systematic conduct of OR (ICP 1982). The CSP demonstrated the importance of farmers' and systems' perspectives and the use of cropping systems research sites to represent different agroecological domains to generate technologies. The other important aspects of the CSP were the development of standard procedures to select research sites, identify problems, and plan and implement programs; use of a multidisciplinary approach, inclusive of social scientists; and emphasis on strong research-extension linkages (ICP 1982,1985).

Farming Systems Research and Development Division (1985-90)

Along with the establishment of an apex research body in 1985, the National Agricultural Research and Services Center now the Nepal Agricultural Research Council (NARC), the Central Farming Systems Research and Development Division (FSRDD) and the Socioeconomic Research and Extension Division (SERED) were established during the period 1985-90 with the support of the USAID-funded Agricultural Research and Production Project (ARPP). The ARPP ran its outreach activities--observation nursery, farmers' field trials, preproduction verification trials, seed multiplication, and minikits in

14 pilot areas (Schillinger 1989). Although confined to a small area, it had a strong OR component and it emphasized research-extension linkage.

Central Farming Systems and Outreach Research Division (1990-92)

The outreach functions of the research stations were administratively brought under FSRDD by reorganizing it into the Central Farming Systems and Outreach Research Division (CFSORD) in 1990, establishing an outreach cell in all the research stations. Following this, the farming systems research sites were handed over to the respective research stations. These sites were gradually phased out and replaced with new OR sites (Gurung and Floyd 1992).

Coordination and Special Project Division (1992-94)

In 1992 came the biggest blow. The two division chiefs of CFSORD and SERED were terminated and the divisions were dissolved and amalgamated to form a new Coordination and Special Projects Division (CSPD). OR was restructured as a unit within CSPD. The OR unit had no experienced personnel due to the departure of the staff.

Central Outreach Research Division

The CSPD was dissolved in 1994 to form a new Central Outreach Research Division (CORD) supported by the USAID-funded Agro-enterprise and Technology Systems Project (ATSP) (1991-97). The ATSP, although an agribusiness improvement project working through private entrepreneurs, also emphasized the strengthening of NARC and OR programs. The ATSP support was primarily to five Terai research stations, later extended to include two hill research stations. After the inception of ATSP, OR seemed to have regained some momentum.

Agricultural Extension and Research Project (1981-93)

The Agriculture Extension and Research Project (AERP) funded by the World Bank was primarily geared to reorganize the extension services of the Terai districts and secondarily to strengthen and reorganize the applied and adaptive research for technological backstopping (Shrestha and Bimoli 1992). The AERP affected only the four regional research stations in the Terai. Research-extension linkage was good and regular research-extension bimonthly and biannual workshops were held. However, impact was low after project termination due to breakdown of the system and low staff morale.

Outreach research methodology

Outreach research process

Most of the procedures of OR were borrowed from the cropping and the farming systems research methodologies. The following steps are followed in the conduct of OR:

1. Domain demarcation. Domain demarcation has been largely agroecologically based, except for the command area agroecological maps of five research stations funded by the ATSP and the use of cluster analysis technique at the Lumle Agricultural Research Center (LARC). The domain demarcation had always been defined from the researchers' perspective. Agroecological domains in use are Terai, inner-Terai, the Shiwalik hills, river basins, low hills, middle hills, high hills, mountains, and the *Himal* (Himalayas). These domains are largely based on altitude which, to a large extent, determines the terrain, land type, crops grown, access to road/market, temperature, animal-rearing systems, animal feed sources, and animal species composition.

2. Site selection. OR site selection, largely based on agroecology, is done by a multidisciplinary team of researchers using *Samuhik Bhraman* technique, a form of rapid rural appraisal developed at the LARC in the early 1980s and adapted to Nepalese conditions. It literally means traveling together.

3. Diagnosis. A site diagnosis is done, largely in a single farmer group, using rapid or participatory rural appraisal (R/PRA) method. The Pakhribas Agricultural Center (PAC) uses wealth-ranking technique using household food self-sufficiency as a proxy indicator to categorize households for diagnosis. Farm household surveys are also conducted. These two form the basis for describing the features of the site, identifying problems and opportunities, and determining the baseline to gauge changes and impacts in future.

4. Planning. Information gathered during the diagnosis is used to plan and design on-farm research. After R/PRA, a village-level workshop (VLW) is conducted, where problems are discussed and priorities set. In subsequent years, the VLW is held yearly before the annual station planning. The VLW results are discussed at the station, then at the regional station, and finally in a central-level panel meeting for approval by NARC.

   OR staff report the researchers' lack of interest to include their research trials in the OR program, except for breeders whose interests are largely to conduct farmers' field trials, a prerequisite for variety release. A drawback in research planning is the lack of consideration given to socioeconomic differences among farmers, which is an important factor in targeting research.

5. Experimentation. Approved OR activities are implemented in farmers' fields with the help of the designated OR staff and locally hired site recorders. PAC designs different

sets of experiments based on farmer category, but such categorization is not practiced at other NARC stations. Although LARC endeavors to encompass a cross-section of the farmers, the OR program of other regional and agricultural research stations emphasizes testing technologies with the medium-rich farmers and excludes the poorer farmers. Research designs are conventional, complex, and beyond farmers' comprehension at times.

6. Assessment. Assessment of the OR trials is done by the OR staff. At LARC, in addition to other visits, multidisciplinary monitoring teams visit the sites toward the crop ripening period. This visit allows the researchers to obtain firsthand farmers' feedback and identify experimental trial plots not suitable for inclusion in the analysis. General emphasis is on collecting quantitative data and performing statistical analysis of yield parameters. Socioeconomic and postharvest evaluations are inadequate, yet they appear as a strong feature of the methodology. Another problem with data analysis is the scientists' lack of knowledge on statistical analysis of qualitative data.

7. Recommendation. Recommendations are formulated in national workshops. Regional stations' research results are presented in summer or winter crop workshops, discussed, and recommended (if found suitable) as technologies for verification. The Variety Release Committee releases the varieties. There are, however no release procedures for management-related technologies.
   Verified technologies are recommended to the extension agencies through the research-extension coordination meetings, which are not regularly held. Compilation of recommendations (at LARC only) is done through workshops and publications. LARC and PAC publish the recommendations through a quarterly Nepal bulletin targeted for the grass roots-level extension workers and leading farmers.

8. Dissemination. Information is the first step to technology dissemination. A central-level Information and Documentation Division exists in the research and extension organogram, but it is absent at the regional stations or at district levels. Due to a lack of clear implementation policy, the information component receives less emphasis in research than in the extension system.

Agricultural technology in Nepal largely comes in the form of new crop varieties. The input delivery system, in both formal and private sectors is very weak. Input delivery agencies operate largely in accessible areas, and availability of improved inputs is generally poor. Availability of at least the seed of recommended varieties is important. To achieve this, the regional stations produce foundation seeds to supply the extension agencies and nongovernment organizations (NGOs) to support their certified seed production programs. A few of the ARS (e.g., LARC) have also entered into formal agreements with NGOs to conduct OR and disseminate tested technologies.

Types of OR trials and activities

A summary of the types of outreach trials and activities is presented in Table 1 and described in subsequent subsections.

Farmers' field trials

Traditionally, farmers' field trials (FFTs) are varietal trials conducted in farmers' fields as the last step of the variety development procedure. With an emphasis on OR and initiation of the concept of regional research stations in 1990-91, different types of FFTs have emerged: coordinated national, regional, or station-level FFTs. These are research-designed and farmer-managed on-farm trials. They have a plot size of 100 m^2 $treatment^{-1}$, 3-6 treatments set^{-1}, and 5 replications OR $site^{-1}$.

Scientists' visits to the field sites and interaction with farmers are poor. Not considering farmers' selection criteria, use of conventional research designs and cultural practices (line planting), overemphasis on quantitative yield parameters, poor postharvest and socioeconomic evaluation decrease the utility of FFTs in improving farmers' adoption of recommended varieties and technologies.

Preproduction verification trials

Preproduction verification trials (PPVTs) were designed by the researchers and implemented through the extension services with the primary purpose of promoting tested and verified technologies at the on-farm research sites. The PPVTs allow farmers to use the recommended practices on two-thirds of a test plot and their own regular practices on the remaining third (Kayastha et al 1989). The original PPVT research design has undergone changes. In 1984, PAC replaced this design with a 2 x 2 factorial (farmers' and improved variety and management) trial (Green 1987). In cereals, inclusive of rice, plot size is 500 m^2, equally divided into four treatments. Although PPVTs are not a common feature of the other NARC stations, they are a strong feature of LARC and PAC. The PPVT is simple in design, easily understood by farmers, has large plot size allowing better postharvest evaluations and saving of seeds, and, best of all, demonstrates interaction effect between the two factors at two levels. They are popular among the farmers first, for its simple design, and second, for the improved inputs that come free with the trial.

Component technology trials

Component technology trials are on-farm trials conducted to determine the optimum level of a given factor or the reduced number of factors where interaction between them is known to occur (Hawkins 1987). They investigate varieties, fertilizer rates or levels of some other inputs, and planting methods in randomized block design, with two or more

Table 1. Types of outreach trials and activities performed by research stations and the various agricultural research-related projects. [a]

Particular	NRRP	Stations (Outreach)	LARC	PAC	ICP	ARPP	AERP
Farmers' field trials							
Coordinated FFTs	Y	Y	Y	Y	Y	Y	Y
Regional FFTs	-	Y	Y	Y	-	Y	Y
Station FFTs	-	-	Y	Y	-	-	-
PPVTs	-	Y	Y	Y	Y	Y	Y
Component technology trials	-	Y	Y	Y	Y	Y	Y
Cropping pattern trials	-	Y	-	-	Y	Y	Y
Minikits	-	Y	Y	Y	Y	Y	Y
Informal research development	-	-	Y	Y	-	-	-
Participatory variety selection	-	-	Y	-	-	-	-
Household acceptance research	-	-	-	Y	-	-	-
Demonstrations	-	-	-	-	-	-	Y
Pilot production programs	-	-	-	-	Y	Y	Y
Foundation seed multiplication	Y	Y	Y	Y	Y	Y	Y

[a] Y = yes, - = no. See text for acronyms of organizations.

replications per farm. They are complex research trials with complex experimental designs (e.g., split-plot and factorial) with a large number of treatments (Upraity and Hawkins [1987] report of 16 treatments in a fertilizer trial) which are beyond farmers' comprehension and are perhaps of little or no relevance.

Cropping pattern trials

The cropping pattern trials are designed to develop cropping systems technologies suitable to the resource-poor farmers (Shrestha et al 1987). They test both farmers' patterns and improved practices that are usually more intensive with additional crops. Agronomic practices come from the component technology trials, and emphasis is placed on agronomic and economic analysis. Almost no attention is given to the socioeconomic dimensions despite available on-farm advantages. Given the highly risk-averse nature of the subsistence Nepalese farmers, it is unlikely that they will adopt the pattern and practices as a package. LARC's experience tells us that the best strategy is to conduct trials on component technology trials and let the farmers change their patterns according to their circumstances.

Minikits

Minikits were initiated as part of the national crop commodity programs during the Integrated Cereals Project period. The minikits started with wheat in 1977, followed by rice, maize, and soybean a year later (APROSC 1983). The minikits were introduced with dual objectives of research and extension--to receive feedback from farmers and to overcome delay in transfer of varietal technology through a farmer-to-farmer seed exchange network. Minikits, distributed free to cooperating farmers, are composed of small quantities of seeds of new crop varieties (released and prerelease), with planting instructions and feedback cards. An evaluation study (APROSC 1983) concluded that the minikit program was the right approach in disseminating improved crop varieties; however, it failed to achieve its research objectives. Initially, minikits were distributed through government extension services; now distribution of minikits by R/ARSs may be considered erratic because of the following reasons:

- Because of budgetary constraints, regional stations are not able to develop minikits for wide distribution.
- There is poor response or not all feedback cards are returned.
- Researchers neglect following up, analyzing, and utilizing the available information, and researchers do not have a strict need for minikits as a research tool.
- Since 1990-91, research stations have operated their own OR sites; thus, they distribute minikits in OR sites within their own command areas.
- There is duplication of effort by government extension agencies who also compose and distribute minikits and seed extension kits.

Informal research and development

LARC first adopted the informal research and development (IRD) method in Nepal in 1989 with agronomic crops as a complement to formal research to help farmers test new materials themselves. PAC also uses a similar tool known as informal research kit. Under IRD, new crop varieties, fruit varieties or species, and animal species are distributed free to farmers at research sites, for their own testing. One of the notable achievements of IRD at LARC is the distribution of rice variety IR28218 which was not promoted beyond FFT but was preferred by farmers when distributed as part of the IRD packets. In contrast, rice variety IR13155 was promoted up to the PPVT stage; it was, however, rejected by farmers when distributed as IRD packets due to threshing problems.

Participatory variety selections)

From previous experience with the IRD and FFTs, the participatory variety selection (PVS) method evolved at LARC in 1995. The method has discarded the conventional experimental design of FFTs and has increased plot size with emphasis on socioeconomic (as against biological) evaluation. In PVS, about 1-2 kg seeds of new crop varieties are given to rich, medium, and poor farmers to test under their own management practice. The method is very new and still being evaluated.

Household acceptance research

Household acceptance research was developed at PAC in 1993 based on the observation that technology acceptance is a decision of the household rather than of a single participating farmer and that it is a process rather than an event which takes place at one point in time (Bhattarai and Ohja 1996). It has two components--household acceptance test and outreach kit. In the household acceptance test, which includes a limited number of farmers (two sets per category), attempts are made to collect more qualitative information. The outreach kit involves more farmers (30-35 per category), and mostly quantitative information is collected to complement information obtained from the household acceptance test (Shrestha et al 1994). The research is designed so that a wide range of socioeconomic information is collected. One problem associated with this is analyzing the information appropriately.

Pilot production programs

The pilot production program (PPP) is a methodology to demonstrate technologies tested at the field research sites before recommending it for block production program. Large blocks of pilot tests easily convince farmers. PPP is not used now, but, some regional research directors see its usefulness and argue for its revival (Dhital et al 1996).

Demonstrations

Regional research stations in their OR sites demonstrate new commodities and technologies to promote them.

Seed multiplication program

Farmers are keen to adopt released and prerelease crop varieties. The formal seed agency, however, multiplies seeds of released varieties only. The formal seed delivery system is weak, and informal seed exchange is the general rule. Extension agencies and nongovernment organizations (NGOs) multiply seeds of both prerelease and released varieties and help disseminate promising prerelease varieties as well. To meet the foundation seed demands of the formal seed company, extension agencies, and the NGOs, research stations produce foundation seeds of both released and prerelease varieties in research farms, but LARC uses on-farm research sites to produce foundation seeds.

Institutional arrangements and linkages

Outreach research in the NARC organogram

After the Central Outreach Research Division (CORD) was formed as a result of the 1994 reorganization, OR activities of all regional research stations were administratively brought under the CORD. CORD's major role is to coordinate, monitor, and evaluate the OR programs. Under the new structure, each station has an outreach cell with an OR coordinator, and a few core and field staff for the OR sites. Scientists of regional stations are also required to participate in the OR programs. The OR coordinator coordinates the scientists and the OR core staff help in the formulation and implementation of OR programs with support from the regional station chiefs (NARC 1996b). However, actual implementation is in jeopardy.

Organization of rice research

The National Rice Research Program (NRRP) in Parwanipur in the Terai is mandated to coordinate all rice improvement work in Nepal. It is responsible for all aspects of rice research for the Terai with subtropical to tropical climate. The Central Agriculture Botany Division in Katmandu has the NRRP mandate for hill rice research and is involved only in rice breeding research (NARC 1996a). Figures 1 and 2 show the organization of rice research in Nepal and the rice research network. The NRRP and the Central Agriculture Botany Division conduct national-level coordinated research in varietal development work only. The other aspects of rice research are carried out by other scientists located in the central disciplinary divisions, and/or in the regional stations in their respective command areas.

Forums and meetings for internal linkages

Regional annual budget and program planning meeting. The meeting is convened by regional stations, about 6 mo before the start of the next fiscal year. It is attended by regional researchers, scientists from central disciplinary divisions, NARC directors, and extension officers from the command area districts.

Central panel meeting. Disciplinary panels are constituted in the NARC for final approval of the annual work programs forwarded by the regional stations. The panel typically consists of 5-7 members representing commodity programs, central disciplinary divisions, senior management members of the NARC, regional stations, the extension department, and others as required. Rice-related panels are Varietal, Agronomic/Management, Plant Protection, and Outreach Research.

Working group meeting. Before developing each annual work plan, each commodity program coordinates and convenes working group meetings to discuss proposed work programs on a commodity to rationalize and eliminate duplication; secure needed support from regional stations and disciplinary divisions; and identify areas of collaborative work among the commodity programs, disciplinary divisions, and the regional stations. Meetings on rice started in 1984 and were discontinued in 1990 for no reason at all. The revival of the meeting is important to rationalize and coordinate program development.

National summer crop workshop. This workshop is organized annually by the director, Crops and Horticulture, on a national basis. Held in the winter months, it is attended by regional researchers, central disciplinary scientists, NARC directors, and representatives from extension and other related agencies. The dominant feature of the meeting is the presentation of research results, inclusive of OR, followed by a discussion on technology recommendations and identifying problems and constraints through working groups. Some serious concerns raised by scientists are that the workshop is becoming a routine and that there is no follow-up on management recommendations presented in the workshops.

Outreach research workshop. Three OR workshops have been held, the first in January 1990, the second in June 1994, and the third in March 1996. Participant composition is similar to that in any national workshop. The first workshop emphasized the role of different disciplinary divisions, the experiences and issues related to OR, and the OR strategies in different disciplines. Other workshops were heavily dominated by presentation of OR trial results, and only a few papers dealt with concepts and issues in OR. Despite this deficiency, workshop quality has improved because there are more papers dealing with OR concepts and issues.

Linkages between extension and research

How to effectively link research and extension is a major problem in many developing countries, and Nepal is no exception. The linkages in externally funded projects were good, but they broke down after the projects ended. This reflects the lack of institutionalization of OR and research-extension linkages. Some of the major reasons for the poor research-extension linkages are the following:

- Out of necessity, OR was developed by researchers and not by extension agents; therefore, there is a lack of commitment from the extension group.
- There is likewise lack of commitment on the part of NARC to effectively implement the defined policy and operational guidelines for further improvement of the OR.
- The OR component is not given adequate importance in the research structure.
- Outreach is not a mandatory function of the Department of Agriculture and there is no outreach component in its structure.
- There is no memorandum of understanding at the apex level between the NARC and the Department of Agriculture.
- There are disparities in the daily travel allowance rates between NARC and the Department of Agriculture personnel—i.e., extension staff members do not have enough incentive to participate in OR programs.
- Extension personnel lack OR orientation.

Training of collaborators and farmers

LARC and PAC have training programs for its junior research staff, extension collaborators, and farmers, but NARC and the regional stations do not have training units in their structure. The training facilities created at the regional stations during the externally funded project periods remain abandoned. Training of staff, collaborators, extension workers, and farmers should be a key feature in all regional stations to enhance the overall effectiveness of OR programs.

Farmers' participation in outreach research

Farmers' participation in OR is inadequate in Nepal (Galt and Mathema 1986, Gurung et al 1996). Many researchers misunderstand the concept of farmer participation and they take it to mean that farmers only make their land available for experiments. It is largely "contractual" and "consultative," and lacks "collaborative" and "collegial" modes of participation (see Biggs [1989] for modes of participation).

Farmer participation in FSR is largely consultative in nature (Galt and Mathema 1986). With the inception of ATSP, the quality of farmer participation has improved and the OR has also recovered some of the lost impetus. Apart from rural appraisal methods, the other features of farmers' participation in OR are village-level workshops, farmers' field days, farmers study tours of research stations/farms and OR sites, and OR site farmers' exchange visits (Gurung and Amatya 1992, Kadayat 1993, Jaiswal 1994). Farmers hesitate to participate in research trials due to lack of compensation for lost yields; only LARC addresses this problem (Gurung et al 1996).

Future considerations

Effective implementation of OR policy and guidelines

Despite the formulation and issuance of OR policies and guidelines, their implementation has been very weak. Staff allocations to the OR cells depend on the decision of the station chief. The OR staff positions approved for the regional stations are not filled and some stations do not even have an OR coordinator. The NARC management should improve the implementation of OR policies and guidelines.

Research and extension linkage

The basic reason for weak research-extension linkage is largely institutional. Although NARC has a mandatory OR component in its organogram, the OR function is not mandatory for the Department of Agriculture. To improve research-extension linkages and the OR program, OR should be institutionalized as a priority in both research and extension through organizational restructuring from the central to the regional and district levels. It is recommended that NARC adopt the organogram proposed by the NARC Strategy Formulation Task Force in 1996 to further strengthen its OR component.

Outreach research staff cadre

In the NARC organization, there is no provision for an OR staff cadre. The needed staff are drawn temporarily from other disciplinary cadres, resulting in frequent staff turnover and inadequate commitment to the OR program. A separate OR staff cadre with adequate career development opportunities should be created in the NARC organogram to enhance OR effectiveness.

Financial autonomy to OR coordinator

In the NARC structure, the budget is managed by the station chief, although budgets are prepared on the basis of programs of the various units within it. The OR coordinators at the stations do not have the authority to manage the OR budget, and often the OR budgets are used for other purposes. Decentralization of budget management is vital to implement OR programs effectively.

Biometrics services

A biometrician is needed at the regional station level to backstop and train researchers on simple statistical designs for field trials and familiarize them with tools and methods for analyses of quantitative and qualitative data collected from OR.

Reviving rice working group meetings

The rice working group meeting provided a multidisciplinary forum to discuss and formulate collaborative research programs. The meeting should be revived and made mandatory and should be coordinated by the rice program, emphasizing an interdisciplinary approach to developing research programs.

Enhancing farmers' participation and orientation in OR programs

Farmers' participation and orientation is an issue of central concern. Recommendations for enhancing farmers' participation and orientation (Gurung and Amatya 1992, Gurung et al 1996) include the following:

- Collaborative and collegial modes of farmer participation should be emphasized and should aim at strengthening the farmers' own research system.
- Simple research designs and large plot sizes which are comprehensible, easy to manage, and statistically valid should be used.
- The use of complex research designs to study complex interactions on farmers' fields should be avoided.
- The "wealth-ranking technique" should be adopted as an OR methodology for categorizing farmers in order to develop appropriate technologies for different farmer groups.
- Frequency of researchers' visits to OR sites should be increased for firsthand interaction and rapport building with farmers.
- A policy of compensating farmers for lost yields in trials should be instituted.
- Orientation briefing to farmers on OR concepts and programs should be given.

Inclusion of training and information component in the organogram

To have an effective flow of research information and technologies from research to extension and farmers, training and information units must be created at the regional research stations.

Exploiting NGOs and agroveterinary input entrepreneurs

The NGOs are emerging as a potential technology disseminators. It is important that the formal system join hands with NGOs to target and serve the farmers better. Agro-veterinary entrepreneurs are a source of improved inputs and information, but they have poor knowledge on the appropriate use of improved inputs and technologies. NARS and the Department of Agriculture should develop strategies to exploit this channel through OR programs.

Conclusions

The basic problem associated with OR in Nepal is institutional. Efforts made in the past two decades through the various donor-assisted projects have not been able to adequately institutionalize OR. These projects affected only some sectors or parts of the research and extension systems, and thus were not effective in institutionalizing the OR. To improve OR and extension-research linkages, OR should be made mandatory to both research and extension agencies through structural reorganization. For this, high-level commitment and action are vital.

In addition, NARC should effectively implement its OR guidelines and operational plan and develop an OR staff cadre; provide training and information component, at least at the regional station level; give financial autonomy to the OR coordinators; ensure interdisciplinary alliances including biometrics; provide training to research and extension staff on OR concepts and methods and farmer participatory approaches; and develop strategies to exploit the potentials of NGOs and agroveterinary entrepreneurs to help accelerate the dissemination of technologies.

Acknowledgements

The LARC is funded by the Overseas Development Administration (ODA) of the British Government and works in collaboration with His Majesty's Government of Nepal. The support of both governments is gratefully acknowledged.

The authors are also thankful to Drs. C.N. Floyd and P.E. Harding for reviewing an early version of the paper.

References

APROSC--Agricultural Projects Services Center (1983) Evaluation of minikit performance (rice, wheat, maize, and soybean). APROSC, Katmandu, Nepal.

Bhattarai A N (1990) Needs for effective research outreach program in Nepal. Pages 1-18. in: Proceedings of the First National Research Outreach Conference, 23-25 Jan 1990. Nepal Agricultural Research and Services Center, Katmandu, Nepal.

Bhattarai R, Ojha R (1996) Household acceptance test- a useful tool for outreach research: the experience of Pakhribas Agricultural Center. Paper presented at the 3rd National Outreach Research Workshop, 21-23 May 1996, Pakhribas Agricultural Center, Dhankuta, Nepal.

Biggs S D (1989) Resource-poor farmer participation in research: a synthesis of experiences from nine national agricultural research systems. OFCOR Comparative Study Paper No. 3. International Service for National Agricultural Research, The Hague, Netherlands.

Dhital B K, Shrestha P K, Subedi P P, Gurung D B (1996) An overview of the outreach research programs of some major agricultural research stations. LARC Discussion Paper No. 96/2. Lumle Agricultural Research Center, Kaski, Nepal.

Galt D L, Mathema S B (1986) Farmer participation in farming systems research. SERED Report No 3. Socioeconomic Research and Extension Division, Khumaltar, Nepal.

Green T (1987) Diamond trials in on-farm research: the experiences of Pakhribas Agriculture Center. Pages 28-44 in Proceedings of the Workshop on Trial Design for On-farm Research in Nepal. Farming Systems Research and Development Division and Socioeconomic Research and Extension Division, Katmandu, Nepal.

Gurung H B, Amatya L K, Kadayat K B, Jaiswal J P, Gurung D B, Dhital BK (1996) Declining farmers participation over time in on-farm research: causes and suggestions. LARC Seminar Paper No. 96/15. Lumle Agricultural Research Center, Kaski, Nepal.

Gurung H B, Floyd C N (1992) Farming systems research and development: experiences of Lumle Agricultural Research Center. Pages 52-65 in: Farm management notes for Asia and the Far East. No. 15, April 1992. Secretariat, FAO, RAPA, Bangkok, Thailand.

Gurung H B, Amatya L K (1992) Farmers participation in hill farming systems research and extension: experiences of the Lumle Regional Agricultural Research Center. LARC Seminar Paper No. 92/20. Lumle Agricultural Research Center, Kaski, Nepal.

Hawkins R (1987) Design of on-farm trials: an overview. Pages 1-11 in Proceedings of the Workshop on Trial Design for On-farm Research in Nepal. Farming Systems Research and Development Division and Socioeconomic Research and Extension Division, Katmandu, Nepal.

ICP--Integrated Cereals Project (1982) Annual report, Oct 1981-Sep. 1982. Ministry of Agriculture, His Majesty's Government of Nepal.

ICP--Integrated Cereals Project (1985) The integration of research and extension in farmers' fields. Terminal report of the Integrated Cereals Project. Department of Agriculture, Ministry of Agriculture, His Majesty's Government of Nepal.

Jaiswal J P (1994) Increasing the relevance of farmers' participation in generating appropriate technology for the hill areas of Nepal. MS dissertation, University of Reading, UK.

Joshi S K (1996) Growth, institutionalization, and sustainability of outreach research in national agricultural research systems. (unpubl.)

Kadayat K B (1993) Research outreach: an approach for the hills of Nepal. MS dissertation, University of Reading, UK.

Kayastha B N, Mathema S B, Rood P (1989) Organization and management of on-farm research in the national agricultural research system. OFCOR-Case Study Map. 4 International Service for National Agricultural Research, The Hague, Netherlands.

NARC--Nepal Agricultural Research Council (1996a) Status of rice research and production in Nepal. Khumaltar, Lalitpur, Nepal.

NARC--Nepal Agricultural Research Council (1996b) Nepal Agricultural Research Council's agricultural research strategy. Strategy Formulation Task Force Report. (unpubl.)

NARSC--Nepal Agricultural Research and Services Center (1987) NARSC policy guidelines. Khumaltar, Lalitpur, Nepal.

NARSC--Nepal Agricultural Research and Services Center (1988) Operational plan for national outreach program. Khumaltar, Lalitpur, Nepal.

NARSC and DOA--National Agricultural Research and Services Center and Department of Agriculture (1988) Research outreach job descriptions for NARSC and DOA Personnel in the Agricultural Research and Production Project (AERP) Command Area. DOA-NARSC Research Outreach Collaboration Paper No. 1. Katmandu, Nepal.

ORD--Outreach Research Division (1995) Guidelines for outreach research. Outreach Research Division, Nepal Agricultural Research Council, Khumaltar, Lalitpur, Nepal.

Rai S K (1992) Experiences in implementing outreach activities. Pages 38-40 in Proceedings of the First Regional Outreach Workshop, 19-24 Jan 1992. Pakhribas Agricultural Center. Dhankuta, Nepal.

Sah R P (1996) National rice research report. Paper presented at the Nineteenth Summer Crops Workshop, 27-29 Feb 1996, National Rice Research Program, Parwanipur, Nepal.

Schillinger W F (1989) Research outreach: the ARPP experience. An end of tour report. Agricultural Research and Production Project, Winrock International, and Ministry of Agriculture, Department of Agriculture, and National Agricultural Research and Services Center, Katmandu, Nepal.

Shrestha P M, Galt D L, Saran K C (1987) Cropping pattern trials: on-farm trials under farmers' circumstances. Pages 12-27 in PrOoceedings of the Workshop on Trial Design for On-farm Research in Nepal. Farming Systems Research and Development Division and Socioeconomic Research and Extension Division, Katmandu, Nepal.

Shrestha S S, Bimoli B P (1992) Strengths and weaknesses of research outreach programs in the AERP command area. Pages 19-27 in Proceedings of the First National Research Outreach Conference, 23-25 Jan 1990. Nepal Agricultural Research and Services Center, Katmandu, Nepal.

Shrestha S S, Saran K C, Rood P G, Gurung B D, Gautam S R (1994) Farmers back to farmers: the farmers paradigm in agricultural research, the experience of Pakhribas Agriculture Center. Pages 45-53 in B.K. Acharya et al (Eds.) Proceedings of the Second National Khumaltar, Workshop on Outreach Research, 29-31 Mar 1994. Coordination and Special Projects Division, Nepal Agricultural Research Council, Khumaltar, Nepal.

Upraity V, Hawkins R (1987) Component technology trials: experiences in Nepal. Pages 45-51 in Proceedings of the Workshop on Trial Design for On-farm Research in Nepal. Farming Systems Research and Development Division and Socioeconomic Research and Extension Division, Katmandu, Nepal.

South-South cooperation: Indonesian apprenticeship program for African rice farmers

A. S. MARTAAMIDJAJA[1] and H. ANWARHAN[2]

[1]*Agency for Agricultural Education and Training (AAET), Ministry of Agriculture, Jakarta, Indonesia*

[2]*Central Research Institute for Food Crops (CRIFC), Bogor, Indonesia*

Key words: farmer-to-farmer learning, learning-by-doing, structured training, innovative technical assistance, demonstration farm, apprentice-farmer, host farmer, rice farming, cultural interchange

Abstract

The success of Indonesia in achieving rice self-sufficiency in 1984 motivated Indonesian farmers to share the fortune with their fellow African farmers who suffer from drought and food shortage. A fund amounting to an equivalent value of 100,150 t of rice was raised and donated to needy farmers in several African countries. The donation was at first given in cash through the Food and Agriculture Organization and later on, beginning 1990, was allotted in the form of technical assistance comprising an apprenticeship program in Indonesia and expert assistance to African countries. The goal of the apprenticeship program is to help African countries increase food production, particularly rice, by improving farmers' capabilities in applying proper cultivation technologies and sound farm-business management. As an educational process, this program provides selected African participants with practical learning experience through interaction with Indonesian host-farmers who serve as facilitators of learning in rice farming activities. In addition, the participants are given practical knowledge and skills through a structured training to complement the farmer-to-farmer learning process. Since 1990, eleven groups comprising a total of 163 apprentice farmers from 14 countries have participated in the program. Many of them have successfully applied in their own farms what they have learned from Indonesian farmers. In the future, more farmers from many African countries will be invited to join the program, either in Indonesia or in the outreach training facilities to be established by the program in Tanzania and The Gambia. Broad-based evaluation showed that most participants regard their experience in both technical and social aspects as meaningful and useful. They also expressed their willingness to apply the technologies they have learned as well as share the experience with fellow farmers. The importance of thorough assessment of training needs, careful selection of host farmers and field facilitators, clear information on the objectives and nature of the program to both prospective participants and host farmers, and intensive supervision are some lessons learned by the organizers from 6 yr of experience. The Ministry of Agriculture of the Republic of Indonesia and the Indonesia Agricultural Society Foundation as organizers of the apprenticeship program pay serious attention to the quality improvement and the continuity of the program for the benefit of African farmers and the enhancement of South-South cooperation.

Introduction

Increasing food production, particularly rice, has long been the priority of agricultural development in Indonesia. Through an innovative national rice intensification program known as BIMAS (which stands for *bimbingan massal* that literally means mass guidance), the government has guided and encouraged rice farmers to improve their farming practices and to organize themselves toward a better and more productive rice farming business. The BIMAS program not only has introduced to farmers new rice-farming technologies and group farm-business management but also provided them with subsidized farm inputs, low-interest credit, and improved village-level marketing facilities and infrastructure. With active participation and leadership of the farmers, the BIMAS program has successfully helped farmers increase their productivity. As a result, in 1984, Indonesia became self-sufficient in rice for the first time since its independence in 1945.

V. Balasubramanian, J. K. Ladha and G. L. Denning (eds.), Resource Management in Rice Systems: Nutrients, **269–280.**

As an expression of happiness and thankfulness to God for this successful achievement, in July 1985, the Indonesian farmers, represented by the National Group of Outstanding Contact Farmers, decided to share their good fortune with the farmers of Africa who often suffer from drought and food shortage, through a grant equivalent to 100, 150 t of rice as a humanitarian donation. During 1986-89, the donation was handed in cash to the African farmers through the Food and Agriculture Organization of the United Nations. Since 1990, the remaining funds have been used to provide technical assistance to African countries through an agricultural apprenticeship training program in Indonesia and technical assistance to selected African countries. Through this program, Indonesian farmers and experts are able to share with their fellow farmers from Africa their knowledge, skills, and experiences in achieving self-sufficiency in rice.

In December 1993, the Indonesian Agricultural Society Foundation or *Yayasan Amal Masyarakat Pertanian Indonesia* (YAMPI) was established to manage the remaining fund particularly to implement the agricultural apprenticeship program (Fig. 1). The first apprenticeship program, organized in 1990, was a 1-mo training with 10 participants from The Gambia, Tanzania, and Zambia. The number of participants in the following years increased to 20 per group. By the end of 1996, 11 apprenticeship groups comprising 163 participants from 14 countries had been trained (Table 1).

Program structure and organization

The apprenticeship program is basically a practical training based on the principles of "learning-by-doing" and "farmer-teaches-farmer" that combines structured training and on-farm training with host farmers.

The main educational objectives of the program are

1. to improve the knowledge, attitudes, and skills of apprentice farmers on cultivation techniques and business management of food production, rice in particular;
2. to exchange experience, ideas, and information on agricultural development between apprentice farmers and Indonesian farmers;
3. to develop creativity, self reliance, and entrepreneurial capacities of both apprentice farmers and Indonesian farmers;
4. to strengthen cooperation, friendship, and brotherhood between apprentice farmers and Indonesian farmers; and
5. to introduce Indonesian culture to African apprentice farmers.

A typical 2-mo apprenticeship program incorporates the following activities:

1. First structured training, 1-2 wk
2. On-farm training with host-farmer, 4 wk
3. Study tour, 1 wk
4. Second structured training, 1 wk

During the first structured training conducted in a research or training center, the apprentice farmers learn basic theory and practices of food production, particularly that of rice. During on-farm training, the participants stay with the host farmers and work with them in the field, on a working schedule prepared by the host farmers and agreed upon by the participants. A maximum of four participants are supervised by one host farmer assisted by a field extension worker and/or a subject matter specialist.

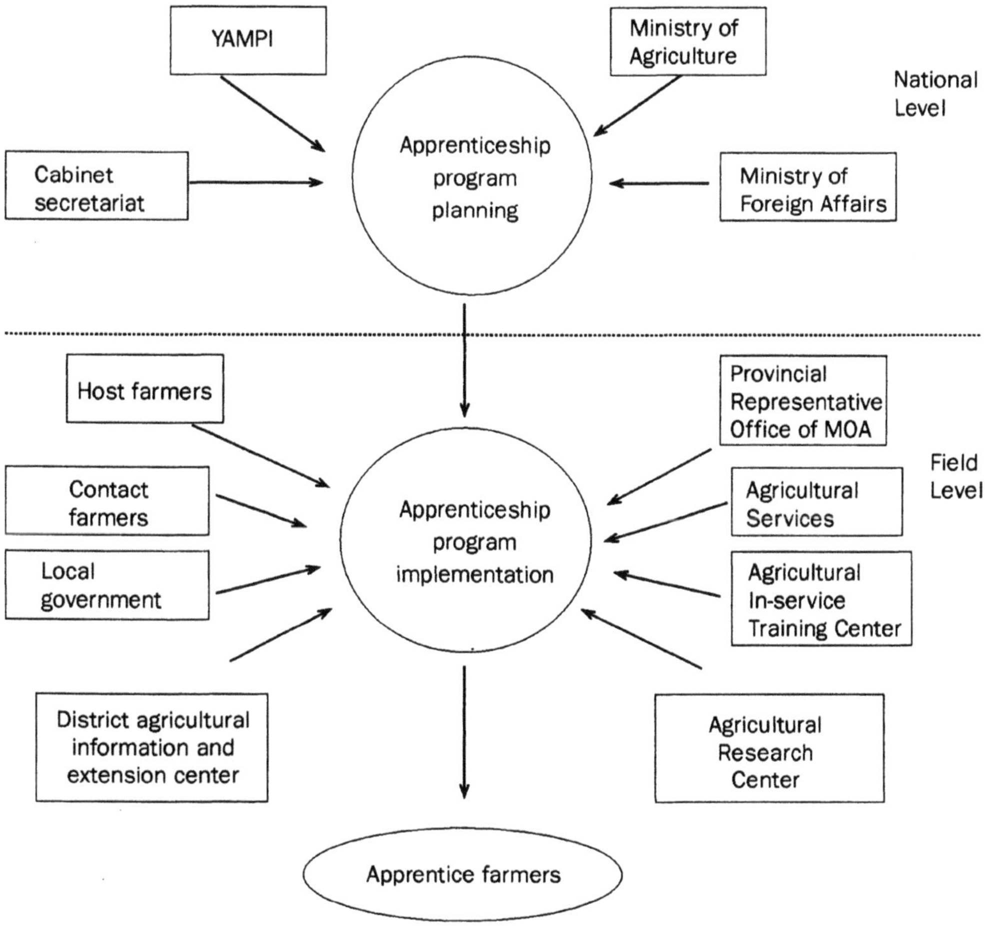

Figure 1. Cooperative undertaking of apprenticeship program.

Table 1. Number of apprenticeship participants, by batch and country (various years).

Batch	Period	Country	Participants[a]						Total	Grand total
			Youth		Farmers		Officials			
			M	F	M	F	M	F		
I	Oct-Nov 1990 (4 wk)	Zambia	-	-	1	1	-	-	2	10
		Tanzania	-	-	2	2	-	-	4	
		The Gambia	-	-	2	2	-	-	4	
II	May-Sep 1991 (17 wk)	The Gambia	5	-	5	5	2	-	17	17
III	Oct 93-Jan 94 (12 wk)	Tanzania	2	-	3	3	1	1	10	10
IV	Jun-Aug 1994 (8 wk)	Tanzania	-	-	4	4	2	-	10	17
		Madagascar	-	-	1	1	-	-	2	
		Papua New Guinea	-	-	2	2	1	-	5	
V	Oct-Dec 1994 (8 wk)	Madagascar	2	-	3	3	1	1	10	17
		Nigeria	1	-	1	1	1		4	
		Senegal	-	-	1	1	1		3	
VI	Mar-May 1995 (8 wk)	Papua New Guinea	4	-	3	3	1	1	12	12
VII	May-Jul 1995 (8 wk)	Senegal	4	-	2	2	2	-	10	16
		Ghana	-	-	1	1	1	-	3	
		Cote d'Ivoire	-	-	1	1	1	-	3	
VIII	Sep-Oct 1995 (8 wk)	Ghana	3	1	2	2	2	-	10	17
		Sudan	1	-	1	1	1	-	4	
		Ethiopia	-	-	1	1	1	-	3	
IX	Jun-Aug 1996 (8 wk)	Surinam	3	-	-	-	1	-	4	17
		Uganda	-	-	-	2	1	-	3	
		Ethiopia	3	-	3	3	1	-	10	
X	Oct-Nov 1996 (4 wk)	Papua New Guinea	1	-	3	3	1	-	8	10
		Iraq	-	-		-	2	-	2	
XI	Nov-Dec 1996 (4 wk)	Gambia	10	7		-	3	-	20	20
Total			39	8	42	44	27	3	163	163

[a]M=male, F=female.

By combining the theory and practice of rice cultivation techniques and agribusiness management, the apprentice farmers are expected to gain meaningful knowledge and skills from both the experts and the farmers. The farming activities cover the practice and management of farm business (production, harvesting, postharvest handling, processing, and marketing) and the management of farmers' institutions (farmer group, farmer association, and farmer cooperatives). In addition, the participants are given ample opportunities for interacting with the rural community. This broad-based learning experience characterizes the uniqueness of the apprenticeship program as an educational medium for technology transfer and cultural interchange.

To ensure the effectiveness of the teaching-learning process, the following principles are applied:

- The subject matter taught is determined based on the general needs of the apprentice farmers as indicated by a training needs assessment;
- Adequate adjustment is made for the specific needs of the participants;
- Adequate opportunity is provided for the participants to observe, experience, and solve problems of their interests;
- A pleasant learning environment is provided throughout the program; and
- A climate of understanding and tolerance between apprentice farmers and host farmers is developed to mutually respect cultural, religious, and customary differences.

The participants are categorized into three groups: adult farmers, farm youth, and agricultural officials. The farmers, both adult and youth, are the main participants in the program; the agricultural officials accompany them and act as facilitators. In general, for every five farmers, one official is invited. Married couples are encouraged to participate together, in the hope that a more positive impact will result when they return to their home countries.

The apprenticeship program is organized by the Ministry of Agriculture, Republic of Indonesia, through the Agency for Agricultural Education and Training, as the executing agency, in collaboration with YAMPI, Ministry of Foreign Affairs, and the Cabinet Secretariat.

At the field level, cooperation is developed with the Agency for Agricultural Research and Development through its research centers, local government, agricultural services, farmer groups, and the Group of Outstanding Contact Farmers. Besides researchers and trainers, subject matter specialists and field extension workers are actively involved as facilitators in the field, particularly during on-farm training.

The cooperative undertaking of the program is shown in Figure 1.

The program funds are provided mainly by YAMPI. The Ministry of Agriculture contributes some funds for domestic expenses, while the Cabinet Secretariat through its Technical Cooperation Among Developing Countries Program contributes some expenses for international airfare of participants from certain countries.

The domestic expenses provided by YAMPI and the Ministry of Agriculture cover living allowance, local transportation, study tour, health insurance, training materials, pocket money, and other expenses related to the administration of the program. All expenses incurred during the preparation of participants' in their own countries, as well as participants' daily allowance during the international travel, are shouldered by the participating African countries.

Training needs assessment (TNA)

Before the participants leave for Indonesia, a TNA is conducted in their countries to ensure the suitability of the program to their actual needs. Information for the TNA is collected mainly from two sources: the Indonesian Embassy in the participants' country and a direct field observation by a TNA team from Indonesia. The TNA team usually consists of a researcher, an extension worker, and a contact farmer. The team conducts the TNA about 6 mo before the program is expected to start. The main objectives of the TNA are

1. To have a clear picture of strengths, weaknesses, and opportunities available for possible implementation of a food crop intensification program in the participating country; and
2. To obtain data on the knowledge and skills needed by the prospective farmers, as well as their interest in learning, which are necessary for developing a suitable curriculum for the agricultural apprenticeship program.

During the last 3 yr, the TNA team visited 8 countries--Tanzania, Madagascar, Senegal, Ghana, Ethiopia, The Gambia, Sudan, and Uganda. In 1996, another team visited Tanzania, Swaziland, Lesotho, Zimbabwe, Guinea Bissau, and The Gambia to assess first hand the priority needs, including training needs, for agricultural development in those countries.

Program implementation

The training curriculum is developed through a series of meetings involving various people such as researchers, trainers, prospective host farmers, subject matter specialists, field extension workers, contract workers, agricultural staff, and other related officials who will be involved in teaching the apprentice farmers. Two curricula are usually prepared: one for the structured courses and the other consisting of scheduled activities for on-farm training during the participants' stay with the host farmers. The curriculum of the structured training comprises topics related to cultivation techniques of rice and secondary crops. Other topics are also incorporated upon the participants' request or if required as a result of the TNA. With some variations for different groups, the following topics are usually incorporated in the structured training:

A. Rice and secondary crop cultivation

- Land preparation
- Seed selection and treatment
- Nursery preparation and maintenance
- Fertilizer application
- Transplanting and direct seeding
- Water management

- Weeding
- Integrated pest management
- Harvesting
- Postharvest handling and processing
- Marketing

B. Other topics

- Cow and goat raising
- Fish farming in ricefield
- Vegetable and fruit cultivation

During the structured training in the Research/Training Center, the theory and practice of rice and/or secondary crop cultivation are given both in the classroom and in the field. Participatory and self-observation methods are applied, with emphasis on individual practical work. Sufficient time is allocated for discussion and exchange of experience and ideas. Some prospective host farmers are frequently invited to the discussion. Often, the instruction, which is given in English, has to be translated into the participants' local language by one of the group members.

After the first structured training is completed, the participants are transferred to their designated host-families. They usually stay for 2-4 wk with the host farmers from whom they will learn day-to-day farming activities. To facilitate better learning and easy communication, a field extension worker and/or a subject matter specialist is assigned to help both participants and host farmers. In addition, the trainers or researchers from the center make periodic visits to supervise the activities and to help with more information if problems or questions arise. Although a curriculum is already prescribed, adjustment of schedule is always possible upon the participants' request. A once-a-week field visit, usually on Saturdays, to nearby farms or agricultural enterprises is scheduled, so that the participants can have a wider learning experience. Staying with a host family in the middle of a rural community gives the participants greater opportunity to understand and take part in various social and cultural activities. This exposure to rural life makes learning easier and richer. At the end of the field practicum, all participants and their hosts go on a study tour, visiting interesting development projects as well as tourist places. This activity is designed primarily to promote Indonesia's achievements in national development programs.

A second structured training is organized as a reinforcement session to review and reiterate the lessons the participants have learned from the field practicum. Upon request, additional new courses may also be inserted in the training. The last day of the training is usually for entertaining feedback and suggestions from the participants. At the end of the program, the participants are given some samples of different high-yielding rice seeds to bring home for trials. Some sample farming tools are provided free on request.

The following impressions were expressed by most of the participants during the feedback sessions:

1. Although oral language is the ultimate barrier to communication with the host farmers, participants feel comfortable living with their hosts and can understand well what they observe and learn by using body language and other means of communication.
2. Participants have no problem adapting to the food, social customs, and cultural practices of the Indonesian rural community.
3. Participants appreciate the structured training program as complementary to the farmer-to-farmer apprenticeship training. The combination of theory and practice helps them understand "the why's" of what the farmers are doing.
4. The apprenticeship training is very useful and applicable, and the experience of living with the Indonesian farmers is worthwhile and enjoyable.
5. The study tour is important and exciting, especially the visits to industry-related places.
6. Participants are willing to try what they have learned in Indonesia on their own farms and to teach their fellow farmers the new knowledge and skills they have acquired.

By the end of 1996, 11 groups totaling 163 persons from 14 countries, 11 of whom are from Africa, had participated in the program. The number includes 133 farmers and 30 agricultural officials. The African participants were from Zambia, The Gambia, Tanzania, Madagascar, Nigeria, Senegal, Ghana, Cote d'Ivoire, Sudan, Ethiopia and Uganda, whereas the other participants came from the Non-Aligned Movement countries of Papua New Guinea, Surinam, and Iraq. Fifty-five (34%) of the participants were female.

Follow-up program

The increasing demand of African countries to join the program has necessitated the Government of Indonesia and YAMPI to undertake a follow-up program that can produce multiplying effects to more African countries. The idea is to organize apprenticeship programs in one or more African countries, which will be managed by the alumni. This African intercountry apprenticeship program will not only be cost-effective but also ensure high rate of post-training applicability due to the relative similarity of agroclimatic and farming conditions of participating countries. Tanzania for the eastern and The Gambia for the western part of Africa have been chosen as two outreach centers for promoting African intercountry apprenticeship training program. Indonesia will assist the two countries by

- providing a short-term technical assistance (3-4 mo) to help the alumni adopt rice cultivation techniques on their own farms using technologies and methods they learned in Indonesia and using them as a demonstration farm for educational and extension purposes;
- establishing a farmers' training center in a location where trained alumni are living. The center will be either fully financed by Indonesia or funded through a joint-financing scheme with other donor agencies/countries;

- inviting an additional group of prospective farmers living near the farmers' training center to join a special apprenticeship program in Indonesia, to prepare them to become facilitators in the farmers' training center; and
- providing long-term experts (1-2 yr) to help the center organize and run the apprenticeship programs and to assist the Ministry of Agriculture to promote national rice intensification programs using the farmers' training center as one of the agricultural human resources development centers.

This follow-up program was started 1995. It is hoped that by February 1997, the construction of the training center will have been completed in The Gambia. A group of farmers who will become facilitators of the center is now undergoing apprenticeship training in Indonesia. Two long-term experts will assist the Gambians in organizing the apprenticeship training and the national rice intensification program.

In Tanzania, physical facilities for the farmer's training center have been constructed with financial support from the Government of Japan. The Government of Tanzania has informally requested Indonesia to assist in equipping the center with educational facilities and agricultural machinery needed for running farmer training courses as well as apprenticeship programs. The request is now being considered by YAMPI and the Ministry of Agriculture. Meanwhile, a long-term technical assistant will go to Tanzania by mid-1997.

Program impact : the cases of Tanzania and The Gambia

Tanzania

Three groups comprising 20 rice farmers and 2 extension workers took the apprenticeship program in Indonesia in 1990, 1993, and 1994. The majority of them came from the district of Morogoro. For this reason, the Indonesian technical team (comprising a contact farmer, a subject matter specialist, and a rice researcher) was sent to Morogoro in early 1995. With the assistance of the technical team, 8 alumni at Mkindo Village established, from February to June 1995, a 3.2-ha demonstration farm, supported by Tanzanian researchers from the Dakawa Research Development Center. Three rice varieties were used in the demonstration farm: Dakawa Line (DL) 88 and DL 85, both local varieties, and IR64 (brought from Indonesia). Beside using high-yielding varieties, a full application of rice cultivation technologies was followed for land preparation, fertilizing, row planting, IPM, and water management. It was reported that production of the demonstration farm was 6.2 t ha^{-1} (milled rice), whereas average local production was only around 2.5 t ha^{-1}. The Ministry of Agriculture officials were impressed by the positive impact of the demonstration farm on surrounding farmers and expressed their eagerness to conduct similar demonstrations in other areas.

The technical team had also motivated the alumni to establish farmer groups and to organize agricultural cooperatives and provide training to other farmers of the nearby villages.

The Gambia

The 19 farmers and 2 extension workers who attended the apprenticeship program in Indonesia in 1990 and 1991 lived in different villages. Therefore, the technical experts who went there in January 1995 decided to select a location for the demonstration farm not based on alumni domicile but on agroclimatic suitability. The farmers who will later be involved in the demonstration farm (18 farmers and 2 officials) are now taking the apprenticeship program in Indonesia (November-December 1996).

Two demonstration farms, one in Tabanani Village representing rainfed area and the other in Waddu Village representing tidal-swamp area, were established. Each farm covered 3 ha and was managed by 12 farmers. Full rice cultivation technology was applied on both farms.

High rice yields achieved in the demonstration farms have motivated the Government of Gambia to promote a national rice intensification program starting 1996-97. Upon the request of the Minister of Agriculture, another technical assistance team from Indonesia was sent to The Gambia in early 1996 to help them plan their national rice intensification program.

A farmers' training center will be established in one of the demonstration farm villages to train farmers of West Africa. A long-term expert team from Indonesia will be dispatched to assist in establishing the center and implementing the apprenticeship training, as well as in promoting the national rice intensification program.

Lessons learned

Although no big constraints and problems were encountered, experience in organizing the apprenticeship program over the past 6 yr indicates that

1. Training needs assessment is one of the important aspects of any apprenticeship program. A thorough analysis of information regarding the characteristics and farming activities of prospective participants is the key to a successful apprenticeship program.
2. Clear and complete information regarding the conditions and requirements of the program should be given to the authorities of the invited country to avoid nomination of unqualified persons or misconception of the nature of the program.
3. Prospective host farmers and participants must be carefully selected and oriented to the objectives and nature of the program to avoid unnecessary social problems. Similar attention should be given to the recruitment of supporting field staff.
4. Organizers must conduct periodic supervision throughout the training period to solve teething problems promptly.
5. A special staff should be assigned to take care of personal problems of participants to create a pleasant climate for learning.
6. Married couples generally make better participants. They usually go along easily with all members of the host family and are eager to learn more about farming as well as household activities.

7. In most cases, nonfarming and social activities attract participants as much as those related to agriculture.

Conclusions

Indonesia's apprenticeship program for African farmers, organized in 1990, provides an educational medium for exchange of knowledge, skills, and experience among farmers of African countries and those of Indonesia on food production technologies, rice particularly.

The program provides participants with practical learning experience on both technical and sociocultural matters. This will ensure not only the transfer of technology but also the enhancement of cultural appreciation among both apprentice and host farmers. The results have been significant, as indicated by the reported successful application of farming technologies and increased rice yields in demonstration farms run by alumni. The follow-up program to establish farmers' training centers in Tanzania and The Gambia for the purpose of organizing intercountry apprenticeship training in Africa will provide greater learning opportunity for more African farmers. Indonesia will assist the governments of Tanzania and The Gambia by equipping the training center, providing a starter fund to initiate the training, and technically assisting the African trainers in the development of curriculum and methodology for the training. Additional information on the Indonesian apprenticeship training program can be obtained from materials listed in Annex 1.

Annex 1. Materials for further reading on the Indonesian apprenticeship program for African and non-aligned countries.

Agustinus P, Tjiptono, Anwarhan H (1996) Priority need assessment in agriculture of six African countries. Unpublished paper, Ministry of Agriculture, Republic of Indonesia.

Bol D, Horst L (1985) Report of evaluation mission, July/August 1985. Smallholders Irrigation Development Unit, Morogoro-Tanzania.

Food and Agriculture Organization (1989) Indonesians promote self-sufficiency--helping farmer-to-farmer. FAO, Rome, Italy.

Ministry of Agriculture (1990) Rice self-sufficiency : the Indonesian experience. Jakarta, Indonesia.

Ministry of Agriculture (1992a) Manual on apprenticeship program for farmers of developing countries in Indonesia. Jakarta, Indonesia.

Ministry of Agriculture (1992b) Apprenticeship program on food for farmers of developing countries Report of the Expert Group Meeting on Cooperation in Food Production of G-15. Jakarta, Indonesia.

Ministry of Agriculture (1996a) Guidelines on the 9th Agricultural Apprenticeship Program for African/Non-aligned Countries in Indonesia. Jakarta, Indonesia.

Ministry of Agriculture (1996b). Indonesian technical assistance in agriculture to the Republic of Gambia. Unpublished paper. Jakarta, Indonesia.

Ministry of Agriculture (1996c) Food donation of Indonesian Agricultural Society to Africa. Jakarta, Indonesia.

Ministry of Agriculture (1996d) Agricultural apprenticeship program for African/non-aligned movement farmers in the Republic of Indonesia. Jakarta, Indonesia.

Technology evaluation, adaptation, and dissemination for efficient fertilizer use in rice systems in India--a cooperative's experience

B.P. GOVIL and V. KUMAR
Indian Farmers Fertilizer Cooperative Ltd. (IFFCO), 53-54, Nehru Place, New Delhi 110019, India

Key words: fertilizer demonstration, farmers' meeting, field day, crop seminar, minikit, lab-to-land program, integrated plant nutrient system

Abstract
Although fertilizers have played a pivotal role in increasing the productivity of rice in India, the utilization efficiency of applied plant nutrients has been a matter of concern. The Indian Farmers Fertilizer Cooperative Ltd. (IFFCO), the largest fertilizer cooperative in India, is helping farmers in identifying relevant fertilizer use technology and its field application. Field demonstrations have, by and large, proved to be the most successful activity, followed by village-level farmers' meetings, field days on demonstration plots, and crop seminars before the start of the season. Distribution of minikits containing critical inputs to small and marginal farmers in adopted villages has immensely helped farmers. IFFCO's lab-to-land program adopted in selected rice growing blocks of Uttar Pradesh has been successful. IFFCO's trials on urea supergranules (USG) have yielded useful information. The USG technology is on the threshold of adoption. Likewise, soil fertilizer trials, integrated plant nutrient system (IPNS) trials, maximum yield research trials, and biofertilizer demonstrations conducted under "on-farm" conditions have convinced a large group of farmers of the importance of soil testing and applying plant nutrients on the basis of the IPNS approach.

Introduction

Rice contributes about 42% to total foodgrain production in India. Rice production has increased fourfold from about 20 million t in 1950-51 to 82 million t in 1994-95, with only a 40% increase in area during the same period. Yield increase is largely due to increase in the use of high-yielding varieties and fertilizers. India's 1879-kg-per-ha yield of milled rice, however, is far below the world average. Two important reasons for the low yield are inefficient utilization of applied plant nutrients, particularly N, and deficiency of nutrients like Zn and S. Nitrogen losses through volatilization, denitrification, leaching, and runoff reduce N use efficiency on lowland rice. Balanced application of plant nutrients through correct methods will improve yields in all rice-based cropping systems, both under irrigated and rainfed conditions. IFFCO's experience shows that there is a good scope for increasing rice yields through transfer of improved technologies at the farmers' level.

In the cooperative sector, fertilizers are promoted by KRIBHCO (Krishak Bharti Cooperative Ltd. and IFFCO is a major promoter), but experience related to rice fertilization is very limited.

IFFCO is the largest producer and distributor of urea, diammonium phosphate (DAP), and NPK complex fertilizers in India. Transfer of agricultural technology with emphasis on promotion of balanced and efficient use of fertilizers at the farmers' level is one of the main objectives of IFFCO. To achieve this objective, IFFCO conducts a variety of fertilizer demonstration and farmer education programs.

V. Balasubramanian, J. K. Ladha and G. L. Denning (eds.), Resource Management in Rice Systems: Nutrients, 281–295.

Field programs for technology dissemination

During 1985-94, IFFCO conducted 10,260 two-plot crop demonstrations, 21,396 farmer meetings, 7,915 field days, 1,494 crop seminars, 8,815 sales point personnel training, 11,804 agricultural campaigns, and 4,418 social campaigns. IFFCO also adopted 2,250 villages and distributed 154,347 minikits containing critical inputs. The message of balanced and efficient fertilizer use was successfully promoted through these programs. During the same period, IFFCO also distributed 25.27 million t of high-analysis fertilizers equivalent to 13.14 million t of N, P, and K nutrients. Currently, IFFCO produces and markets about 3.0 million t of fertilizers like urea (46% N), NPK complete (10-26-16, 12-32-16), and DAP (18-46) annum^{-1}.

Field demonstrations

Conducting crop demonstrations in farmers' fields is the most effective method of convincing the farmers of the utility of agricultural technologies to improve crop yields and fertilizer use efficiency. In field demonstrations, farmers do all operations from sowing until harvesting with their own resources, supervised by IFFCO's field staff. The farmers get convinced and gain confidence about the introduced technologies, following the principles of "learning by doing" and "seeing is believing." The types of crop demonstrations conducted include two-plot demonstrations, block demonstrations, and critical-input-package demonstrations.

Results of 896 two-plot demonstrations on rice conducted during monsoon (kharif) and winter (rabi) of 1992, 1993, and 1994 are summarized in Table 1. The averages over six seasons indicate that rice yields can be increased by about 28% over farmers' practice through the adoption of the recommended or soil test-based fertilizer application. In these trials, farmers used 72-32-11 kg ha^{-1} of N-P_2O_5-K_2O against the recommended dose of 95-48-32, indicating less use of P and K by farmers. The response was better during the monsoon than during winter, presumably because farmers used a lower fertilizer dose during the monsoon season.

Similarly, results of 38 crop demonstrations on rice conducted during by KRIBHCO during 1994-95 demonstrated that rice yields can be increased up to 40% by adopting recommended doses and methods of fertilizer application.

Reclamation of coastal saline soil in West Bengal

The Sunderban area of 24 Parganas District alone constitutes about 94% of salt-affected area in West Bengal. Eighty percent of the total rainfall of 1750 mm is received duringJune, July, and August, causing serious waterlogging and salinity. Cultivation of high-yielding varieties is practically not feasible during monsoons. The lack of water precludes rice cultivation during winter and summer. Based on the technology generated by the Canning substation of the Indian Council of Agricultural Research, IFFCO executed a project to manage these coastal saline soils on 138 farms.

Table 1. Results of two-plot fertilizer demonstrations on rice in India, 1992-95.[a]

Season	Demonstrations	Nutrients (kg ha^{-1}), FP			Nutrients (kg ha^{-1}), RP			Grain yield (kg ha^{-1})		
	(no.)	N	P_2O_5	K_2O	N	P_2O_5	K_2O	FP	RP	% increase in yield
Kharif 1992	268	54	19	9	88	47	36	3420	4520	32.2
Kharif 1993	148	73	25	8	95	46	33	4150	5160	24.3
Kharif 1994	282	77	29	5	98	49	26	3200	4398	37.5
Rabi 1992-93	51	87	42	28	108	53	53	3900	4850	24.4
Rabi 1993-94	81	84	27	16	96	47	29	4700	5250	11.7
Rabi 1994-95	60	100	50	32	111	50	37	4200	5050	20.2
Av for kharif	6698	67	24	7	94	48	31	3485	4606	32.2
Av for rabi	198	87	38	24	100	48	37	4200	4927	17.3
Av for 6 seasons	896	72	27	11	95	48	32	3643	4676	28.4

[a]Kharif=monsoon season, rabi=winter season, FP=farmers' practice, RP=recommended practice.

Soil reclamation included excavation of ponds 3 m deep on 20% of the area of each holding. These ponds collected rainwater during the monsoon, for later irrigation to winter crops. The ponds were also used for fish culture. The level of the remaining 80% of the holding was raised to a height of 0.5 m with excavated soil. A field bund 1.20 m high around the plot prevented waterlogging. The entire reclaimed area was planted to high-yielding varieties with recommended doses of fertilizers to achieve high yields (Table 2). Farmers have switched from local/traditional varieties like Soda Mota and Kala Mota to improved cultivars such as Punkaj and CMM25. With the availability of water for supplementary irrigation, wheat, mustard, potato, and green vegetables were introduced as winter crops. Pisciculture was practiced in excavated ponds and poultry and ducks provided additional income.

Lab-to-land Program in Uttar Pradesh

Normally, there is a wide communication gap and time lag between the generation of modern agricultural technologies by scientists and their adoption by farmers. IFFCO, in its Lab-to-Land Program, selected clusters of 3-5 villages in Gauriganj and Misafirkhana blocks in Sultanpur District and Singhpur Block in Raebareli District of Uttar Pradesh (UP). during 1985-86. In each village, 10-25% of the total cultivated area was covered by 0.2-ha demonstration plot in each selected farm. More than 50% of the farm families were covered in selected villages. The following measures were taken to transfer improved technologies from research stations to farmers' fields.

- Cereal - legume crop rotations (e.g., rice - wheat - green gram) were followed.
- Seeds of high-yielding varieties like Jaya and Saket 4 (rice) and HD2329 (wheat) were supplied to collaborating farmers.
- Fertilizers, macro as well as micro nutrients, were applied to the crops.
- Agrochemicals like 2,4-D, isoproturon, and dimecron were used for crop protection.
- Agricultural implements were supplied to farmer cooperators.
- Various extension methods like regular farm visits, farmers' meetings, field days, exhibitions, and soil testing were followed to educate the farmers on improved crop cultivation techniques.

Table 2. Average crop yields (kg ha^{-1}) in Coastal Saline Soil Management Project, Sunderban area, 24 Parganas District, West Bengal, India, 1984-90.

Crop	1984-85	1985-86	1986-87	1987-88	1988-89	1989-90
Rice	1,240	3,410	-	3,500	3,800	3,900
Mustard	-	1,125	-	1,200	1,400	1,450
Wheat	-	-	2,600	2,600	-	-
Green vegetables	-	6,000	6,700	6,800	9,000	10,000
Potato	-	-	27,000	-	-	-

This continuing activity is the most practical method to effectively transfer technologies to farmers. During 1991-92, 300 rice demonstrations were laid out in each block during the monsoon and 300 on wheat during winter in Tajwapur and Pallia blocks of UP. In these demonstrations, wheat yields ranged between 4 and 5 t ha^{-1} and rice yields, 5 and 6 t ha^{-1}. The selected villages became the nucleus for propagating the benefits of the improved package of practices to others nearby.

Soil test-based trials

To evaluate the comparative efficacy of soil test-based fertilizer recommendations, IFFCO conducted 206 on-farm trials on rice from 1992-93 to 1994-95: 60 in UP, 25 in Punjab, 24 in West Bengal, 21 in Andhra Pradesh; 19 in Bihar, 14 each in Gujarat and Haryana, 8 in Orissa, 7 each in Madhya Pradesh and Tamil Nadu, 5 in Himachal Pradesh, and 2 in Rajasthan (Table 3). The experiment was laid out in randomized complete block design with four replications and five treatments--control; farmer's practice; soil test-based dose of N, P, and K; recommended dose of N, P, and K; and 150% of the recommended dose.

In all states except Gujarat and Tamil Nadu, the soil test-based recommendation gave higher yield than the generally recommended dose (Table 4). In Gujarat and Tamil Nadu, the yields were marginally higher in the recommended dose treatment than in others. Soil test-based recommendation proved better in 4 of the 12 states, i.e.,. Madhya Pradesh, Punjab, Rajasthan, and West Bengal. In other states, yields were higher in the 150% of the recommended dose treatment.

The soil test-based recommendation was the most profitable (Table 5). Both benefit-cost ratio (BCR) and response were highest for this treatment except in Orissa where BCR and response were highest for the recommended dose. IFFCO conducted farmers' meetings and field days to educate farmers on the utility of soil testing. Results of the soil test-based trials will be published for wider circulation. Simple leaflets/brochures will be published in various regional languages. IFFCO has already prepared 14 films in different regional languages on balanced fertilization, with emphasis on soil testing.

Maximum yield research trials

Maximum yield research (MYR) trials seek to achieve the highest possible yields through scientific crop management. MYR considers factors such as crop rotation, sowing date, seeding rate, control of insect pests and diseases, irrigation management, and use of farm implements to achieve maximum productivity.

Table 3. Results of soil test-based trials conducted on farmers' fields in various states of India, 1992-93 to 1994-95.

State	Trials (no.)	Treatment[a]				
		A	B	C	D	E
Andhra Pradesh	21	-	73-28-14*	90-53-30	80-49-30	120-74-46
		3670	4680**	5420	4970	5560
Bihar	19	-	51-14-5	76-44-28	80-40-20	120-60-30
		1980	2660	3380	3380	3560
Gujarat	14	-	172-54-0	93-72-0	116-58-11	174-87-16
		2920	2660	4590	4650	4860
Haryana	14	-	160-52-4	130-51-18	139-62-62	209-92-92
		3000	5540	5890	5880	6330
Himachal Pradesh	5	-	59-0-0	131-40-36	108-40-40	162-60-60
		3060	3870	5120	4800	5150
Madhya Pradesh	7	-	48-27-14	101-75-35	86-57-29	129-86-43
		2290	3250	4630	4220	4350
Orissa	8	-	67-32-34	89-44-47	76-39-39	114-58-58
		3070	4270	4710	4650	5160
Punjab	25	-	152-31-0	127-15-6	125-30-29	188-45-43
		4290	5880	6340	6250	5840
Rajasthan	2	-	109-58-0	111-50-14	120-40-30	180-60-45
		1630	2040	2220	2210	2080
Tamil Nadu	7	-	105-58-0	102-38-63	125-53-53	188-79-79
		4180	2040	5380	5690	5610
Uttar Pradesh	60	-	87-18-3	112-61-32	120-60-29	180-90-44
		2740	3750	4860	4770	5220
West Bengal	24	-	108-51-45	85-42-47	88-44-44	131-66-66
		3460	5010	5060	5030	5040

[a]For each state, figures on the first line are nutrient rates ($N-P_2O_5-K_2O$ in kg ha^{-1}), figures on the second line are yields (kg ha^{-1}).
A=control, zero nutrient; B=farmers' practice; C=soil test-based recommendation; D=recommended dose; E=150% of the recommended dose.

Table 4. Benefit-cost ratio (BCR) and response in soil test-based trials conducted in farmers' fields in various states of India, 1992-93-1994-95.

State	Treatment B		Treatment C		Treatment D		Treatment E	
	BCR	Response	BCR	Response	BCR	Response	BCR	Response
Andhra Pradesh	3.4	8.6	3.9	10.2	3.1	8.1	3.0	7.9
Bihar	4.4	10.6	4.2	10.2	4.0	9.8	3.1	7.5
Gujarat	2.7	6.8	3.7	10.1	3.6	9.5	2.7	7.0
Haryana	4.9	12.0	6.1	15.1	4.7	11.0	3.6	8.5
Himachal Pradesh	6.8	14.0	4.3	10.1	4.0	9.3	3.3	7.6
Madhya Pradesh	4.7	11.8	5.0	12.3	4.5	11.3	3.3	8.1
Orissa	3.5	9.0	3.5	9.1	4.1	10.3	3.6	9.0
Punjab	3.7	8.9	6.7	14.8	4.7	10.7	2.5	5.6
Rajasthan	0.9	2.4	1.5	3.8	1.3	3.0	0.7	1.6
Tamil Nadu	2.1	4.8	2.7	6.0	2.9	6.6	1.8	4.2
Uttar Pradesh	4.5	10.0	4.3	10.7	4.0	9.7	3.2	7.9
West Bengal	3.1	7.8	3.8	9.3	3.8	9.3	2.5	6.1

IFFCO, in collaboration with the Potash and Phosphate Institute of Canada (India Program) in Gurgaon, initiated the MYR trials under on-farm conditions following rice - wheat rotation in Haryana, Punjab, and UP. Five farmers were selected in each village, and three nutrient treatments each for rice and wheat were compared. The treatment details, along with location and yield data, for Haryana, Punjab, and UP are given in Tables 5, 6, and 7.

In Haryana and UP, green manuring + 120-30-30 kg ha^{-1} of N-P_2O_5-K_2O gave the maximum rice yield. The higher dose of 180-30-30 was only marginally better than the normal recommended dose of 120-30-30. In Punjab, however, the 180-30-30 dose proved better in two of three villages, and green manuring + 120-30-30 in one village. Thus, in Punjab, there is need to increase nutrient dose to obtain maximum yields. It is possible to harvest 10 t ha^{-1} of rice plus wheat grain under conditions of Haryana and UP and 12 t ha^{-1} in Punjab. For wheat, green manuring + 180-90-45 kg ha^{-1} of N-P_2O_5-K_2O gave the highest grain yield in all three states.

Promoting nutrient application through IPNS approach

Realizing the growing imbalance in fertilizer use and the importance of incorporating organic nutrient sources to achieve sustainable productivity, IFFCO initiated the integrated plant nutrient system (IPNS) demonstration program to demonstrate and communicate relevant technologies to farmers for adoption. The Food and Agriculture Organization of the United Nations (FAO) collaborated with IFFCO by sponsoring two technical missions in 1995-96. In this regard, IFFCO is promoting

- fertilizer application based on soil test values to ensure balanced fertilization,
- green manure and leguminous crops for inclusion in cropping systems, and
- improved methods of manure and crop residue management.

Table 5. Grain yield (t ha^{-1}) of rice and wheat in maximum yield research trials, Haryana, India, 1990-92.

$N-P_2O_5-K_2O$ (kg ha^{-1})	Season[a]	Crop	Sohna, Ambala	Mohidinur,Karnal	Khundewala, Yamunanagr	Girdharpur, Kurukshetr	Pasinakhur, Panipat	Mean
120-30-30	Kharif 1990	Rice	6.1	4.9	6.1	4.9	5.9	5.6
	Kharif 1991	Rice	5.9	5.2	6.1	6.0	5.9	5.8
120-60-30	Rabi 1990/91	Wheat	3.3	4.2	3.7	4.7	4.1	4.0
	Rabi 1991/92	Wheat	3.7	4.5	4.5	4.4	4.3	4.3
GM+120-30-30	Kharif 1990	Rice	6.5	5.7	6.8	5.7	6.4	6.2
	Kharif 1991	Rice	6.4	5.9	6.7	6.5	6.5	6.4
GM+180-90-45	Rabi 1990/91	Wheat	4.4	4.8	5.4	5.2	4.8	4.9
	Rabi 1991/92	Wheat	4.1	5.0	5.1	5.0	4.8	4.8
180-30-30	Kharif 1990	Rice	5.8	5.2	5.7	5.1	5.5	5.5
	Kharif 1991	Rice	5.8	6.2	5.6	5.8	5.8	5.8
180-90-45	Rabi 1990/91	Wheat	4.0	4.4	4.7	4.9	4.2	4.4
	Rabi 1991/92	Wheat	3.9	4.7	4.7	4.8	4.5	4.5

[a]Kharif=monsoon, rabi=winter.

Table 6. Grain yield (t ha^{-1}) of rice and wheat in maximum yield research trials, Punjab, India, 1990-92.

$N-P_2O_5-K_2O$ (kg ha^{-1})	Season[a]	Crop	Dhapai/ Kapurthala	Kangjagir/ Jalandhar	Gunika/ Patiala	Mean
120-30-30	Kharif 1990	Rice	6.4	6.2	7.4	6.7
	Kharif 1991	Rice	6.7	6.2	8.1	7.0
120-60-30	Rabi 1990-91	Wheat	5.5	5.6	5.0	5.3
	Rabi 1991-92	Wheat	4.8	4.4	6.0	5.1
GM+120-30-30	Kharif 1990	Rice	5.8	6.0	8.1	6.6
	Kharif 1991	Rice	6.7	6.5	8.2	7.1
GM+180-90-45	Rabi 1990-91	Wheat	5.8	6.2	5.3	5.7
	Rabi 1991-92	Wheat	5.0	4.6	6.1	5.2
180-30-30	Kharif 1990	Rice	6.4	6.3	7.2	6.6
	Kharif 1991	Rice	6.7	6.7	8.1	7.2
180-90-45	Rabi 1990-91	Wheat	6.3	6.0	4.8	5.7
	Rabi 1991-92	Wheat	5.0	4.6	6.0	5.2

[a]Kharif=monsoon, rabi=winter.

Table 7. Grain yield (t ha^{-1}) of rice and wheat in maximum yield research trials, Uttar Pradesh, India, 1990-92.

$N-P_2O_5-K_2O$ ($kg\ ha^{-1}$)	Season[a]	Crop	Rajpurinwada/ Bareilly	Usmannagar/ Moradabad	Baghelwala/ Nainital	Mean
120-30-30	Kharif 1990	Rice	5.0	5.5	6.4	5.6
	Kharif 1991	Rice	4.3	5.4	6.4	5.4
120-60-30	Rabi 1990-91	Wheat	4.0	3.9	4.4	4.1
	Rabi 1991-92	Wheat	-	3.8	5.1	4.4
GM+120-30-30	Kharif 1990	Rice	5.2	6.0	6.7	6.0
	Kharif 1991	Rice	4.5	5.9	6.8	5.8
GM+180-90-45	Rabi 1990-91	Wheat	4.4	5.7	5.1	5.1
	Rabi 1991-92	Wheat	-	5.6	5.2	5.4
180-30-30	Kharif 1990	Rice	5.0	5.6	6.6	5.7
	Kharif 1991	Rice	4.5	5.6	6.3	5.4
180-90-45	Rabi 1990-91	Wheat	4.1	5.7	4.2	4.7
	Rabi 1991-92	Wheat	-	5.4	4.8	5.1

[a]Kharif=monsoon, rabi=winter.

The IPNS studies were initiated in the monsoon (kharif) of 1994 in eight villages countrywide. In each village, seven farmers are participating in a 3-yr demonstration program, on a 4000-m^2 plot. Half of the plot is cultivated under farmers' practices, the other half under the IPNS. Soil fertility and crop yields are monitored in both plots. The same plots will be used for 3 successive years. A sample nutrient balance based on the inputs and outputs of nutrients is established for N, P, and K. The in-out balances provide a preliminary view of the sustainability of the farming systems under study.

Kanech, a village in Ludhiana District of Punjab, was analyzed to develop nutrient balance from the study. All plots are under a rice - wheat rotation. Results for monsoon and winter for 1994-95 showed that rice and wheat are equally profitable. The BCR values for both crops are similar. The current cropping system in Kanech Village is not sustainable because soil is mined of its N and K reserves. The main concern is the decline in soil N and soil organic matter status. Small quantities of organic matter are added to the system to maintain satisfactory levels of organic matter.

The results for seven farms in Bhulai Ka Pura Village in Allahabad District of UP are presented in Table 8a, b. The IPNS practice definitely gave much higher yields than farmers' practice. There was a negative nutrient balance for N and K in all cases under both practices. Phosphorus was positive under IPNS practice in all farmers' fields; it was only marginally negative in farmers' practice plots.

The results for Singechak village in West Bengal are presented in Table 9a, b. In terms of grain yield, IPNS practice was far better than farmers' practice. Negative nutrient balance was very prominent for K followed by N. Although the negative balance of P was not common, it was more prominent during winter than during monsoon.

Table 8a. Integrated plant nutrient system on-farm trials with rice in Bhulai Ka Pura Village, Allahabad District, Uttar Pradesh, India, 1994 kharif (monsoon).

Farmer	Nutrients applied[a] ($kg\ ha^{-1}$)		Yield ($kg\ ha^{-1}$)		Nutrients removed ($kg\ ha^{-1}$)	
	FP	IPNS	FP	IPNS	FP	IPNS
R. A. Vishvakarma	92-0-0	150-75-37 +13[b]	2960	4440	129-5-63	158-7-90
R. A. Yadav	69-0-0	121-60-30 +40	4440	6350	182-17-75	223-9-127
N. K. Mishra	117-9-18	151-60-30 +40	3100	4290	127-16-79	157-10-115
K. S. Yadav	69-0-0	121-60-30 +40	4130	5990	178-15-108	212-6-109
G. Yadav	92-0-0	121-75-37 +40	3050	3600	131-8-58	164-12-102
Sushila Devi	69-0-0	150-75-37 +40	2860	4260	121-12-71	152-11-93
B. I. Pal	69-0-0	121-40-30 +40	3040	4830	104-6-55	170-13-117

[a]Nutrients=N-P_2O_5-K_2O. FP=farmers' practice. [b]Zinc sulfate ($kg\ ha^{-1}$).

Table 8b. Nutrient balance and benefit-cost ratio (BCR) of IPNS practice in farmers' fields in Bhulai Ka Pura Village, Allahabad District, Uttar Pradesh, India. 1994 kharif (monsoon).

Farmer	System[a] and nutrient balance				Cost of nutrients	Cost of produce	BCR
	System	N	P_2O_5	K_2O	($Rs\ ha^{-1}$)	($Rs\ ha^{-1}$)	
R. A. Vishvakarma	FP	-37	-5	-63	644	10,071	16
	IPNS	-8	68	-53	2253	15,093	7
R. A. Yadav	FP	-114	-17	-75	483	15,106	31
	IPNS	-102	51	-98	1812	21,576	12
N. K. Mishra	FP	-10	-7	-61	1047	10,540	10
	IPNS	-6	50	-85	2022	14,572	7
K. S. Yadav	FP	-109	-15	-108	483	14,032	29
	IPNS	-91	54	-86	1769	20,352	11
G. Yadav	FP	-39	-8	-58	644	10,353	16
	IPNS	-42	63	-65	2050	12,223	6
Sushila Devi	FP	-52	-12	-71	483	9,724	20
	IPNS	-2	64	-56	2253	14,494	6
B. L. Pal	FP	-35	-6	-55	483	10,346	21
	IPNS	-49	47	-87	1812	16,436	9

[a]FP=farmers' practice, IPNS=integrated plant nutrient system.

Table 9a. IPNS demonstrations in farmer's fields in Singhchak Village, Midnapore District, West Bengal, India, 1994 kharif (monsoon).

Farmer	Season[a]	System	Nutrients applied ($kg\ ha^{-1}$)			Yield ($kg\ ha^{-1}$)	Nutrients removed ($kg\ ha^{-1}$)		
			N	P_2O_5	K_2O		N	P_2O_5	K_2O
S. C. Samanta	Kharif	FP	126	22	22	2123	46	17	29
		IPNS	65	33	25	2830	67	25	37
	Rabi	FP	128	42	42	4856	107	40	59
		IPNS	125	63	65	5054	119	45	67
P. Jana	Kharif	FP	39	64	42	2022	45	18	30
		IPNS	75	23	25	2829	67	26	42
M. Samanta	Kharif	FP	29	27	30	1922	42	17	25
		IPNS	65	33	25	2931	73	27	39
M. C. Khatna	Kharif	FP	101	33	33	2224	48	19	33
		IPNS	75	33	25	3032	75	30	47
	Rabi	FP	118	62	62	4549	97	40	68
		IPNS	125	63	63	5055	126	50	80
T. Mahji	Kharif	FP	36	18	20	1519	31	12	19
		IPNS	43	20	20	1921	44	36	23
	Rabi	FP	78	35	35	4955	95	39	57
		IPNS	83	40	40	5155	114	95	61
M. Mandal	Kharif	FP	38	18	20	1619	37	14	23
		IPNS	43	20	20	1921	48	18	28
	Rabi	FP	82	24	40	4346	96	38	58
		IPNS	83	40	40	4943	123	45	71
S. Jana	Kharif	FP	35	20	20	1519	32	12	18
		IPNS	45	23	23	1921	41	16	22
	Rabi	FP	70	40	40	4854	98	36	54
		IPNS	85	43	43	5659	119	46	63

[a]Kharif=monsoon, rabi=winter.

Extension of IPNS

To further demonstrate the system to farmers, IPNS studies have been extended to 50 locations in the country, in two farmers' fields at each location, from the monsoon of 1995. Trials will continue for four seasons. The IPNS technology has proved beneficial in terms of sustainability and economics.

Table 9b. IPNS demonstrations in farmer's fields in Singhchak Village, Midnapore District, West Bengal, India, 1994 kharif (monsoon).

Farmer	Season[a]	System	Nutrient balance ($kg\ ha^{-1}$)			Cost of nutrients (Rs)	Cost of produce (Rs)	BCR
			N	P_2O_5	K_2O			
S. C. Samanta	Kharif	FP	80	5	-7	1,304	7,140	5
		IPNS	-2	8	-12	1,038	9,520	9
	Rabi	FP	21	2	-17	1,701	16,320	10
		IPNS	6	18	-2	2,095	17,000	8
P. Jana	Kharif	FP	-6	46	12	1,364	6,970	5
		IPNS	8	-3	-17	978	9,520	10
M. Samanta	Kharif	FP	-13	10	5	739	6,460	9
		IPNS	-8	6	-14	1,038	9,860	9
M. C. Khatna	Kharif	FP	53	14	0	1,340	7,480	6
		IPNS	0	3	-22	1,108	10,200	9
	Rabi	FP	21	22	-6	2,015	15,300	8
		IPNS	-1	13	-17	2,083	17,000	8
T. Majhi	Kharif	FP	5	6	1	609	5,100	8
		IPNS	-1	-16	-3	684	6,596	10
	Rabi	FP	-17	-4	-22	1,217	16,660	14
		IPNS	-31	-55	-21	1,347	17,340	13
M. Mandal	Kharif	FP	1	4	-3	623	5,440	9
		IPNS	-5	2	-8	684	6,528	9
	Rabi	FP	-14	-14	-18	1,132	14,620	13
		IPNS	-40	-5	-31	1,347	16,660	12
S. Jana	Kharif	FP	3	8	2	628	5,100	8
		IPNS	4	7	1	756	6,596	9
	Rabi	FP	-28	4	-14	1,256	16,320	13
		IPNS	-34	-3	-20	1,419	19,040	13

[a]Kharif=monsoon; rabi=winter.

Promotion of biofertilizer for rice among West Bengal farmers

From the 1995kharif (monsoon), IFFCO, with the cooperation of the Agricultural University in the state, has launched a special scheme to promote biofertilizers for rice among farmers of West Bengal. The University has agreed to prepare biofertilizers like phosphate solubilizing bacteria (PSB), azotobacter, and rhizobium. Culture packets were distributed free to farmers of Birbhum, Midnapore, Hooghly, and Malda districts. An extra yield of 2-11% has been reported (Table 10). PSB seems to be gaining popularity among farmers.

Table 10. Response to biofertilizers[a] in farmers' fields, West Bengal, 1995 monsoon (kharif).

Particular	Birbhum	Hooghly	Midnapore	Malda
No. of villages	4	1	3	5
Name of biofertilizer	PSB	PSB	Azotobacter	Rhizobium
No. of packets distributed	8,000	400	8,000	12,000
No. of beneficiaries	1,318	130	1,021	4,000
Crop	Rice	Rice	Rice	Black gram/ Moong
Yield in treated plot (kg ha^{-1})	4,836	5,200	4,328	1,000
Yield in untreated plot (kg ha^{-1})	4,733	4,940	4,256	900
% increase	2	5	2	11

[a]Source of biofertilizers: BCKVV (Bidhan Chandra Krishi Vishiva Vidyalaya), Kalyani, West Bengal.

Evaluation of urea supergranules vis-à-vis prilled urea

IFFCO, in collaboration with ICAR, agricultural universities, and state departments of agriculture, has been responsible for initiating work on urea supergranules (USG) in India since 1980. The USG study was conducted in rice lowland and upland conditions. The USG on-farm trials consisted of three treatments:

a) Urea (S): PU applied at 60 kg N ha^{-1} in 3 splits--1/3 each at planting, tillering, and panicle initiation;
b) USG (P): USG at 60 kg N ha^{-1} placed at a depth of 5-10 cm between 4 hills of rice within 7-10 d of transplanting; and
c) USG (B): USG at 60 kg N ha^{-1} broadcast and incorporated at puddling.

Plot size was 250 m^2. Plant spacing was 20 cm x 20 cm. Two USG granules of 1 g each were placed at the center of four hills of rice. The dose works out to 125 kg urea or 57.5 kg N ha^{-1} for 62,500 four-hill groups in 1 ha. The remaining 2.5 kg N ha^{-1} was broadcast and incorporated in the field to raise the dose to 60 kg N ha^{-1}. The entire quantity of P and K was applied before transplanting through IFFCO NPK complex (12-32-16) at the rate of 40 kg P_2O_5 and 20 kg K_2O ha^{-1} plus 15 kg N ha^{-1}. Thus, the level of N became 75 kg ha^{-1} in all three treatments.

The results are summarized in Table 11. The USG (P) treatment was more effective in increasing the grain yield of rice than USG (B) and urea (S) by 6-18%. Next to USG (P), USG (B) was also effective in increasing rice yield in all the states except in Andhra Pradesh and Maharashtra where yields were 4.6 and 5.8% lower than yields for urea (S). Thus, placement followed by broadcast application of USG gave higher grain yields than urea three-split application.

All available data on the USG experimental results on rice in India have been compiled in IFFCO's publication *Soil fertility and fertilizer use*.(IFFCO 1989). As a result of sustained efforts, USG has been included in the "fertilizer control order".

Table 11. Grain yield of rice as influenced by urea supergranules (USG) in farmers' fields in India, 1981-85 monsoon (kharif).

Year	Trials (no.)	Grain yield (kg ha^{-1})			% increase / decrease in grain yield		
		Urea (S)[a]	USG (P)[b]	USG (B)[c]	USG (P) over Urea (S)	USG (B) over Urea (S)	USG (P) over USG (B)
1981	25	3848	4539	4171	18	8	9
1982	169	4957	5799	5249	17	6	10
1983	134	4899	5339	5015	9	2	6
1984	31	4401	4927	4504	12	2	9
1985	27	5341	5654	5304	6	-2	7

[a]Urea (S)=prilled urea applied in 3-splits, 33% each at basal, mid-tillering, and panicle initiation.
[b]USG (P)=USG deep placement at 5-10 cm depth in alternate four hills, 7-10 d after transplanting.
[c]USG (B)=USG broadcast and incorporated at the time of puddling.

The product, however, has not seen the market yet because of some technical problems. Recently, upon request of the Indian Council of Agricultural Research, IFFCO supplied 20 t of urea tablets to be used in a demonstration using an Indonesian prototype hand applicator. One hundred and fifty demonstrations have been laid out in farmers' fields in Bara Gaon Village in Karnal District of Haryana State during kharif (monsoon) 1996.

Conclusions

Trials and demonstrations conducted by IFFCO in farmers' fields have clearly shown combined yield levels of 10-12 t ha^{-1} under rice - wheat system by adopting a green manure plus the normal fertilizer recommendation. Soil test-based fertilizer recommendations economized nutrient use without sacrificing yield. The MYR trials showed higher productivity with increasing nutrient application. USG deep placement showed 6-18% yield advantage over prilled urea broadcast application. The benefits of biofertilizer use appeared marginal. Productivity increase due to reclamation of saline soils was demonstrated in 24 Parganas District, West Bengal.

Farmers can obtain higher grain yields, nutrient use efficiency, and BCRs by suitably modifying their nutrient management. The general fertilizer recommendations need to be fine-tuned at the farm level, keeping in view the crops in the system and the likely residual effects of nutrients applied to a crop. The IPNS trials have proved that in most conditions, nutrient balance is negative for N and K, but positive for P. However, the BCR of IPNS was invariably lower than that of farmers' fertilizer rates. Thus, under intensive cropping systems, proper adjustment of fertilizer recommendations to maintain nutrient balance at economic cost would be desirable.

IFFCO is adopting a farmer-friendly approach for technology adaptation and dissemination. Field demonstrations have proved to be the most beneficial, followed by village-level farmer meetings, field days on demonstration sites, and crop seminars before the start of the season. Distribution of minikits containing critical inputs like fertilizers and rice seeds of high yielding varieties to small and marginal farmers and laying out small demonstration areas in their fields have immensely helped this category of farmers. IFFCO's lab-to-land program adopted in selected blocks of central UP is also promising.

Reference

IFFCO--Indian Farmers' Fertilizer Cooperative Ltd. (1989) Soil fertility and fertilizer use. Vol. 3. Urea supergranules for increasing nitrogen use efficiency. IFFCO. New Delhi. India.

Grass roots organization for promoting sustainable agriculture and food security: the experience of Grameen Bank in Bangladesh

M. HOSSAIN[1] and MD. S. ALAM[2]
[1]*International Rice Research Institute (IRRI), P.O. Box 933, Manila, Philippines*
[2]*Grameen Krishi Foundation, College Road, Alamnagar, Rangpur, Bangladesh*

Key words: microcredit, land-poor household, social development, women empowerment, poverty alleviation, Krishi (Agricultural) Foundation, food production, natural resource management

Abstract
The Grameen Bank of Bangladesh has developed a target group-oriented financial institution to provide credit to land-poor households so that they could generate self-employment in economic activities of their choice. Initially, the Bank basically financed nonfarm rural enterprises. But with increases in income, the land-poor borrowers of the bank got access to land through the tenancy market and became involved in crop production activities and contributed to growth in foodgrain production. The Grameen Bank attempted to increase food production by establishing an independent Krishi (Agriculture) Foundation for developing an efficient irrigation management system to promote modern agriculture by providing inputs in kind and management advice, in exchange for a share of the harvest. More than 95% of Grameen Bank borrowers are women. Since the Bank has already mobilized these socially disadvantaged groups in village-level associations, it will be easier to extend natural resource management and mechanical technologies to conserve natural resources, add value to rice production, and reduce the drudgery of women.

Introduction

Bangladesh has been experimenting with a rural credit program for the low-income households, through the Grameen Bank. Grameen means rural, but Grameen Bank is different from rural branches of commercial banks or specialized financial institutions. It is a target group-oriented credit institution set up to provide loans to households that own less than 0.2 ha of land. Initiated in 1976 by a professor of economics from Chittagong University (M. Yunus), the bank started as an action research project in a village to test the hypothesis that if the poor are provided access to working capital, they can generate self-employment and organize economic activities that can ultimately alleviate poverty without external assistance. After a 3-yr experimental period in the Chittagong District, the project was replicated in five districts of Bangladesh during 1979-82, with financial assistance from the State Bank of Bangladesh and the International Fund for Agricultural Development. A government ordinance in September 1983 transformed the project into a specialized credit institution owned by its borrower-members. Since then, the Bank has expanded at a rapid rate in Bangladesh, and is now being replicated in other countries, as a model of poverty alleviation through microcredit.

The success behind the Grameen Bank was the setting up of a strong village-level organization for all-round social development, in which credit is an entry point. Initially, the focus was on financing self-employment in rural nonfarm activities, in which the landless have traditionally been involved; removing illiteracy and other social ills; promoting nutrition and good sanitary practices; and encouraging smaller family size. The food security of the poor was sought by generating additional income for the landless in

V. Balasubramanian, J. K. Ladha and G. L. Denning (eds.), Resource Management in Rice Systems: Nutrients, 297–311.

the rural nonfarm sector. But it has gradually moved in the direction of increasing food production by exploiting underutilized land and water resources (including programs for utilization of fallow land, crop diversification, restoration of idle irrigation equipment and underutilized government fisheries), and supplying agricultural inputs on credit and directly participating in farming activities, by setting up an independent nongovernment organization called Grameen Krishi Foundation (GKF).

This paper describes the grass roots organization that the Grameen Bank has developed, assesses the evolution of the Grameen Bank's involvement in agricultural development activities, and outlines the efforts of the GKF in introducing new crops and promoting the adoption of new technologies and improved farming practices.

Organizational development

Premise

The basic premise behind the development of Grameen Bank is the consideration that prevailing government institutions that follow the traditional top-down approach of delivery of services are ineffective and incapable of solving poverty in Bangladesh. It is, therefore, imperative to build grass roots organizations that can nurture people's creativity, initiative, and entrepreneurship, for full utilization of their potential to alleviate poverty. Grameen Bank is an attempt toward that end. Through trial and error during the first 6 yr of its experiment, it evolved into a nonconventional institution that is able to organize the poor, particularly the women, and bring them into mainstream economic activities (Hossain 1984, 1988, 1993; Funglesang and Chandler 1987, 1993).

The Grameen Bank is better described as an organizational development for participation (Funglesang and Chandler 1993). Dr. Mohamed Yunus, founder of the bank, explains the philosophy behind the Grameen Bank concept: "The most essential element I would emphasize in any development strategy is its focus on the human being. An integral part of the focus would be particular attention to the structure of institutions and organizations and the processes keeping them operational. I would promote organizational forms which help people achieve their potential. In doing so, I would focus particularly on the poorest, the bottom 50% of the population. The upper 50% are usually the beneficiaries of the existing institutions, practices, and processes. Creation of new institutions will either take some benefits away from the upper half or they will not see any new benefit in them. I would lay emphasis on creating local self-government institutions at the village level. The smaller the local government territory is, the better the chance for the poor to participate in decision making. No giveaways, no hand-outs. I would be tough in negotiating "prices" each community or individual should pay for services they receive from the institutions." (cited in Funglesang and Chandler 1993).

Credit as a key development input

The main constraint to increasing employment and income for the poor is that they lack access to capital for organizing economic enterprises. Generation of self employment requires working capital and basic skills in reading, writing, and arithmetic. The poor are constrained in taking up such employment opportunities by lack of financial resources and necessary education and skills. The financial capital required may be small, but at very low levels of income, it is difficult to accumulate even small amounts. Although development of human capital is a long-term program, working capital can be provided to the poor through financial institutions.

But credit is rarely available to the poor at reasonable interest rates. Commercial banks fail to cater to the credit needs of the poor for three reasons. First, these banks require collateral, which the poor find difficult to provide. Second, their procedures and formalities for obtaining loans are too cumbersome for the illiterate poor. And third, commercial banks prefer handling large loans rather than the petty loans that the poor need.

The poor, however, have access to informal sources of loans: professional moneylenders known as *mohajons*, well-to-do-farmers, friends, and relatives. But these sources charge very high interest rates, usually 10% per month. The high interest rate severely limits the production activities that can be undertaken with the loans. Therefore, high-interest credit from informal sources basically finances consumption, for which the poor are forced to take very short-term loans when their physical existence is at stake.

Faced with shortage of capital, many nonfarm operators often borrow raw materials in kind from the moneylenders against the stipulation of compulsory delivery of output to the lender. Such transactions involve higher prices for raw materials and lower prices for output, compared with the market, and the implicit interest charges claim a large proportion of the potential income of the borrower.

Also, lack of capital sometimes forces the poor with necessary skills to organize nonfarm enterprises in "partnership" with the rich whose only involvement in the activity is the contribution of capital. The income obtained by the poor in such enterprises may not be higher than the imputed value of the family labor at the prevailing wage rate. Supply of capital at "reasonable" rates of interest could thus make a significant dent in the life of the rural disadvantaged groups.

An organization for appropriate credit delivery system

The Grameen Bank now follows standard operational procedures for delivery of credit, which evolved through experimentation during 1976-83 in a process of continuous interaction with the poor (Hossain 1988). A Grameen Bank branch normally covers about 15-20 villages located in a union, or at most two unions, in the vicinity of the branch. A person from a household that owns less than 0.2 ha of cultivated land, or assets with a value equivalent to less than 0.4 ha of medium-quality land, is eligible for a loan. Preference is given to a woman member of the household. Because most members of the target group are illiterate and are shy about coming to banks, the banking services are brought to their doorsteps. The size of the branch and the area covered are determined by the bank workers who travel daily to maintain contact with the target group.

Interested persons are asked to form groups of five like-minded people of similar economic standing who enjoy mutual trust and confidence. The bank has settled on a group of five through trial and error. Male and female members form separate groups. Each group elects a chair and a secretary (these positions rotate among members yearly so that all members have the learning experience that accompanies the responsibilities of these positions). The chair is responsible for discipline in the group and for supervision of loan utilization by the members. Members conduct business with a bank worker, through the chair, at weekly meetings which all members must attend.

A number of groups from the same village are federated into a center, and the weekly meetings are held at the center level. Again, groups of male and female members form separate centers. The group chairs elect a center chief and a deputy center chief, who hold office for 1 yr only. Center chiefs ensure attendance at weekly meetings, payment of loan installments, and overall discipline; and conduct the program of the meetings. The bank worker attends the center meetings, and all bank business is conducted openly in the meeting, in front of the members.

Loans are given to individual members without any collateral. A borrower may use the capital in any productive activity of his choice, but the loan has to be used immediately, and the principal repaid in 50 weekly installments, each covering 2% of the loan amount plus interest. Initially, the bank charged 16% as the annual interest rate, the same rate other commercial banks in Bangladesh charged. The rate of interest was, however, raised in 1991 to 20% to cover the high cost of operation of this small-scale credit program.

Disbursement of loans is not a simple matter. When a group is formed, it is observed closely for a month by the bank worker to see if members are conforming to the rules of the Grameen Bank. Prospective borrowers are obliged to participate in a group training program for a minimum of 7 d of continuous instruction by the bank worker. The training is intended to make the members thoroughly conversant with the rules and regulations of the bank.

The group is accorded formal recognition when all members are found to be well versed in the bank rules and procedures. Two members of the group then receive loans; their loan-repayment behavior is observed for 2-3 wk. If they pay the weekly installments regularly, the next two members become eligible for loans. The group chair is the last member to receive a loan. No repeat loan is approved for any member until the previous loans are cleared.

The group ensures mutual accountability. The individual is kept in line by considerable pressure from other members of the group. The existence of the group thus acts as the collateral for the bank loans. The credibility of the group as a whole and its future benefits in terms of new loans are in jeopardy if one member breaks the discipline and defaults on loan repayments. Groups sometimes decide to fine or to expel a member who fails to attend a meeting and willfully defaults on payments. Other members of the group also extend financial support to a member in times of genuine difficulty when the member cannot pay. A member may leave the group when the loan is fully repaid. If a member leaves earlier, the responsibility for paying the balance falls to the remaining group members. When an entire group defaults, the responsibility for repayment falls to the center.

One of the regulations of the Grameen Bank is that each member save one Taka every week, in addition to 5% of the loan amount, which is kept aside at the time of disbursement. This amount is accumulated in the account called the Group Fund, which is managed by the group on a consensus basis. When a member leaves the group, he or she is entitled to withdraw the savings accumulated in the Group Fund. Any group member can borrow from the Group Fund for any purpose, investment or consumption, at terms and conditions decided by the group. It is argued that the poor value credit from informal sources more highly than they value credit from formal financial institutions because they perceive the informal lenders as a permanent and reliable source available to finance consumption as well as production expenses. By creating the Group Fund, the Grameen Bank has ensured the same flexibility in its credit services as in the informal markets.

Credit as an entry point for social development

In 1984, the bank introduced a social development program called "Sixteen Decisions" to imbue members with discipline, unity, and hard work and to improve their living standard. These decisions include certain codes of conduct that members are encouraged to follow in their daily life, such as production of fruits and vegetables in kitchen gardens, improvement of housing and education of children, use of latrines and safe drinking water for better health, and rejection of dowry in marriages. Physical training and parades for both men and women are held at weekly center meetings, and the "Sixteen Decisions" are chanted as slogans. The bank management says that observance of these decisions, including participation in physical training and parades, is not compulsory. In the field, however, their observance has become a requirement for receiving a loan.

With the introduction of "Sixteen Decisions", the Grameen Bank has moved into an overall development program for the poor. Credit has been an entry point and now serves as a catalyst in the development process. The bank views the social development program as a key to the sustainability of the credit program. In the long run, better housing and sanitation, and increased consumption of vegetables and fruits, improve the health of the borrowers, increase their productivity and income, and hence ensure better recovery of loans. Kitchen gardens provide a regular source of income from which the borrower can partly finance the weekly loan repayments. Dowry constitutes the greatest risk of default in repaying loans, as the poor borrower is compelled to use capital to finance it. The bank is working hard to eliminate this social ill and the attendant economic burden on its members. Children's education is being promoted through active help in establishing center schools, distributing textbooks, and setting up children's savings funds, since basic skills in reading, writing, and arithmetic are required for increasing the productivity of labor.

Progress of the Grameen Bank

Within a decade since its establishment in 1983, the Grameen Bank has become a major development institution in Bangladesh. By March 1996, the bank has established 1,056 branches in almost every corner in the country, covering 35,644 villages, nearly 41% of the total villages in the country (Table 1). The bank has reached nearly 2.5 million house-

Table 1. Progress of Grameen Bank as of March 1996 (*Source:* Grameen Bank 1996a).

Indicator	
No. of bank branches	1,056
No. of villages covered	35,644
No. of households covered (000)	2,062
Amount of loans disbursed (US$ million)	1,373
Amount of loans recovered (US$ million)	1,164
Amount outstanding with borrowers (US$ million)	209
Amount saved in Group and Emergency Funds (US$ million)	110
Coverage of women	
Percent of members	94

holds. That constitutes 14% of all rural households in Bangladesh, and 28% of the households with less than 0.2 ha of land, who are the Bank's target group.

Grameen Bank has gradually turned into a bank for rural women. During the initial period of experimentation, only about 30% of the Grameen Bank borrowers were women (Hossain 1984). With experience, Grameen Bank has found women as more credit-worthy than men: they are careful in proper utilization of the loan and are regular in repayments. Also, access to capital provides women control over the household economic enterprises that are conducted with the assistance of other family members. The financial control contributes to the improvement of women's status in the household and empowers the women to participate in the household decision-making (Rahman 1986). That is the reason Grameen Bank has followed a policy of giving preference to women in extending credit, although in many cases the loan could be used by other members of the household to run family enterprises. In March 1996, about 94% of the Grameen Bank members were women who tapped 92% of the loans extended by the Bank.

The Grameen Bank has set Tk 5000 (US$125) as the maximum amount for an individual loan. But a larger amount may be provided to a group or center for joint activities or collective enterprises for acquisition of technologies that require larger investment for increasing labor productivity and reaping economics of scale. The average size of loan received by the borrowers increased from Tk 1759 in 1984 to Tk 2845 in 1992. The small size of loan and the intensive supervision inherent in the credit delivery and monitoring system led to a high cost of operation which contributed to a large proportion of bank branches operating at a loss (Hossain 1988, Khandaker et al 1995). To increase the size of the business and thereby reduce operation costs, the Grameen Bank introduced in 1992 a seasonal loan program for agricultural operations with a larger ceiling, which can be repaid after harvest of the crop. The new policy contributed to an increase in the average size of the loans in 1994 to Tk 3554 for female borrowers and Tk 4120 for male borrowers.

The Bank has been successful in recovering more than 96% of its loan within due time. The key to the excellent recovery are (a) establishment of the mutual accountability system and peer pressure through the small informal groups, (b) financing the genuine credit need of the borrower for profitable income earning opportunities as perceived by the borrower, (c) provision of loan for activities that generate regular incomes, (d) the procedure of collecting repayments in small amounts in weekly installments, and (e) development of an intensive management information system that keeps the management at different layers aware of the operations of the branch in the field.

The expansion of the Grameen Bank in terms of disbursement and recovery of loans, and accumulation of the members savings in Group Fund is shown Table 2. The cumulative amount of loans disbursed up to March 1996 reached US$1.47 billion, of which US$1.21 billion has been repaid by the borrowers. The annual disbursement of loans increased almost fourfold during 1991-93, mainly due to the introduction of the seasonal loans and the conscious effort of the bank to increase the size of business per branch to reduce its operation costs. The annual disbursement during 1995 reached US$340 million, much higher than the total amount of rural credit delivered by all financial institutions in the country. The accumulation of savings in the Group Fund reached US$108 million by the end of 1995, nearly 50% of the amount of outstanding loans with the borrowers. Thus, the Grameen Bank is able to finance a large part of its operations by utilizing the savings of its members in the Group Fund.

Involvement in crop production activities

The Grameen Bank members are drawn mostly from the landless and the marginal landowners. Early evaluation of the bank's operation showed that it was successful not only in confining its services to its target groups, but also in mobilizing the extreme poor within the target group (Hossain 1984). Only 3% of the member households owned land more than 0.2 ha in size, and 60% owned no more than the homestead land. The cultivation of land is thus marginal to the economic enterprises of the Grameen Bank borrowers. Also, the requirement that the members should repay the loans in regular weekly installments provides a built-in incentive to use the loans in activities that generates regular incomes.

During the early stage of the development of the bank, the loans were used mainly for rural processing and manufacturing, transport, and trading activities. With the expansion in the coverage of women borrowers, livestock and poultry raising (using resources within the homestead) became popular, as weekly installments could be paid from the sale proceeds of milk and eggs. Many members also invested the loan for fattening cattle, for sale during the Eid Festival when cattle are purchased by high-income households for sacrifice, and hence the market remains strong. The sluggish market for rural industrial products and the slow growth of rural incomes, however, led to a gradual decline in the proportion of loans taken for industrial and trading activities.

Table 3 shows changes in the pattern of the disbursement of Grameen Bank loans for activities grouped into broad sectoral categories. There was a dramatic increase in the share of crop production activities. During 1992 nearly a quarter of the loans were used for crop production activities; the share further increased to about 35% in 1994 after the introduction of the seasonal loans, which are taken mostly for agricultural activities. The Grameen Bank annual reports list 466 activities for which loans are taken by its members.

Table 2. Expansion of Grameen Bank (million Tk), 1983-95 (*Source:* Grameen Bank 1995).

Year	Loans disbursed	Loans recovered	Members' savings
1983	99	59	19
1984	304	199	44
1985	428	380	84
1986	542	466	137
1987	810	654	221
1988	1280	1014	352
1989	1768	1495	536
1990	2262	1989	779
1991	2640	2325	1075
1992	5204	3620	1499
1993	10622	7624	2336
1994	13912	12186	3372
1995	13664	13318	4330

Table 3. Changes in the pattern of disbursement of Grameen Bank loans, by sector, 1984-94 (% of total loans) (*Sources:* Grameen Bank, various years).

Sector	1984	1992	1994
Crop production	1.6	26.8	35.0
Livestock and fisheries	30.6	31.9	26.1
Trading and shopkeeping	32.9	21.4	22.4
Processing and manufacturing	22.2	18.8	15.5
Others	12.7	1.1	2.0
Total	100.0	100.0	100.0

The top 10 activities, however, account for nearly two-thirds of the total loans. The changes in the portfolio of these top 10 activities can be seen in Figure 1. In 1984 most of them were from the livestock, processing, and trading sectors. The list did not contain any crop production activity. But in 1994, the list included four crop production activities: rice cultivation, nonrice crop cultivation, land leasing, and tubewell irrigation. In 1994, the Grameen Bank issued US$122 million for agricultural activities, of which US$91 million was taken for those four activities. About 486,000 households borrowed money for rice cultivation, 307,000 for operation of shallow tubewells, 109,000 for leasing land, and 60,000 for purchase of fertilizers. The amount of loans for the purchase and operation of agricultural machinery is shown in Table 4.

How can the landless and the marginal landowners have the absorptive capacity for such huge investment in the crop sector which is supposed to be marginal to their economic enterprises? An in-depth impact evaluation study of long-term borrowers (Gibbons and Todd 1993, pers. commun.) suggests that Grameen Bank members can improve their access to land through purchase of small parcels and wider participation in the tenancy market. In Bangladesh, sharecropping is the predominant tenancy arrangement. Under share tenancy, the higher the yield of the crop, the higher the rent

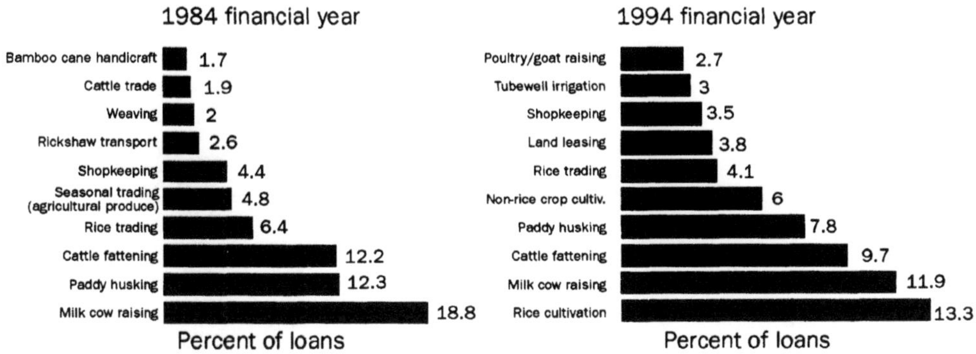

Figure 1. Changes in the portfolio of activities financed with Grameen Bank Loans, 1984 and 1994 (Source: Grameen Bank Annual Reports 1984 and 1994).

received by the landowner. To maximize the rental income, the landowner rents out land to households that have some land of their own and maintain a farm establishment, i.e., a pair of draft animals. This ensures that (a) the tenant has the necessary experience of efficiently managing the rented land, and (b) the tenant would apply labor and draft power to the maximum because of the low opportunity cost of underutilized capacity of the fixed establishment (family labor and draft animals). Thus, it is the small farmers who get preference over the landless in obtaining land from the tenancy market. The Grameen Bank has changed this pattern. Most Grameen Bank member households now own draft animals due to widespread use of the loan for livestock raising. The introduction of the modern varieties requires heavy investment in irrigation and chemical fertilizers. With improved liquidity, the Grameen Bank members are in a better position to make such investments than the small landowners. So, in the tenancy market, the land is now redistributed from small landowners to landless and marginal farmers who are Grameen Bank members. Also, many of the marginal landowners previously used to rent out their nonviable tiny holdings and work as agricultural laborers. Being Grameen Bank members, and having improved access to tenancy markets, they now have resumed self-cultivation of the holding. In this way, the Grameen Bank has made a great impact on associating the land-poor households in farming activities.

The Grameen Krishi Foundation

The Grameen Bank has also been making a conscious effort to extend modern agricultural practices to increase food production and achieve food security. This has been done by setting up an independent institution in 1991, known as the Grameen Krishi Foundation (GKF).

Table 4. Use of Grameen Bank loans for acquiring and operating of agricultural machinery and equipment, 1994. (*Source:* Grameen Bank 1995).

Machinery and equipment	No. of borrowers	Loans (000 US$)
Shallow tubewells	306,742	26,490
Deep tubewells	822	155
Power pump	1,698	36
Hand tubewells/treadle pump	9,303	446
Power tillers	2,136	364
Threshing machines	202	37
Rice mills	2,259	272

The idea behind the setting up of GKF started with Dr. Yunus's experiment in the mid-1970s of introducing a three-share tenancy system in a village near the Chittagong University in place of the traditional 50:50 crop sharing tenancy. Under the three-share farm, the output is distributed as follows: the management committee responsible for inputs of water, fertilizer, and seeds is entitled to one-third of the harvest, the farmer gets one-third of the crop for his labor and draft power, and the landowner gets one-third of the crop for his contribution of land. The basic objective is to relieve the farmer of the direct management problem particularly in adopting new innovative technologies, the burden of mobilizing resources for investment in water and fertilizer which often acts as a constraint to the adoption of modern varieties; and to distribute the risk of crop failure among the tenant, the landowner, and the management committee, so that the adverse impact of natural disasters on the poor farmer is minimized.

The opportunity of testing this innovative land tenure arrangement on a larger scale came in 1986 when the government decided to de-invest the inactive government-owned deep tubewells to the Grameen Bank. This move led to the establishment of the Grameen Agricultural Project for managing the deep tubewells and fish farms bought from the government. To de-link its involvement in agricultural production from the credit delivery function and to develop an independent nonprofit organization in agriculture, the GKF was registered under the Companies Act in December 1991. The mandate of the Foundation is to organize the landless, the assetless, and the poor women to enable them to gain access to resources for their productive self-employment, and to encourage them to undertake income-generating activities to alleviate poverty and enhance their quality of life. The Foundation operates in the northwestern region of greater Rangpur, Dinajpur Bogra Tanguil, and Jamalpur, one of the most economically depressed regions in the country, and centers around the tubewells purchased from the government.

The main objectives of the Foundation are to (Rahman 1993)

- develop an efficient and planned irrigation infrastructure around a deep tubewell;
- reorganize agricultural system by working directly with the farmers;
- help small and marginal farmers by supplying modern agricultural inputs;
- bring fallow land under cultivation and contribute to more efficient utilization of land;
- introduce nontraditional crops and promote crop diversification;

- introduce modern technology and management system in agriculture;
- produce high-quality seeds and encourage farmers to use them;
- encourage farmers to practice sustainable agriculture, particularly through the adoption of integrated pest management;
- involve the Grameen Bank members, particularly women, in crop production and encourage them to adopt improved technologies;
- develop an efficient grain marketing system to alleviate the adverse impact of the fluctuation in foodgrain prices on poor farmers; and
- ensure food security of the vulnerablė groups.

Given the existing socioeconomic structure in rural Bangladesh, the Foundation realizes that the irrigation system cannot be managed selectively only for the poor. Therefore, it accepts as a principle that the beneficiaries of its irrigation program would be all categories of farmers. The only change from the prevailing practices is that access to water will be provided more equitably, and the costs of other material inputs will be borne equitably. The Foundation mobilizes financial resources for agricultural investment and provides them to farmers in exchange for a share of the harvest. By taking over the responsibility of financing the modern agricultural inputs, and the management issues such as at what ratio different inputs have to be combined, the Foundation takes the risk of introducing a new technology out of the farmers, and eliminates the risk aversion factor that constrains the adoption of modern technology.

Irrigation-based modern agriculture is technology-centered and needs precise knowledge of the characteristics of natural resources and effective coordination among farmers for efficient operation. Practical operational questions are:

- How do we organize a large number of farmers who own land in small and scattered plots within the command area of a deep tubewell?
- How do we establish a trusting relationship among farmers so that they can participate in developing a joint crop production plan for efficient utilization of the resources?
- How do we develop a water management system that ensures equity among participating farmers and efficiency in the utilization of this scarce natural resource?
- What crops and varieties best suit the specific soil type and quality of the heterogeneous parcels of land under the command area of the tubewells?
- What should be the optimum quantity and timing of the use of seeds, fertilizers, and pesticides?
- What should be the allocation of land to different crops that would maximize farm income in view of the expected change in market conditions and relative prices?

These are momentous practical tasks requiring a variety of skills, knowledge and experience, and capacity for facilitation. These are the challenges that the Foundation has been attempting to address (Funglesang and Chandler 1993, Rahman 1993, Alam and Ali 1996).

The base of the Foundation's agricultural program is primary farm (PF). It consists physically of the land serviced by a deep tubewell, which is around 50-60 ha. The social organization to be formed for collaboration and collective decision-making within the PF is conditioned by the incidental physical layout of the landholdings around the farm. The collection of people in a PF is highly heterogeneous (unlike in the organization of the Grameen Bank Credit Program), consisting of the elements in the entire social spectrum from big landowners to landless sharecroppers. They are brought together by the incidence of land layout, but maintaining cohesion within the group is difficult, as these traditionally diverse socioeconomic groups have conflicting interests and hidden social agendas that aim to benefit one at the expense of the other. This heterogeneity makes it extremely difficult to plan and implement an efficient agricultural development program.

The PF is managed by a farm manager who is an employee of the Foundation. The manager assists the farmers in preparing a farm plan through group discussions; prepares contractual arrangements with farmers for crop production, stipulating the assistance they need from the Foundation and the terms and conditions, and implements them; and encourages farmers to cultivate nontraditional crops, by leasing land to establish a demonstration plot. The manager is also in charge of developing and managing the irrigation infrastructure.

The farm manager is supervised by a unit office covering 8-16 PFs. The unit office approves crop production plans, mobilizes financial resources, procures inputs, and assists the farm manager in maintaining the irrigation equipment. The unit office is supervised by a regional office, which, in turns, reports to the GKF head office in Rangpur.

The GKF is still going through a learning process and has yet to demonstrate the kind of success that the Grameen Bank normal credit program has achieved.

Experiments are going on regarding the appropriate sharing arrangement of crops for providing inputs to farmers on credit. Before the establishment of the Krishi Foundation in 1991, the Grameen Bank used to provide inputs only for the cultivation of boro rice, and collected one-fourth of the produce for providing irrigation and one-third of the produce when the farmer also received seeds and fertilizers on credit. The Bank soon found out that under this arrangement, farmers underreported the yield and deprived the Grameen Bank its due share. The share rental was increased in 1992 to 30 and 40% respectively for providing irrigation only and irrigation + seed + fertilizers. Inputs on credit were also extended to other crops and the sharing arrangement was fixed in consultation with the farmers. Recently the GKF moved away from share rent to fixed-rent arrangement for single and multiple inputs (Alam and Ali 1996). Thus, the burden of crop failure has now been shifted entirely on the farmer. For the cultivation of boro rice, the rental share is as follows:

Seed	:	3 kg of rough rice for 1 kg of seed
Fertilizer	:	2 kg of rough rice for 1 kg of urea 2.75 kg of rough rice for 1 kg of TSP or potash
Irrigation	:	1/15 of rough rice harvested from irrigated area per season
Seed and fertilizer	:	1/6 of rough rice for 50 kg of seed, 210 kg of urea, 128 kg of TSP and 69 kg of muriate of potash per hectare
Seed, fertilizer, and water	:	2/3 of rough rice per hectare when the farmer receives all the inputs in the specified quantity

Considering that the average yield of boro rice is only about 6 t ha^{-1}, the rental share charged by the GKF appears too high to provide adequate incentives to farmers. For tenants, who have to pay 50% of the gross produce as rent to the landowners, this leaves only 10% of the produce as the share of his labor and draft power. The relatively low earning for the tenant's labor is a major constraint to accepting the services from the GKF. Farmers also complain about the high charge of irrigation and untimely supply of inputs (Rahman et al 1996).

Since 1995 the GKF has started a credit program for small farmers with landholdings of up to 1.2 ha. Already 30,000 members have been mobilized and loans amounting to Tk 100 million have been disbursed. Eighty percent of the members are women.

The GKF has had limited success in promoting crop diversification. It attempted to introduce new crops such as soybean, hybrid maize, and hybrid potato, but had difficulty convincing farmers to adopt them on a large scale. By 1995, only 4% of the land cropped under contract with the GKF could be brought under these new crops (Table 5). Nearly 90% of the land cropped was under rice and wheat. The main constraint to crop diversification has been the limited size of the market.

A recent evaluation of the impact of the GKF (Rahman et al 1996) showed that the GKF farmers reaped only marginal benefits. Table 6 summarizes the findings on costs and returns for the two major crops, aman and boro rice, for GKF and non-GKF farmers. The irrigation cost was higher for the GKF farmers, because GKF charges full cost of irrigation to the farmer, while those operating outside benefit from the subsidized irrigation under the government-operated systems. Providing credit for financing investment in chemical fertilizer did not make any impact at all. The fertilizer application rates were very high for both groups of farmers. The only positive impact was in the adoption of integrated pest management. The GKF farmers incurred 17-22% lower costs on pesticides than the control group. The GKF group had only marginally higher yield compared with the control, mainly due to the difference in labor use. The net returns from

Table 5. Allocation of land to different crops (% of cropped area) in Grameen Krishi Foundation compared with all Bangladesh (*Sources*: Alam and Ali (1996), BBS 1996).

Crop	Grameen Krishi Foundation (1994-95)	Bangladesh (1993-94)
Rice	85.4	72.2
Aus	0.0	12.0
Aman	29.6	41.4
Boro	55.8	18.8
Wheat	5.0	4.5
Maize	2.4	0.0
Sugarcane	2.4	1.3
Mustard	0.7	2.5
Soybean	1.1	0.0
Potato	0.5	4.2
Banana	0.2	0.3
Others	2.2	15.0
Total	100.0	100.0

Table 6. Costs and returns in the cultivation of modern rice varieties, GKF-supported villages compared with non-GKF area, Bangladesh, 1995-96 (*Source:* Rahman et al 1996).

Particular	Aman (wet season) rice			Boro (dry season) rice		
	GKF	Non-GKF	Difference (%)	GKF	Non-GKF	Difference (%)
Seed (kg ha^{-1})	676	677	0	733	685	7
Irrigation (Tk ha^{-1})	213	79	170	5,459	3,993	37
Fertilizer (kg ha^{-1})	222	222	0	348	369	6
Pesticides (Tk ha^{-1})	83	100	-17	195	251	-22
Labor (d ha^{-1})	148	144	3	173	163	6
Yield (kg ha^{-1})	3,022	2,928	3	4,797	4,677	3
Gross return (Tk ha^{-1})	17,418	16,877	3	24,482	23,120	6
Total costs of production (Tk ha^{-1})	8,913	9,298	-5	16,504	16,069	3
Farm operating surplus (Tk ha^{-1})	8,505	7,579	12	7,988	7,051	13

the cultivation of the crop was, however 12% higher for the GKF farmers compared with the control, partly due to higher prices of output received for the products. This benefit may have accrued from participation in the GKF's crop marketing program, and the reduced compulsion for selling the produce immediately after the harvest when the prices remain low.

Conclusions

Food security has two dimensions: (a) producing enough food to meet the growing demand, and (b) creating adequate purchasing power through productive and remunerative employment particularly for the low-income households to have access to the food (Sen 1981). The Grameen Bank tried to achieve the second component by developing a target group-oriented financial institution to provide access to credit to the land-poor households so that they could generate self-employment in economic activities of their choice. It developed an organization at the grass roots level to make the credit delivery system effective and to ensure the efficient utilization of the credit for sustainable income growth. Initially, the Bank financed nonfarm rural enterprises. But with the increases in income, the land-poor household accessed land through the tenancy market and started crop production activities, thereby indirectly contributing to food security. In 1994, 1.35 million households received US$122 million from the bank to invest in crop production activities, 486,000 used the loan for rice cultivation, 307,000 for operation of irrigation equipment, and 109,000 for leasing land.

The Grameen Bank attempted to increase food production by establishing an independent Krishi Foundation for developing an efficient irrigation management system around the underutilized deep tubewells in the private sector. The Foundation has also been trying to promote modern agriculture by providing inputs in kind and management advice, in exchange of a share of the harvest. The basic objective is to take out of the farmer the risk involved in adopting new crops and farming methods, and input-intensive technologies. But so far, it has had limited success in this endeavor. The Foundation is in a learning process of how to make it effective.

There is a vast potential of using the grass roots organization that the Grameen Bank and Krishi Foundation have developed to extend the knowledge-intensive natural resource management technologies to farmers: direct-seeding methods, efficient fertilizer use, appropriate production and postharvest mechanization, and by product utilization.

References

Alam M S, Ali M H (1996) Food security. Grameen Krishi Foundation, Rangpur, Bangladesh.

BBS--Bangladesh Bureau of Statistics (1996) Statistical bulletin of Bangladesh, August, 1995. Bureau of Statistics, Dhaka, Bangladesh.

Funglesang A, Chandler D (1987) Participation as process--what we learn from Grameen Bangladesh, Norwegian Ministry of International Development.

Funglesang A, Chandler D (1993) Participation as process--process as growth--what we can learn from Grameen Bank, Bangladesh. Grameen Trust, Dhaka, Bangladesh

Grameen Bank (1995) Annual report 1994. Mirpur, Dhaka.

Grameen Bank (1996a) Monthly statements, April 1996. Mirpur, Dhaka.

Grameen Bank (1996b) Policies for sharing arrangements in crop cultivation. Grameen Krishi Foundation, Rangpur.

Grameen Bank (various years) Annual report. Mirpur, Dhaka.

Hossain M (1984) Credit for the rural poor: the Grameen Bank in Bangladesh. Research Monograph No. 4. Bangladesh Institute of Development Studies, Dhaka.

Hossain M (1988) Credit for alleviation of rural poverty: the Grameen Bank in Bangladesh. IFPRI Research Report No. 65. International Food Policy Research Institute, Washington, D.C.

Hossain M (1993) Credit for alleviation of rural poverty: the Grameen Bank in Bangladesh. Tropicultura 11(3):115-121.

Khandaker S R, Khalily B, Khan Z (1995) Grameen Bank: performance and sustainability. World Bank Discussion Paper No. 306. World Bank, Washington, D.C.

Rahman M S (1993) Annual report 1993 of Grameen Krishi Foundation. Grameen Krishi Foundation, Rangpur.

Rahman R I (1986) Impact of the Grameen Bank on the situation of poor rural women. Grameen Bank Evaluation Project Working Paper No. 1. Bangladesh Institute of Development Studies, Dhaka.

Rahman R I, Saha B K, Quasem M A (1996) Impact of Grameen Krishi Foundation on the socioeconomic condition of rural households. BIDS Working Paper New Series No. 17. Bangladesh Institute of Development Studies, Dhaka.

Sen A (1981) Poverty and famines: an essay on entitlement and deprivation. Clarendon Press, Oxford.

Increasing the impact of public-sector agricultural engineering research, development, and transfer

B. DOUTHWAITE

International Rice Research Institute (IRRI), P.O. Box 933, Manila, Philippines

Key words: agricultural engineering, agro-mechanical technologies, technology transfer, impact, stripper-harvester, axial-flow thresher, hydrotiller

Abstract

Public-sector agricultural engineering has a vital role in compensating for the externalities and market failures that cause the poor supply of agro-mechanical technologies to small-area rice farmers in Southeast Asia. A review of literature suggests that public-sector agricultural engineering is constrained in meeting farmers' needs because of the "top-down" and linear approach it adopts to the research, development, and transfer (RDT) of technology. This paper analyzes the RDT processes for three technologies which the public sector has been involved. For the three case studies, the RDT processes are clearly nonlinear, viz. iterative, participatory, and evolutionary. However, this is not usually reflected in the public sector planning of RDT, which is assumed linear. As a result, public-sector participation is inadequate or missing at critical intervention points. A rethink of the way the public sector carries out the RDT of agro-mechanical technology is needed, based on a good understanding of the processes of technology change in South East Asian ricefields. The paper shows that analysis and documentation of RDT process can contribute to this understanding.

The need for public-sector intervention in the development of agricultural equipment

Farmers in some regions of Asia are experiencing labor shortages and rising labor costs. This is because the productivity of agricultural workers has been growing at less than 2% a year, which is lower than that of other sectors. This means the agricultural sector has not been able to increase its workers' pay as fast as other sectors and as a result, labor has been leaving agriculture (Koppel and Hawkins 1994), (Fig. 1). Labor shortages are most acute in countries that have experienced the most rapid rates of economic growth. In Thailand, for example, wage increases have been at 15-20% over the last 5 years, fueled by a growth of 9% per year in industrial output. Thailand legalized imports of unskilled labor for the first time in 1996 (Economist Intelligence Unit 1996) because flows of unskilled labor away from the agricultural sector could no longer meet the demand.

Faced with increasing labor costs, farmers' only option is to increase their labor productivity by adopting appropriate technologies. The option of increasing the price of rice to pay for higher labor costs is closed to them because the price they can command is linked with the world price. Indeed, the world price of rice has halved in real terms between 1980 and 1994 (IRRI World Rice Statistics Database). If rural labor productivity is not increased to be in line with other sectors, the agricultural sector will stagnate and rural poverty will increase (Koppel and Hawkins 1994).

Are appropriate mechanical technologies available to rice farmers in Southeast Asia? Farmers in Europe, Australia, and North America have increased their labor productivity with mechanization, so perhaps can a transfer of "Western" farm machinery solve Asian farmers' problems? Unfortunately not, because most of the ricefields in South East Asia

V. Balasubramanian, J. K. Ladha and G. L. Denning (eds.), Resource Management in Rice Systems: Nutrients, 313–337.

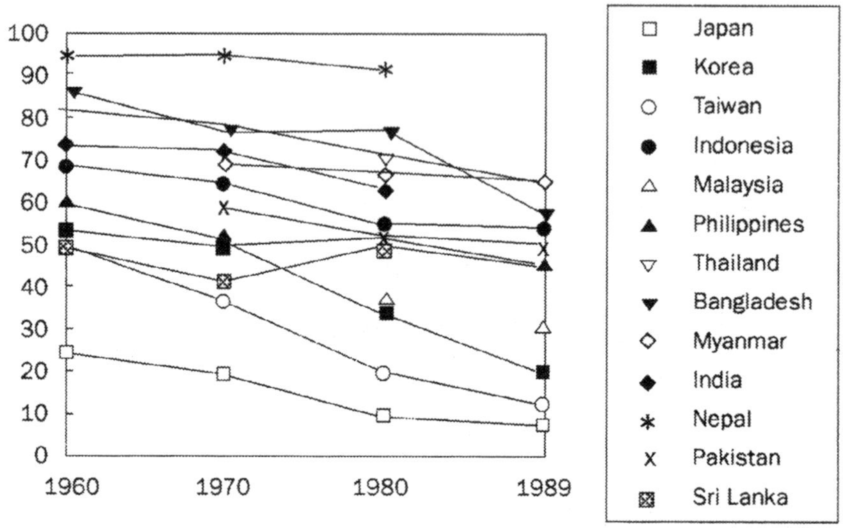

Figure 1. Percentage of labor force involved in agriculture (Koppel and Hawkins 1994).

are small (<0.5 ha) and the farmers' incomes are low relative to farmers of developed countries. As a result, equipment developed in the West is usually too big, complicated, and expensive to operate in South East Asian rice fields, and the infrastructure to support it is often lacking.

More suitably sized equipment commercially available from Japan, South Korea, or Taiwan tends to be too expensive for South East Asian farmers, unless purchased second-hand, because it was developed and priced for rice farmers who get up to six times the world price for their rice. China produces a large amount of small-field equipment, but market acceptance of Chinese equipment in some countries (e.g., Thailand) is low because of a reputation for poor quality, and because the equipment may not be adapted to local conditions.

Although some technologies have been developed where they were needed by manufacturers in South East Asia, the amount of private R&D activity tends to be low, partly because the mostly small- and medium-scale local firms cannot afford the cost of extensive R&D. In a survey of agricultural machinery manufacturers in the Philippines, Mikkelsen (1984) found that only 9% of firms had a turnover of above $500,000 per year (1980 prices) and the median turnover was between $25,000 and $50,000. Only 36% of firms reported any R&D. Gilbert (1991) states that it is difficult for the agricultural machinery industry to capture the surpluses from farmers to invest in agricultural machinery innovation because agricultural activity in developing countries is marginal. This may explain why multinational agricultural machinery companies that have resources to invest in R&D have generally ignored the small-farm sector in SE Asia.

To fill the gap, public-sector agricultural engineering research attempts to identify and adapt needed technologies and hands them over to the local private sector. This practice has stimulated local production of machinery. For example, Mikkelsen (1984) concluded that the Philippine agricultural machinery industry was stimulated by the release in 1972 of a simple hand tractor design by IRRI. Mikkelsen also found that the provision of free innovations by IRRI had "augmented private invention without reducing it, and has possibly even increased the level of invention". Unfortunately, IRRI's success has been the exception rather than the rule. A survey of the agricultural engineering components of the Consultative Group for International Agricultural Research (CGIAR) system concluded that:

> *In most developing countries the national agricultural engineering research and extension organization is weak and its link with local industry very poor.....Significant adoption of equipment designs by CGIAR engineers has been achieved only where it is supported by a programme of industrial extension.* (NIAE 1985)

Since the report was written, agricultural engineering research at the CGIAR centers has dwindled. Only IRRI has an agricultural engineering division but it has lost its industrial extension program. Industrial extension is now seen to be in the domain of national agencies, but there is little evidence that national agricultural research systems' (NARS) links with manufacturers have strengthened since the NIAE report.

Does a constraint in the supply of equipment suitable for small rice fields matter in the medium- and long-term, given that the trend is toward larger field and farm size (Cassman and Harwood 1995), which ought to make it possible for farmers to use commercially available technologies developed elsewhere?

The answer is yes for three reasons. First, the transition from labor-intensive to capital-intensive mechanized farming systems usually happens gradually. In Central Luzon in the Philippines, for example, nearly all harvesting is still being done by hand, 20 years after the widespread adoption of mechanical threshers developed by IRRI. The mechanical threshers used today in Central Luzon (capacity of 2 ha /day) can still be described as small-field equipment. These threshers and other small-farm equipment supplied by local manufacturers allow farmers to adopt mechanization in increments, causing less disruption to the existing farming system and labor markets.

The second reason is that an indigenous agricultural machinery industry is important to provide rural nonagricultural employment and reduce urban migration. Much of the indigenous-agricultural-machinery-manufacturing industry in Southeast Asia is based in the main rice-growing areas. The need for employment generation in rural areas of Asia is underscored by Edgren and Muqtada (1987):

> *Against the backdrop of existing levels of unemployment, underemployment, poverty, economic inequality, and landlessness in several countries in the Asia-Pacific region, the need to generate rural employment on a substantial scale can hardly be exaggerated.*

Third, small-field equipment will always be needed in areas where field size is constrained by topography, hydrology, land tenure patterns, and government legislation, and where flooding in the wet season precludes growing anything but rice. The future of rice farming in these less favorable areas, which is important to reduce urban drift and help provide the 55% more rice the world will need by 2020 (Cassman and Harwood 1995), depends on the availability of suitable small-field equipment.

Given the importance of small-field equipment and the sector that builds it, how can the public sector become more effective in generating and supplying new technologies to local manufacturers and farmers?

Public sector methodology for evaluating and transferring equipment technologies

One of the main reasons suggested in the development literature for the low impact of public-sector agricultural engineering research on farm mechanization is the way the research, development, and transfer (RDT) of technology are attempted. There is a consensus in the literature that public-sector agricultural research adopts a linear and hierarchical conceptualization of the technology development and transfer process (Chambers and Jiggens 1986, Biggs 1989, Clarke 1994) irrespective of the type of technology (Kaimowitz et al 1989). This conceptualization sees knowledge flowing through a pipeline that has basic research activities at one end and knowledge embodied as useful products at the other. Innovations are seen to flow sequentially down the pipeline, with different participants responsible for different parts of the process. Chambers and Jiggens (1986) called this conceptualization the Transfer-of-Technology (TOT) model, while Biggs (1989) called it the central (central source of innovation) model.

Figure 2 shows an adaptation of the model for equipment technology. The model sees CGIAR centers and other advanced research centers at the beginning of the pipeline carrying out basic and strategic research that results in new technology building blocks and concept prototypes. In the next stage, these building blocks are passed on to the research component of NARS for assembly or refinement into locally relevant technology options. The technology is then transferred to the extension component of the NARS, which extends it to manufacturers and farmers.

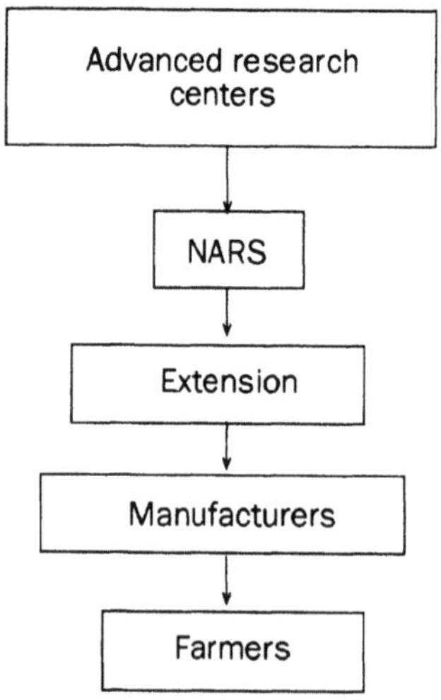

Figure 2. The transfer of technology methodology adapted for equipment technology (Chambers and Jiggins 1986).

Problems with the methodology

There is also a consensus in the literature that the central or TOT model has severe shortcomings because it sees the RDT process as linear and top-down, while in fact, innovation and technology change are iterative, complex, and evolutionary (Mokyr 1990, Clarke 1994). Chambers and Jiggens (1986) say that the model is part of the normal professionalism of agricultural science, i.e., it is the paradigm that professionals in international centers and NARS assume. Biggs (1989) criticizes the model because it sees CGIAR centers as the significant sources of innovations, while information from farmers, manufacturers, and national agencies is seen as "inputs" for generating new technologies, not as significant sources of innovation in their own right. Chambers and Jiggens (1986), Biggs (1989), and Clarke (1994) all go on to suggest that a change is needed in the way that agricultural research is conceptualized and carried out, if research institutes are to have more impact in farmers' fields. Common themes that emerge from their criticism of the central model are

- The research process must consider the nature of the technology change process and be more iterative and flexible;
- Resource-poor areas need a more flexible and iterative approach because they tend to be more complex and diverse than more favorable areas;
- The need for iteration and the two-way flow of information means that research needs should be integrated with technology diffusion over time. For agricultural machinery development, this means links are needed between "upstream" research that takes place at the beginning of the product development cycle, and the "downstream" local adaptation that occurs during technology transfer and diffusion.
- Innovation and knowledge are equally important wherever they arise.

Kaimowitz et al (1989) state that the central innovation model, although applied to all technology types, was developed for, and is most applicable to, seed-based technologies. They go on to say that there is little reason to believe that the model is applicable to other technology types. A main difference between seed and equipment is the difference in the amount of "how to" knowledge that needs to be transferred with a new technology. Growing a new variety usually requires relatively little new knowledge on the farmers' part, whereas using a new machine may require extensive training to gain any benefits. Also, unlike seed, machines do not multiply. New equipment must be manufactured, and the manufacture itself often requires a large transfer of know-how through industrial extension. The need for machine manufacture means that private-sector manufacturers are usually the conduit by which new mechanical technology reaches farmers. In contrast, new seed technology is usually channeled to farmers through public-sector seed testing and multiplication centers (R. Chaudhary, IRRI, Philippines, 1997, pers. commun.).

Most public-sector agricultural engineering units, whether they are at CGIAR Institutes or part of NARS, are usually part of a much wider organizational structure whose main business is the development and transfer of seed-based technology. It is reasonable to suppose that public-sector agricultural engineers plan their research assuming the "corporate" central model which is unlikely to be applicable to mechanical technology that reaches farmers through the private sector. There is no equivalence to industrial extension in the transfer of seed-based technologies. Perhaps this explains the almost complete lack of formal industrial extension in the public-sector agricultural engineering in Southeast Asia.

Methodology of the study

Hypotheses

This paper tested three hypotheses:

1. Successful RDT of mechanical technology occurs in an iterative and evolutionary manner.
2. In contrast, public-sector agricultural engineering RDT is planned linearly, assuming the central model of the RDT process.
3. The discrepancy between the way RDT is planned and the way it happens in practice is a constraint to the supply of needed mechanical technology in Asian ricefields.

The hypotheses were tested by analyzing three case studies of mechanical technologies developed by IRRI. The main role of IRRI's Agricultural Engineering Division in the last 30 years has been to develop and adapt machinery designs that are suitable for manufacture by local workshops, and to provide design drawings and other information free to national agencies and manufacturers.

The success of the RDT process for each technology is assessed according to whether the technology was sustainably adopted. Sustainable adoption means that the machine is manufactured and purchased without any direct or indirect subsidy for more than 3 years.

Current literature is not clear on how the RDT process itself should be analyzed to identify the factors that determine the success or failure of the process. Van der Bliek and van Veldhuizen (1993) concluded in a review of participatory technology development projects that an analysis of process documentation (the documentation of the events during the endeavor in question) is invaluable in understanding the nature of the RDT process, but this documentation is rarely carried out. Indeed, they make no mention of how the process documentation should be assessed.

This study has assumed its own approach. It uses the information available in the process documentation to describe chronologically the changes made to the technology itself (hardware) and the way it is used (software) from initial concept to sustainable adoption in farmers' fields. An attempt is made to identify and explain

- which participants made what changes;
- participants' reasons for making changes to the technology;
- the effect of these changes on the success of the RDT process;
- why the participants' became involved, the constraints they faced, and the resources they committed.

The justification for focusing on the changes made to technology comes from the experience that technologies must be adapted to local conditions -- the local society, its ecology and physical environment, its (agri-)cultural experience, and its socioeconomic structures – for them to be appropriate and successful (Haverkort 1991). Therefore, the quality of the changes that are made to a new technology determines whether it will be sustainably adopted.

The type and quality of modifications are also a measure of the participation of various actors in the RDT process because it shows the level of understanding that the participants have achieved, and the resources the participants are prepared to commit. Haverkort (1991) identified participation as crucial to successful RDT.

The three case studies were chosen on the basis that appropriate process documentation exists. More process documentation exists for the stripper gatherer harvester, so this case study received more in-depth analysis.

Case study 1. The Stripper Gatherer harvester

Stripper harvesting, which involves stripping or combing the grain from the crop while the plants remain anchored, has long attracted the attention of engineers because it is intrinsically simpler than conventional harvesting, and it reduces the volume of material handled during harvesting and threshing.

IRRI began work on stripper harvesting in 1968 and continued until 1975. No design was released, though, because of problems with high loss in wet, weedy, and lodged crop. IRRI restarted the work on stripper harvesting in 1989 after a breakthrough in stripping technology by Silsoe Research Institute (Silsoe) in England. Silsoe appeared to have solved the intrinsic problem of stripper rotors in difficult crop conditions that had killed the earlier IRRI work.

Development of the SG harvester prior to release to national agencies and manufacturers

Initially, IRRI conducted a feasibility study on whether the Silsoe stripper rotor had potentials as the basis of a small-field rice harvester. This work concluded that the Silsoe stripper rotor offered a considerable opportunity to develop a low-cost harvester. The feasibility study, completed by the end of 1990, identified two promising small-field stripper harvesters -- the stripper gatherer (SG) harvester and the stripper thresher (ST) harvester. On the basis of the feasibility study, the Overseas Development Administration (ODA) funded a one-year project to develop the SG and ST harvester concept prototypes for possible commercialization. The time frame proved too short to develop both machines, so the project focused on the SG harvester in the first year. An extension of the project allowed work on the ST harvester.

Of the two, the SG harvester was simpler. The machine consisted of the stripper rotor feeding harvested material back into a receptacle which was changed when full. The ST harvester had on-board rethreshing and straw separation. Although initially most people thought the ST harvester had more potential because it had lower labor requirement than the SG harvester, it was later found that the ST harvester was too heavy and cumbersome, and work stopped in 1995.

The ODA project provided only very limited time and funding for evaluation and transfer of the "commercially ready" SG harvester. It is doubtful whether there would have been any real adoption of the SG harvester without a German government-funded GTZ [Deutsche Gesellschaft fr Zusammenarbeit (GmbH)] project on "Postharvest Technologies for Rice in the Humid Tropics " that began 6 months after the ODA money stopped. Probably the main reason stripper harvesting was included in the GTZ project was the publicity generated by a press release IRRI put out in June 1992, which was picked up by many national Philippine newspapers and the foreign press. Foreign press stories included an article in New Scientist (Home-grown harvester eases rice growers' burden, p. 18, 22 December 1992) and a feature article in Far Eastern Agriculture (Bright future for stripper harvester, p. 29-31, July/August 1992). The awareness created by the press release may have influenced the NARS representatives to choose the SG and ST harvesters for evaluation during the GTZ-IRRI-NARS workshop planning meeting in October 1993.

Figure 3 shows the evolution of the SG harvester. The first "commercially ready" SG harvester prototype (SG800 ver. 1) was officially released in September 1993, less than 3 years after the first concept prototype was built. By this time, seven manufacturers had already started building the design by copying prototypes lent by IRRI and from drawings "in progress". The SG800 ver. 1 was the fifth prototype built, which is indicative of the large number of changes the technology went through in the early stages of development as the design team learned from field testing and feedback from farmers.

After the beginning of the GTZ project, IRRI's official role with respect to the SG harvester changed from technology developer to technical backstopper. Under the terms of the project, IRRI had transferred a "mature" technology--a technology that did not require further major development--to the national agencies. Nevertheless, IRRI continued to make its own improvements in the design and incorporate those of manufacturers and national agencies. In July 1994, IRRI issued drawings of the second version of the SG800 Mk. I. In December 1996, IRRI issued the SG800 Mk. II drawings, although most of the modifications incorporated in the Mk. II had been incorporated in manufacturers' machines for more than a year. IRRI's main concern after 1994 was the production and transfer of the technology "software"--training materials, quality check-lists, consultancy and technology commissioning trips, and information about the factors that affect adoption of the equipment.

Evolution after release in the Philippines

The role of manufacturers. Initially, three manufacturers close to IRRI began to manufacture the SG harvester on orders from IRRI and farmers. The orders for the machines were as a result of the 1992 press release. The first units were completed in April 1993. Manufacturers began building using incomplete drawings, including full-scale plots of critical components, and by copying an IRRI-built prototype. The lack of a complete set of production drawings and the fact that the design had not been finalized

Fig. 3. Evolution of the SG harvester.

(i) Influential work prior to conceptualization of SG harvester		**Chronology**
• Conceptualized to solve mobility problems of conventional combine harvesters by being lighter • Problem of high grain shatter loss never solved	<u>IRRI's first stripper harvester</u> 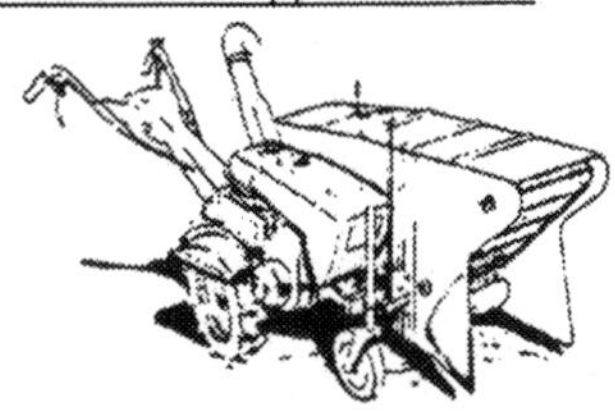	**1968** First concept prototype built (IRRI, 1968) **1973** Six-wheel, ride-on stripper combine harvester prototype built **1975** Project terminated - no commercialization 1978
<u>IRRI-CAAMS mechanical reaper</u> 	• Lightweight, walk-behind mechanical harvester • Leaves harvested crop lying in a windrow in the field • Initially 20 manufacturers make over 450 machines • Sales fall off because of down-turn in economy and reliability problems, at the same time sales of Kubota reaper increase	**1980** IRRI begins working with the Chinese Academy of Agricultural Mechanization (CAAMS) engineers to develop simpler version of Chinese reaper. **1982** MAF-IRRI begins promoting reaper to over 200 manufacturers **1982** Kubota (Japan) launch similar machine **1983** Over 400 IRRI units sold **1985** Sales of IRRI unit fall to almost zero
• Silsoe stripper rotor retro-fitted to conventional combine harvesters in place of cutterbar and reel. • First commercially-successful stripper system in Europe • Increased capacity of combines by up to 100%	<u>Silsoe stripper rotor system</u> 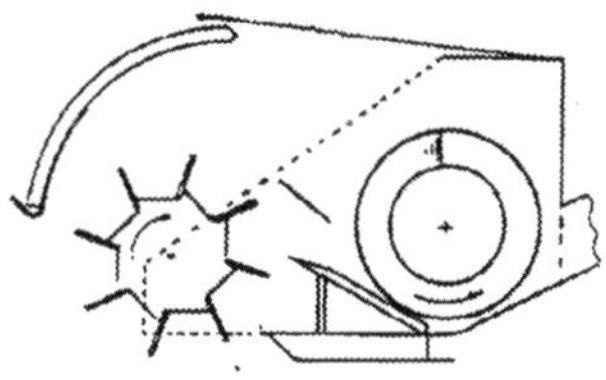	**1984** Silsoe field test first prototype **1987** SRE, a commercial company, field test 7 units **1988** SRE market 25 units
(ii) Development of the SG harvester		**Chronology**
<u>SG450 ver. 1</u> 	• Pushed by operator • Stripped material feeds into sack • 450 mm diameter stripper rotor • 450 mm wide rotor • <100 kg	**10/89** Silsoe stripper rotor transferred to IRRI **10/90** Feasibility study at IRRI on stripper rotor completed **10/90** SG450 ver. I completed
• Ground-drive fitted • Stripped material feeds into collection container • Can only harvest in anti-clockwise because drive wheel protrudes beyond the width of rotor • Machine pulls to the left • Weight 120 kg	<u>SG450 ver. 2</u> 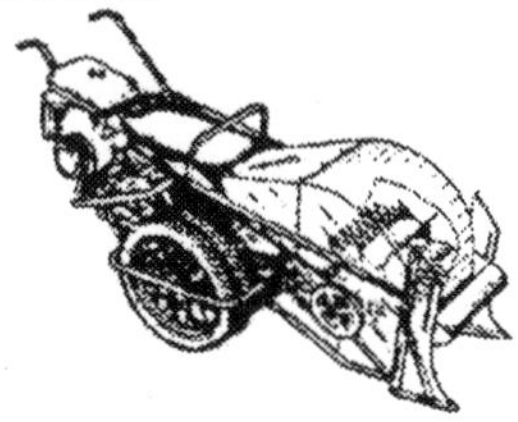	**4/91** ODA funded project begins **6/91** SG450 ver. II completed

Figure 3: Evolution of the SG harvester (cont.)

(ii) Development of the SG harvester (cont.)		**Chronology**
SG600 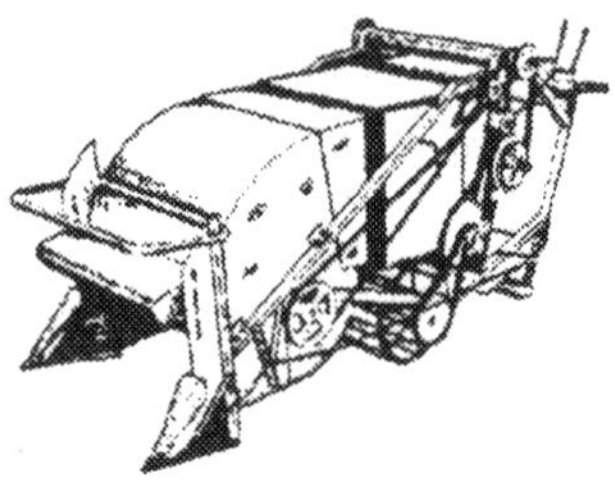	• 540 mm diameter rotor (original Silsoe dimensions) • 600 mm wide rotor • Full-width 420 mm diameter cage-wheel • Can harvest in both directions • Machine travels straight • Performance in wet conditions improved • Weight 160 kg	**9/91** SG600 completed
• 800 mm diameter rotor • Hand tractor transmission used • Star-wheel instead of cage-wheel • Reverse drive fitted, but unreliable • Weight 200 kg	SG800 ver. I 	**3/92** SG800 completed **4/92** ODA project extended for 15 months **6/92** IRRI make press release oı SG800 **2/93** Manufacturer builds first prototype **3/93** IRRI transfers SG800 to PhilRice and AED in Thailan **6/93** ODA project ends; IRRI temporarily funds project salaries
SG800 Mk. I 	• Low profile stripper hood adopted from SRE • Fixed nose position • Improved reverse drive • 540 mm diameter star-wheel • Weight 240 kg	**9/93** 7 manufacturers in production **9/93** Drawings of SG800 Mk. I officially released **11/93** Company A field test pre production prototype **1/94** GTZ project begins **5/94** Company A make 50 SG harvesters
• 600 mm diameter cage-wheel • Larger ground-drive transmission • Modular frame • Non-hinging, wedge shaped containers • Push rod controls replace cables • Weight - 290 kg c/w 240 kg	SG800 Mk. II 	**12/96** Drawings of SG800 Mk. officially released

Sources: IRRI, 1968; IRRI, 1973a; IRRI, 1975; Juarez et al., 1988

At the time of writing, Philippine manufacturers have built more than 100 SG harvesters. The bulk of these have been built by just two manufacturers, Company A and Company B.

Company A found out about the SG harvester from a press article following the 1992 press release. The company's owner and president visited IRRI to assess the technology, after one of his staff saw the SG harvester during a joint IRRI-Philippine Rice Research Institute (PhilRice) field demonstration. He immediately ordered a unit from a Los Baños manufacturer to test and copy.

Because of a communication breakdown, Company A picked up the machine from the Los Baños manufacturer before IRRI project staff could check it. After preliminary testing of the machine by Company A, the company's president was convinced that the technology had a large market potential. Company A wanted to be the first on the market and quickly built its own prototype, basically copying the machine it had bought from Los Baños. Although IRRI staff had since visited the factory and identified some of the problems with the Los Baños machine, a few mistakes were duplicated. Company A also modified the machine to be easier and cheaper to build. Some of the modifications were improvements, but others adversely affected the performance. IRRI assisted Company A in evaluating the company's prototype and subsequently made recommendations on improvements. However, the recommendations were rather late because Company A had already made jigs and fixtures and begun production of 50 units. Although improvements were made, problems remained, particularly with the quality of subcontracted components such as the ground-drive transmission. Most of the transmissions fitted on the first machines sold by Company A broke down because of a host of problems such as misaligned sprockets, substandard and undersized components, and weld slag inside the transmission boxes which subsequently entered the bearings and caused machine breakdown.

Company A, which initially predicted sales of 200 units a year, has to date, sold only about 30 SG harvesters. Altogether Company A has built 100 units in 2 batches. Most of the sales took place in early 1994 when the machine was first released on the market. However, sales quickly went down when it became clear that the machines were breaking down. Company A had to replace one customer's machine 5 times. The company offered a 6-month warranty on its equipment and free reconditioning after the first year. Company A stopped promoting the SG harvester when it realized the liability it was incurring, and that selling a poor quality product would damage the market in the medium term. Since IRRI handed over its industrial extension role to PhilRice, PhilRice has worked with Company A to improve its product. Company A is now promoting the SG harvester again, although less energetically.

A second factor that contributed to the poor sales of the first Company A SG harvesters on the market was the fact that farmers were not taught how to operate the machine properly. The assumption was that an operator experienced with a hand tractor would have no problem with the SG harvester, if shown the basics. However, although the SG harvester is a relatively simple machine, it has proved to be more operator skill-dependent than other ricefield machines in use in the Philippines. Before the SG harvester was launched, Company A had built only mature technology, which farmers already knew how to use and so they did not see the need for training. IRRI was not able to provide adequate training (although it provided some) because IRRI had neither the mandate nor the resources to do it. Since PhilRice began working on the GTZ project, it has fulfilled the necessary training function.

A third factor that contributed to the poor sales was that the SG harvester did not perform technically as well as farmers expected. Farmers wanted a machine that could harvest in soft field conditions in the wet season. Although the SG harvester could operate in flooded fields with a hardpan 10-15 cm deep, in many fields the hardpan was deeper than 15 cm. The machine also had mobility problems in muddy fields: mud stuck to the star-wheel (a single, full-width wheel made in the shape of an eight-pointed star from steel sheet).

Company B was one of the first three manufacturers to build the SG harvester. After building two units which it delivered in May 1993, it stopped producing the SG harvester for more than a year. Company B started again when it was able to hire the craftsman who had built the first SG harvesters and who had a reputation for good quality and innovativeness in improving the design. Between October 1994 and October 1996, Company B produced more than 45 SG harvesters. Unlike Company A, Company B produces machines only on order. Its craft style of production has enabled it to continually improve the design, both from the company's experience and on IRRI's recommendation. The following improvements or ideas for improvements have come from Company B:

- fitting rollers on the container rails for easy fitting and removing of the containers;
- fitting a Hydrotiller-type straw chopper to prevent straw wrapping around the rotor shaft;
- the use of push rods instead of cables to control the forward and reverse ground drives;
- design of a novel wheel to improve mobility in wet and dry fields.

The most important modification in the list is the novel wheel because it is an attempt to solve the biggest factor hindering adoption of the SG harvester - poor mobility in soft fields, common during the wet season harvest (Douthwaite et al 1996). Company B began developing the wheel after losing a sale to a customer when its machine bogged down during a demonstration in early 1996. In each case, manufacturers consulted IRRI engineers about their ideas and often, the ideas were developed together. Without IRRI's support and advice on the practicability of the ideas, it is unlikely the manufacturers would have made most of the changes.

The role of national agencies. PhilRice. IRRI's counterpart institute in the Philippines is PhilRice. IRRI donated a SG harvester to PhilRice in March 1993 and then worked with PhilRice to evaluate the technology in Bataan Province. PhilRice lent the SG system to a cooperative and helped organize demonstrations to interested farmers. PhilRice then sought the reaction and comments of onlookers and operators, who gave the following comments about the machine:

- the forward speed was too fast;

- the SG harvester produced too much dust;
- the operator should be able to ride on the machine (Vasallo et al 1994)

IRRI modified the design to consider the first two points. On the third point, PhilRice felt that modifying the machine so that the operator could ride on it would make the machine heavier, more complicated, and more expensive and thus would be a departure from the design specification for a simple and low-cost harvester with easy field-mobility.

In September 1993, PhilRice held a value analysis workshop on the SG harvester at its headquarters in Muñoz, Nueva Ejica. One recommendation was to redesign the power train on the machine to eliminate the cross-belt, and to use an imported Japanese hand-tractor transmission. The SG harvester was modified, and thus was not available for continued testing in Bataan during the 1993 wet season harvest, as was originally planned. The modified power train was not adopted in the end because it was found to be more complicated than the original, and there were doubts about the availability of the imported transmission. The modified SG harvester was not used for field testing again and the Bataan work petered out.

PhilRice's work with the SG harvester may have ended there, but for the GTZ project. PhilRice attended the planning workshop in September 1993, and was subsequently funded to evaluate the SG harvester.

Under the GTZ Project, PhilRice would verify and adapt the SG harvester to Philippine conditions, with IRRI providing technical backstopping. IRRI, however, was already working with several companies, including Company A and B, to do just this. It could not immediately withdraw from this work. Initially, PhilRice planned to carry out a major redesign of the SG harvester, based on findings from field testing in Bataan, and the Value Analysis workshop (Eng. E. U. Bautista, PhilRice, Philippines, 1994, pers. commun.). One idea was to make the SG harvester convertible into a ST harvester. IRRI felt that a major redesign was not justified and would cause problems with manufacturers who had already tooled up to produce the design that had already been released. The danger of PhilRice and IRRI developing competing stripper harvesters was averted when IRRI, PhilRice, Company A, and some farmers held a joint evaluation of the SG and ST harvesters. The farmers preferred the SG harvester, and PhilRice dropped its plan. IRRI and PhilRice then reached an agreement in which PhilRice undertook to concentrate on improving the design of the existing SG harvester, rather than design a new machine, while IRRI agreed to hand over to PhilRice the responsibility for the technology. IRRI and PhilRice agreed to release joint drawings.

The role of farmers. The focus of the design and development of the SG harvester was to evelop a machine that would be acceptable to farmers. All the SG prototypes were tested in farmers' fields. Nearly all the design changes have been based on farmers' perceptions and recommendations. However, the design has not been modified to meet all farmers' perceptions. For example, farmers often suggest that the SG harvester should rethresh and clean the grain, and then put it in sacks. IRRI has not adapted the machine in this way because it would make the machine too heavy and cumbersome to operate in small fields, which was the main design specification from the outset.

Farmers have adapted the SG harvester themselves since they first started buying the machine. A recent survey of first-adopter farmers undertaken by IRRI found that farmers had been very innovative in developing labor arrangements and harvesting patterns to increase efficiency and work rates. Farmers' success in this is shown by the field capacities they reported. The best farmer was harvesting 1.5 ha d^{-1} with 4 people, while another farmer could manage only 0.5 ha d^{-1} with 5 people. The SG harvester has a rated capacity of 0.7 ha d^{-1} with 4 people (PhilRice 1995).

Current status of adoption of the SG harvester

Philippines. About 60 SG harvesters had been purchased by Filipino farmers by October 1996. Twelve manufacturers are producing or have produced the machine. Two of the biggest and most professional manufacturers have stopped promoting the machine in the wet season harvest until the mobility problem has been solved.

The survey of first adopters found that the SG harvester increased labor productivity fourfold compared with manual harvesting. The machine allowed farmers to reduce their harvest costs by up to 50% and was profitable at their annual use rate. However, utility was constrained by mobility problems in the wet season harvest and frequent machine breakdowns (Douthwaite et al 1996).

Myanmar. The first SG harvester in Myanmar was built by the Agricultural Mechanization Department (AMD) in July 1995 from drawings given by IRRI. It was built to meet a perceived need by the Department of Agriculture (DA) to speed up harvest during the wet season in September and October. The DA had begun encouraging farmers to grow two crops, rather than one, during the wet season.

Within 4 weeks of building and testing the machine, the DA decided to build 2000 units to be ready to go to government tractor stations by September. Although it proved impossible to build this number in one month, AMD's Base Workshop was in full production by September 1995, and the first 100 units were built.

I visited AMD in September 1995 to assist with the production. It was obvious that the factory was under a lot of pressure to build machines quickly so that the machines could be of immediate benefit to farmers. To meet the production quotas left little time for quality control. The machines I checked had a number of quality problems including missing parts, unbalanced rotors, rotors fitted the wrong way around, containers that did not fit, off-centered and misaligned pulleys, and bent shafts. Also, although several sensible design changes had been made, other changes, such as replacing the skid torsion shaft with pipe, meant the machine would not work.

Some improvements in the SG harvester were made based on my recommendations. However, the workshop was unable to significantly improve quality. AMD had built 1000 stripper harvesters by January 1996.

On my visit to Myanmar in November 1996 to follow up on the use of the SG harvester in the tractor stations where the machine had been sent, I found that very few, if any, of the 1000 machines were being used for the following reasons:

- many farmers did not want to use the SG harvester because they need the straw for cattle feed;
- the DA reversed its policy on encouraging farmers to grow two crops in the wet season;
- machines were not working because of poor quality and some design defects;
- operators did not know how to use the machine due to lack of training; and
- there wasn't much feeling of "ownership" for the machines because in most cases, the machines were not requested by the tractor stations.

Other countries. Prototypes of the SG harvester have been shipped to 14 countries. Prototypes have been built by national agencies and manufacturers in Senegal, Indonesia, Thailand, Vietnam, and China, but so far commercial production has not begun.

Case study 2. Axial-flow thresher

Development prior to release to national agencies and manufacturers

IRRI began developing threshers in 1967. Initial efforts concentrated on "hold-on" types of thresher, where the operator holds the stems of the rice plants while the threshing rotor removes the grain (Kahn 1986). IRRI began developing a "throw-in" thresher in 1971, when engineers realized from the commercial failure of their hold-on type thresher design, that farmers wanted a machine with higher labor productivity. IRRI began the development of a tractor-mounted axial-flow thresher (AFT). However, this machine did not meet with much commercial success.

In 1971, IRRI began developing a smaller axial-flow thresher -- the TH3, powered by a 7-hp multipurpose stationary engine. Figure 4 shows how the design evolved, and important milestones in the process. The TH3 was released to manufacturers in June 1973. By 1983, an estimated 19000 AFTs had been produced by 75 manufacturers in the Philippines (Kahn 1986).

Evolution after release in the Philippines

The market was ready for the AFT thresher when it was released to manufacturers in 1973, because of two changes sweeping Philippine rice agriculture:

1. Farmers were rapidly adopting short-duration modern varieties, which enabled them to grow more than one crop a year. However, these varieties were harder to thresh than the traditional ones, and growing more than one crop a year increased the labor requirement for threshing, and the need to thresh quickly.
2. The mechanical threshers available to farmers at the time were mainly large machines imported from the United States owned by the large-land owners. Landlords required farmers to have their crop threshed on these machines so that the landlords could extract rent in kind. In 1973, land reform was leading to the breakup of the large farms (haciendas) and farmers were looking for a smaller mechanical thresher which could thresh in their own fields.

Fig. 4. Evolution of the axial-flow thresher.

Technology development		Chronology
• For 30 to 60 hp 4-wheel tractors • Axial-flow crop movement • Lower peripheral drum velocity than conventional threshers to reduce grain damage • Rotary cleaner adopted from failed table thresher (IRRI, 1969)	PTO-driven thresher 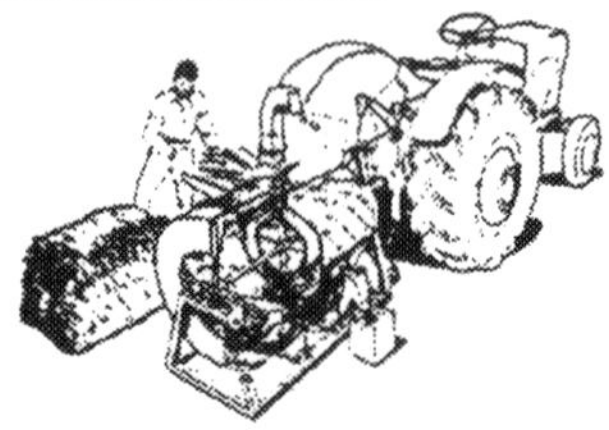	**6/69** Development of PTO-driven thresher begins **1982** Commercially producec in Pakistan
TH3 ver. 1 and 2 	• First prototype has 36 cm wire-loop drum, 127 cm long • Initial data shows drum needs to be 203 cm long • Perforated screen used for concave • Rotary cleaner adopted from PTO-driven thresher	**6/72** Development of low-cost, lightweight axial-flow thresher begins (IRRI, 1972)
• Peg teeth used instead of wire-loops • Concave made from round bar • Full-width feed platform adopted	TH3 ver. 3 	**6/73** Design released to four manufacturers **6/73** Drawings released to NAs in other countries **12/73** Manufacturers' prototypes subjected to intensive field testing by IRRI, and improved
TH3 self-mobile ver. 	• Self-mobile version of TH3 launched but meets limited acceptance with manufacturers	**1974** IRRI sends drawings to Vietnam **1974** Manufacturers adopt oscillating screen **6/75** TH3 produced by 15 manufacturers in 5 countries **1983** 19,000 threshers built by 75 manufacturers in the Philippines
• Manufacturers improve the AFT by replacing the rotary screen with an oscillating one; use open rather than closed threshing cylinder; and fit leaf springs on the wheels to make towing over long distances easier.	TH3 manufacturers' ver. 	**1978** Manufacturers adopt open threshing cylinder **1979** IRRI releases drawing of TH8 with oscillating screen and other manufacturer innovations

Technology development (cont.)		Chronology
TH6 lightweight thresher 	• Developed to give easier access to soft fields • Two models developed - TH6 and TH7. • Weight 105 kg • Fitted with winnowing fan but no cleaning screens	**7/1975** Development of lightweight thresher begins at IRRI **Early 1977** Released in the Philippines
• TH7 fitted with oscillating screen cleaner	TH7 lightweight thresher 	**1978** 3000 units built

Sources: IRRI, 1969; IRRI, 1972; IRRI 1973b; IRRI, 1975; IRRI 1977; IRRI, 1979

Initial response to the TH3 was very positive.

> *"The response from farmers and manufacturers who have observed the machine (TH3) in operation has been very encouraging. The four manufacturers who have received drawings are enthusiastic about its market potential". IRRI, 1973.*

The first orders the manufacturers received were from IRRI, for evaluation trials and demonstrations. Two of these machines were subjected to long-term durability and reliability testing; as a result, several modifications were made to improve machine durability and ease of operation. From experience in manufacturing the machine, modifications were also made to make the TH3 easier to build. These changes were incorporated in the drawings and fed back to the four manufacturers.

Two manufacturers, Kaunlaren and Oberley, established themselves as the main manufacturers by selling good quality products (Kahn 1986). Once farmers became familiar with the new product, manufacturers producing poorer quality machines lost ground.

During the period of rapid increase in production of the AFT, manufacturers came up with some modifications. The most significant were the following:

- The rotary cleaner was replaced with a single oscillating screen, which made the machine more compact and simpler to build;
- The closed threshing cylinder was replaced with an open one which was easier to manufacture, facilitated more convenient peg teeth replacement and straw removal when the thresher was clogged, and reduced blow-back in the operator's face;
- Leaf springs were fitted on the wheels to facilitate road transport.

Manufacturers and IRRI quickly adopted these innovations. In 1979, IRRI released drawings of the TH8 which included these and other changes from the original TH3 design. There is no record of where the idea for most of these changes came from, but we can be certain that a large proportion had come from manufacturers, and from the farmers using the technology.

In 1975, IRRI began the development of two types of portable thresher. The TH6 was designed without a screen grain cleaner to keep the weight down to a minimum. The TH7 had an oscillating screen cleaner but still weighed less than 200 kg and could be carried to the field by four men. This was part of an effort at the time to develop a range of lightweight equipment that could gain access to soft fields far from the road.

Development and commercialization in Vietnam

IRRI sent drawings of the AFT to Vietnam in 1974 to a collaborating manufacturer -- the Vietnam Agricultural Machinery Company (VIKYNO). The drawings were taken to Vietnam by Dr. Phan Hieu Hien, an employee of VIKYNO who was studying at the University of the Philippines in Los Baños. IRRI gave Hien a 6 -week training on the manufacture and operation of its equipment designs, including the AFT and the hand tractor. Hien took a full set of IRRI's drawings to VIKYNO. VIKYNO already had the franchise to build Kubota hand tractors in Vietnam so it decided to concentrate on the AFT. Hien and another engineer were assigned to the project (P.H. Hien, University of Agriculture and Forestry, Vietnam, 1996, pers. commun.). They built a test prototype which was field tested in four provinces in the Mekong Delta. Farmer reaction to the technology was so positive that VIKYNO decided to build a batch of 50 machines (Hien 1991). In 1975, when the government of Vietnam changed to communist, VIKYNO was instructed to stop building "small" equipment like the thresher and to build 4-wheel tractors instead. Hien left the company.

However, almost as soon as the VIKYNO machine appeared on the market, farmer-mechanics started building their own versions for their own use and for sale to their neighbors. As the farmer-mechanics became familiar with the technology, they started making modifications in the design. In 1977, an AFT appeared which used flat plate instead of peg teeth on the threshing drum. The flat plate was made of high carbon steel from car springs, which was much more durable than peg teeth. It was also discovered that the flat plates, if aligned properly, could impart axial crop movement, and these replaced the stationary guides set in the threshing drum cover that induced axial flow in the IRRI design. Gradually the diameter of the core of the threshing drum increased, and the number of teeth fell. The flat bar threshing drum replaced the peg tooth drum to the point where almost no peg tooth threshers are now made in Vietnam.

Another important innovation made by farmer-mechanics was in the cleaning system. As in the Philippines, but independently, the rotary screen on the TH3 thresher was replaced by an oscillating screen. However, in Vietnam, the screen oscillates in the same direction as the axis of the rotor, whereas in the Philippines, the screen oscillates at right angles to the axis of the rotor. The Vietnamese arrangement allows the use of an axial-flow automobile radiator fan, often mounted directly onto the engine output shaft. This is simpler, cheaper, and lighter than the centrifugal fan used in the Philippines. All these modifications were made without any public-sector R&D input.

In 1991, there were an estimated 50,000 AFTs in the Mekong Delta made by hundreds of small-scale manufacturers. Designs range from a modular design with a 0.8 tone-per-hour capacity, to a self-mobile version with a 3 ton-per-hour capacity. The modular design can be quickly knocked down into components which two people can carry into the field along narrow bunds.

Case study 3. HydroTiller

Initial development in the Philippines

The HydroTiller is an adaptation of the Turtle Tiller developed by SV Agro-Industries Enterprises, Inc. based at Jaro, Iloilo Province, in the Philippines. SV Agro-Industries began marketing the Turtle Tiller in 1976 (Villaruz 1986) after 2 years of development work. The Turtle Tiller is a floating tiller with a full-width rotor at the front end of the machine and floats behind on which the engine is mounted. In 1986, IRRI compared the performance of the Turtle Tiller and a number of other floating tillers with hand tractors and water buffalo (Calilung and Stickney 1987). In both normal and waterlogged fields, the Turtle Tiller had a higher field capacity and lower operating cost. However, compared with hand tractors, the Turtle Tiller was hard to maneuver in firmer soils; the single float created wide furrows that required subsequent harrowing; the machine had to be pushed by the operator to maintain mobility in very soft conditions; and, it could not transport itself to the field. IRRI began developing its own version of the Turtle Tiller to overcome these problems. The machine was subsequently called the HydroTiller. The HydroTiller had

- two narrow pontoons instead of a single, wide, elliptical shaped hull which is claimed to almost totally eliminate the furrow problem (Manaligod and Stickney 1991);
- cage-wheel lugs inclined forward 10 degrees to the radial, instead of 0 degrees on the Turtle Tiller, for extra lift and puddling action. This was claimed to improve flotation in very soft fields.

IRRI first released drawings of the HydroTiller in 1987. By 1990, eight firms were producing the HydroTiller in the Philippines (Manaligod and Stickney 1991) and the number has increased since then.

The HydroTiller in Vietnam

In 1991, IRRI donated a HydroTiller to Cantho University in Vietnam. Engineer Bert Manaligod commissioned the technology in 1992 and made a follow-up visit in 1993. In February 1994, Cantho University staff and a manufacturer working with Cantho University reported that

- the HydroTiller proved too heavy to handle to be acceptable to Vietnamese operators who are used to operating walk-behind equipment that, unlike the HydroTiller, has steering clutches;
- the forward speed of the machine was too high and the tillage depth too shallow;
- to work properly, the HydroTiller requires much water.

The manufacturer reported, however, that farmers liked to use the HydroTiller for clearing irrigation channels.

The HydroTiller began to find acceptance among Vietnamese farmers only in January 1995, when Cantho University, with the Center for Agricultural Technology and Extension, demonstrated the HydroTiller in waterlogged fields near Ho Chi Minh City, where most tractors were unable to work (H. Manaligod, IRRI, 1996, pers. commun.). One farmer, Ly Van Bup, was so impressed with the machine during the demonstration that he immediately bought it and began using it in his fields. Other farmers who saw Mr. Bup's machine at work began asking Cantho University for units. By May 1996, Cantho University had sold 21 units and 2 private manufacturers had begun manufacturing. Farmers overcame the difficulty of operation by paying two operators who took turns using the machine, as operators do in the Philippines.

One manufacturer, SAMECO, built a batch of 50 units with a number of modifications in the design, mainly to reduce material cost. Perhaps because of the design changes, in particular the changes in the shape of the pontoons, and the poor quality of manufacture, the machines were not salable and had to be scrapped (H. Manaligod, IRRI, 1996, pers. commun.). One reason suggested for the poor quality and design changes was that SAMECO had made its first prototype from drawings they may have had difficulty reading (V. Balasubramanian, IRRI, 1996, pers. commun.).

The HydroTiller in Indonesia

HydroTillers were sent in the early 1990s to the Centre for Development of Appropriate Technology (CDEAT), a national agency, and a German-funded project "Assistance to the Indonesian Agricultural Machinery Industry" (ATIAMI). CDEAT had the same problems as Cantho University initially experienced, and had not established commercial production of the technology. In contrast, the ATIAMI project, begun in April 1992, had established production of the equipment in 11 workshops who had sold 334 machines by the end of 1995 (K. Steinmann, ATIAMI, 1996, pers. commun.). ATIAMI's success can probably be attributed to three things. First, the project was able to reduce the size and weight of the machine to use the commonly available and cheaper 7-hp engines rather than the 9-or 10-hp engines used in machines in the Philippines. Second, ATIAMI has strong industrial extension and farmer demonstration programs which have helped maintain quality, provided the necessary operator training, and made farmers aware of the technology. Third, the project solved the problem of mobility between fields by fitting rubber wheels on the outside of the cage-wheels, and a removable caster wheel on the back.

HydroTillers in other countries

HydroTillers have been sent to 13 other countries, including China and India. However, no commercial production has been reported, even though the technology is likely to be suitable in at least some of these countries.

Discussion

The nature of the RDT process

The three case studies clearly support the hypothesis that equipment RDT takes place in an iterative and evolutionary manner, not in the linear manner prescribed in the central model. "Upstream" research institutes are not always the source of innovation, as we see with the HydroTiller. IRRI's role with the HydroTiller was to make some modifications in a manufacturer's innovation and then to attempt to disseminate the technology by supplying prototypes, drawings, and some consultancy visits.

Even if the public sector is the source of innovation, the technology does not move down a pipeline, handed on from one participant to another, as assumed in the central model. With both the AFT and the SG harvester, the design was released to manufacturers and national agencies within 2 years after the first prototype was built at IRRI. There then followed a longer evolutionary phase in which the technology was adapted by manufacturers, with assistance from IRRI, so they could use locally available materials and manufacturing techniques, and reduce costs. The use of technology was also adapted by farmers to their local society, field conditions, cultural experience, and socioeconomic structures to improve the performance of the machine.

Some of these changes were major improvements that sharply increased the marketability of the technology, and, in the case of the SG harvester, may prove to make the difference between market acceptance or rejection. Other changes detracted from the performance of the technology. With both the SG harvester and the thresher, IRRI played a "product champion" role during this phase by trying to ensure that beneficial modifications are incorporated in the design, promoting the technology, supplying information, and, where necessary, carrying out specific research and development to overcome problems as they emerged. In the case of the SG harvester, IRRI was able to follow the central model and hand over much of the role of product champion in the Philippines to PhilRice. However, problems during the hand-over period compromised the quality of industrial extension to manufacturers, and the quality of machines and training that farmers received.

Both the SG harvester and the HydroTiller case studies show how problems of poor machine quality and inadequate training can threaten the sustainable adoption of the technology if adequate industrial extension is not provided. Adequate industrial extension can be provided only by an agency that has knowledge of the technology. With a new technology, this may well be the agency that developed the equipment in the first place. This is probably why with all three technologies, the extension part of the NARS did not contribute significantly up to the point of adoption.

Company A's experience with the SG harvester suggests that companies used to manufacturing "mature" technologies may not fully realize that commercializing a new technology has very different requirements, particularly in training of operators.

Both the AFT and the SG harvester were developed at IRRI after unsuccessful attempts to introduce other designs of threshers and harvesters to manufacturers and farmers. If IRRI Agricultural Engineering Division (AED) had avoided extension work, as the central model dictates, then IRRI engineers probably would not have gained the insights and experience from which the AFT and SG harvester arose. Duff (1986) concluded:

> *"An integrated approach involving market assessment, product planning, design, testing, and extension is essential if viable products are to be developed and successfully marketed."*

The way the public sector plans RDT and its effect on impact

The development of the SG harvester was planned assuming the central model. The ODA-funded project that initially supported the development of the SG harvester was for R&D, not for transfer or extension. ODA assumed that if the project was successful, then IRRI would continue using its own resources and its own approach to technology transfer, which had proven successful with the AFT. This assumption might have been justified if IRRI still had its industrial extension program, but IRRI closed the program in 1987 when the donor stopped the funding. IRRI decided that national agencies would take over the industrial extension, as the demarcation of the central model dictates, so at the end of the ODA project, prototypes of the SG harvester were transferred to PhilRice in the Philippines and the Department of Agriculture in Thailand.

Without the GTZ project, the SG harvester would probably have failed to develop further after the ODA funding. PhilRice's first work with the SG harvester petered out and the machine was found unsuitable for Thai conditions. The GTZ project funded PhilRice, and national agencies in Indonesia, Thailand, and Vietnam, to carry out the next steps of the central model -- local evaluation and modification of the SG harvester and the initial commercialization of the technology. The GTZ project provided some funding for IRRI to transfer the technology to national agencies, but not for further R&D or to work directly with manufacturers or farmers. It was assumed that the technology was mature so IRRI no longer had a developmental role to play, and that the national agencies, not IRRI, should work with manufacturers and farmers. In practice though, there was sufficient flexibility for IRRI to continue to interact with manufacturers and farmers and play the role of product champion.

The demarcation of responsibilities that the central model prescribes almost certainly delayed the availability of a good quality machine suited to the needs of Philippine farmers. Before the ODA project started, IRRI began to work with SG harvester manufacturers and with farmers who bought the machine, although with insufficient resources because direct interaction with these groups was outside the scope of the ODA project. After the GTZ project began, responsibility for this work was passed on to PhilRice. However, it took some time for PhilRice to become familiar enough with the technology. The RDT would probably have taken place more efficiently if one team had been responsible for the whole process. This is a lesson being learned in the private sector. For example, Heat and Control Pty., a multinational provider of industrial food processing equipment, has changed its traditional, hierarchical organizational structure to a team-based system where R&D and commissioning are integrated to give better and more efficient service to their customers [Heat and Control (Australia) Pty 1996, pers. commun.].

The experience with the SG harvester in Myanmar shows that although central determination of need and supply of equipment can achieve results quickly, there can be a lack of responsiveness in the system to feedback, particularly negative feedback. One reason is that negative findings appear not to be fed back at all. The SG harvester experience shows that without feedback and flexibility in the RDT system to respond to unforeseen problems, it is possible to invest a lot of resources in developing and extending a technology that does not meet farmers' real needs.

IRRI has sent out prototypes of the HydroTiller to 14 countries but has commissioned the machine only in two countries. IRRI records show that apart from the Philippines, the HydroTiller has been adopted only in Indonesia and Vietnam where commissioning trips were made. This failure to commission a technology as a matter of course stems from the central model perspective that separates R&D from extension. The failure to see the transfer of the HydroTiller to another country as part of a continuum in which IRRI has a role to play has almost certainly limited impact.

There is evidence that not all RDT processes are planned following the central model. The University of Agriculture and Forestry (UAF), Ho Chi Minh City, Vietnam, has been successful in developing and transferring two types of dryer, because the dryer development team is responsible for all stages of the process--R&D, promotion, sales, commissioning, and manufacture (although this is subcontracted). The dryer team retains between 5 and 10% of the purchase price, half of which it reinvests in R&D. The team is accountable for its product and has a profit incentive (P. H. Hien, UAF, Vietnam, 1996, pers. commun.). UAF has adopted the integrated approach that Duff (1986) recommended.

A second example of where research, development, and technology extension are integrated is the ATIAMI project in Indonesia. The ATIAMI project works directly with agricultural-machinery manufacturers in West Sumatra. The same team is responsible for needs assessment, identification of technical solutions, evaluation, local adaptation, industrial extension, and promotion to farmers. The project has successfully increased the farmers' choice of labor-saving mechanical technologies, raised farmers' awareness of their options, improved machine quality, and increased manufacturers' sales.

Conclusions

The case studies clearly support the hypothesis that successful research, development, and transfer of mechanical technology occurs in an iterative and evolutionary manner, even if it is not planned as such. Job demarcation in IRRI and the NARS as dictated by the central model, has meant that RDT has been planned linearly, with different participants taking exclusive responsibility for different parts of the process. This has affected impact. Two exceptions to this model are the ATIAMI project and UAF which have been able to establish relative autonomy from the NARS and take responsibility for most stages in the RDT process. IRRI was also able to effectively integrate research, development, and transfer from 1976 to 1987 when its Agricultural Engineering Division had an industrial extension program. NIAE (1985) concluded that this program had enabled IRRI to have a large impact, compared with other CGIAR agricultural engineering units that did not.

The corollary is that IRRI and the NARS need to adopt a different conceptual model of the RDT process if their agricultural engineering units are to have greater impact. What should be the basis of the new model, and what work is needed to develop it?

This study has shown that successful RDT requires a high level of interaction and participation among the participants in the process. Each participant has a vital role to play based on the knowledge it possesses and the resources it commands. No single participant can develop a technology that will be sustainably adopted on its own. A new approach should, therefore, be based on the public sector identifying ways in which it can catalyze and foster a participatory, interactive, and interdependent RDT process. This will first require a better understanding of the existing degree of participation in the public-sector RDT process, and the constraints to it. This study has shown that this understanding can come from an analysis of the changes that different participants made during the RDT of a new technology, and how and why these changes were made. Further process research is needed looking at other mechanical technologies and comparing their requirements for successful RDT to other types of technology such as seed and resource management.

References

Biggs SD (1989) A multiple source model of innovation of agricultural research and technology promotion. ODI Agricultural Administration Network Paper: 6. London, England.

Cassman KG and Harwood RR (1995) The nature of agricultural systems: food security and environmental balance. Food Policy. Vol. 20, No. 5, pp.439-454.

Calilung EJ and Stickney RE (1987) Comparison of land preparation equipment for small rice farms. Agricultural Engineering Department, International Rice Research Institute, P.O. Box 933, Manila, Philippines.

Chambers R and Jiggens J (1986) Agricultural research for resource poor farmers: a parsimonious paradigm. Institute of Development Studies, University of Sussex, Brighton, UK.

Clarke N (1994) The interactive nature of knowledge systems: some implications for the Third World. Paper presented at a Workshop on the Research/Consultancy Interface, Centre for Development Studies, University of Wales, Swansea, England. 21-23 March.

Douthwaite B, Sexauer J, Lopez RP, Cedillo PB and Quick GR (1996) First adoption of a stripper harvester in the Philippines: Potential and constraints to wider take up. Paper presented at the Conference on Engineering in Agriculture and Food Processing, Gatton College, Toowoomba, Queensland. 24-27 November.

Duff B (1986) Changes in small-farm rice threshing technology in Thailand and the Philippines. IRRI Research Paper Series No. 120. P.O. Box 933, Manila 1099, Philippines.

Economist Intelligence Unit (1996) Thailand country report 1st quarter. Economist Intelligence Unit Limited. London.

Edgren G and Muqtada N (1987) Rural employment creation in Asia and Pacific: an overview, in Rural Employment Creation in Asia and the Pacific. Asian Development Bank, Manila, pp7-12.

Gilbert AL (1991) Modalities of technology transfer - a practical strategy. Paper presented at AGRIMACH '91. Manila, Philippines.

Haverkort B (1991) Farmers' experiments and participatory technology development. In Joining Farmers' Experiments. Edited by B. Haverkort, J. van der Kamp and A. Waters-Bayer. Intermediate Technology Publications, London.

Hien PH (1991) Development of axial-flow thresher in Southern Vietnam. Agricultural Mechanization in Asia, Africa and Latin America. Vol. 22 No. 4 pp 42-46.

IRRI--International Rice Research Institute (1968) Semi-annual substantive report no. 7 - January 1 to June 30, 1969. P.O. Box 933, Manila 1099, Philippines.

IRRI--International Rice Research Institute (1969) Semi-annual substantive report no. 8 - July 1 to December 31, 1968. P.O. Box 933, Manila 1099, Philippines.

IRRI--International Rice Research Institute (1972) Semi-annual progress report no. 15. - July 1 to December 31. P.O. Box 933, Manila 1099, Philippines.

IRRI--International Rice Research Institute (1973a) Semi-annual progress report no. 16. - January 1 to June 30. P.O. Box 933, Manila 1099, Philippines.

IRRI--International Rice Research Institute (1973b) Semi-annual progress report no. 17. - July 1 to December 31. P.O. Box 933, Manila 1099, Philippines.

IRRI--International Rice Research Institute (1975) Annual Report for 1974. P.O. Box 933, Manila 1099, Philippines.

IRRI--International Rice Research Institute (1977) Annual Report for 1976. P.O. Box 933, Manila 1099, Philippines.

IRRI--International Rice Research Institute (1979) Annual Report for 1978. P.O. Box 933, Manila 1099, Philippines.

Juarez F, Te A, Duff B, Crissam L, Stickney RE, Manaligod HT, Salazar GC and Fernandez CP (1988) The development and impact of mechanical reapers in the Philippines. Ag. Econ. Paper No. 88-23. International Rice Research Institute, P.O. Box 933, Manila, Philippines.

Kahn AU (1986) The Asian axial-flow thresher. In Small Farm Equipment for Developing Countries. Proceedings of the International Conference on Small Farm Equipment for Developing Countries: Past Experiences and Future Priorities. 2-6 September. IRRI, P.O. Box 933, Manila, Philippines

Kaimowitz D, Snyder M and Engel P (1989) A conceptual framework for studying the links between agricultural research and technology transfer in developing countries. Linkage theme paper no. 1. International Service for National Agricultural Research (ISNAR), The Hague, Netherlands.

Koppel B and Hawkins J (1994) Rural transformation and the future of work in rural Asia in Development or Deterioration? Work in rural Asia. Edited by B. Koppel, J. Hawkins and W. James. Lynne Rienner Publishers, Boulder, Colorado.

Manaligod HT and Stickney RE (1991) Puddling-type floating power tiller for small-scale rice farms. Agricultural Mechanization in Asia, Africa and Latin America. Vol. 22 No. 4 pp 9-12.

Mikkelsen KW (1984) Inventive activity in Philippine Industry. Ph.D. thesis. Yale University, USA.

Mokyr J (1990) The lever of riches: technological creativity and economic progress. Oxford University Press, Oxford, England.

NIAE--National Institute of Agricultural Engineering (1985) Review of the impact of the agricultural engineering component of the work on the CGIAR institutes. OD/85/2. Silsoe, England.

PhilRice (Philippine Rice Research Institute) (1995) SG800 rice stripper-harvester. Rice Technology Bulletin. No. 11. Maligaya, 3119 Munoz, Nueva Ecija, Philippines.

Van der Bliek J and van Veldhuizen L (1993). Developing tools together: Report of a study on the role of participation in the development of tools, equipment and techniques in AT programmes. No. 3 pp. 39-42. GATE, Eschborn, Germany.

Vasallo AB, Tado CJM and Douthwaite B. (1994) Field evaluation and preliminary study on the adoption of stripper-gatherer system in Bataan. Paper presented at the 10th Annual Scientific Conference of Crop Science Societies of the Philippines, Asia World Hotel, Puerto Princessa City, Palawan, Philippines. May 16-20.

Villaruz MS (1986) The floating power tiller in the Philippines. In Small Farm Equipment for Developing Countries. Proceedings of the International Conference on Small Farm Equipment for Developing Countries: Past Experiences and Future Priorities. 2-6 September. IRRI, P.O. Box 933, Manila, Philippines

Organizing experimenting farmers for participation in agricultural research and technology development

J. A. ASHBY[1], T. GRACIA[2], M. DEL PILAR GUERRERO[2], C. A. QUIROS[2], J. I. ROA[2], J. A. BELTRAN[2]

[1]*IFPRI, Washington, DC, USA*

[2]*Hillsides Agro-ecosystem Program, IPRA Project, CIAT, Cali, Colombia*

Key words: farmers' research committee (CIAL), participatory problems diagnosis, farmer-managed adaptive research, support farmers, training, technology dissemination

Abstract

Experimenting farmers are an important resource in on-farm testing, adaptation, and dissemination of new technologies, especially in resource-poor, highly heterogeneous areas . In Southern Colombia, farmers' research committees (Comites de Investigacion Agropecuaria Local or CIAL) were formed and used to mobilize local leadership among farmers to take responsibility for experimenting with technologies not known in their community. The CIALs were trained in participatory diagnosis and problem definition, planning and evaluation, and, ultimately, monitoring of adaptive technology testing and dissemination. It is possible to institutionalize this responsibility with farmers who often provide "hard-data" from farmer-managed adaptive research. Results show that current costs and coverage of CIAL approach compare favorably with some state or private sector systems, although the basis for comparison is very limited. A favorable cost structure is clearly related to the demonstrated effectiveness of paraprofessionals such as support-farmers, for scaling up and for achieving devolution.

Introduction

Farmers who experiment with new ways of farming are an important resource helping rural communities to solve their farming problems. Yet these experimenting farmers are generally unrecognized, unsupported, and disconnected from the often substantial investment in formal agricultural research. The conventional approach is like a doctor-patient relationship. The researcher and the extensionist are like doctors who are supposed to formulate a prescription for the farmer-patient's ills. When the doctor or scientist can neither diagnose problems correctly nor formulate appropriate prescriptions because the needs are many and diverse, then the approach breaks down. Developing technology suited to the particular, location-specific needs and problems of the 1.5 billion people who depend on complex, diverse, risk-prone agriculture requires a different approach (Chambers 1994).

One solution could be to establish a community-based capacity for conducting adaptive research with farmers participating. An increasing number of experiences involve organizing groups of farmers or working with existing farmer organizations to implement farmer participation (Mattee and Lasalle 1994, Muchagata et al 1994, Mushita 1993, Drinkwater 1994a, Heinrich et al 1991). The strategy of organizing groups of farmers to participate in adaptive technology testing is, in part, a response to concerns about how to reduce the costs of involving farmers in research when this makes heavy demands on the time of salaried professionals (researchers or extensionists). It also addresses the need to 'scale up' farmer participation in research and extension so that technology testing can be

V. Balasubramanian, J. K. Ladha and G. L. Denning (eds.), Resource Management in Rice Systems: Nutrients, 339–355.

conducted in numerous, diverse microenvironments without incurring excessive expenses and compromising the quality of participation (Ashby and Sperling 1994, Bebbington et al 1994, Okali et al 1994).

Many questions have been raised about the viability of institutionalizing an adaptive research role for farmers and the constraints such efforts are likely to face (Bebbington et al 1994). Critics argue that farmers' traditional or folk experimentation is a form of knowledge generation superior to western science. The strength of folk experimentation is in making contingent, sequential adjustments over time to changing circumstances that are unpredictable (Richards 1989, Drinkwater 1994b). This indigenous form of knowledge generation does not readily fit within models of controlled experimentation used by western science.

A more useful analysis draws on understanding the nature of folk experimentation. For example, farmers compare "treatments," but the check or control may be "in the farmer's head" because farmers compare this year's performance with last year's. Another comparison used by farmers contrasts results in a distant field with those in a nearby one; or the results of adding a little bit more fertilizer to one furrow compared with results in the remainder of the field. Folk experimentation involves mainly replication over time, in contrast to replication in space and time used in scientific method. Moreover, farmers recognize confounding effects in folk experimentation. For example, a small amount of seed of a new variety is typically nurtured and multiplied in the more fertile home garden; then the next planting moves the new variety to testing in different types of soil, testing the genotype x environment interaction. Only when performance is assessed in a variety of environments are conclusions drawn about the likely performance of new germplasm in the farmers' environment.

Experience shows that farmers' knowledge generation can draw on both the scientific method of controlled comparison and folk experimentation; it is not an either-or dichotomy (Hardon and de Boef 1993, Berg 1993, Lightfoot 1987).

Another issue is whether creating a special group builds on existing authority structures or creates a parallel nontraditional structure and the extent to which such groups can represent the research agendas of different interest groups within the community or may indeed exclude particular groups (Bebbington et al 1994). Experience with on-farm research showed that when formal criteria for selecting farmers to participate in research were not used, the resultant participants were usually more wealthy and politically active farmers (Merrill Sands et al 1991).

Experimenters, or innovators who can afford to experiment, are likely to be the relatively better-off farmers, who have the skills and resources (including power) to devote to a particular kind of knowledge generation. Past experience with research-minded farmers suggests that it is desirable to purposively select innovators who have the time and interest for experimentation (Abedin and Chowdry 1989, as cited in Merrill-Sands et al 1991; Ashby et al 1987).

The real issue is whether the better off experimenting farmers can be harnessed to achieve a research agenda, defined at the community level, which is also useful to the very poor or to other interest groups, such as women, who may have different priorities.

Could this local capacity, if linked effectively to research agencies, share the costs and expand the coverage of locally relevant adaptive research? Can a self-sustaining capacity and responsibility for promoting farmer participation be created in rural communities? How can linkages among these different actors be managed without increasing the transaction costs to an unwieldy extent? Little systematic work has been done on the costs of creating organizations at the community level to fill this function, and there has not been much empirical assessment of the extent to which such organizations can increase coverage and improve targeting of adaptive research in a self-sustaining way (Axinn 1994).

This paper reports on an effort to provide empirical data on some of these issues from action-research carried out in 1990-94 by the IPRA project of the International Center for Tropical Agriculture with support from the W.K. Kellogg Foundation: devolution of adaptive research responsibilities to committees of experimenting farmers, the effects of scaling up this approach to achieve broad coverage, and its cost.

Methodology

The project's strategy is to implement participatory research methods for adaptive technology testing by forming committees of farmers based in rural communities to carry out technology testing together with public sector agricultural research and extension agencies and intermediate organizations (nongoverrnment organizations [NGOs] and farmer cooperatives). Development of training courses and materials for this purpose is integral to the strategy.

The purpose of the farmers' research committees (Comites de Investigacion Agropecuaria Local or CIAL) is to mobilize local leadership among farmers to take responsibility for experimenting with technologies not known in their community. In this way, the project aims to create "demand-pull" by clients on agricultural research and extension, diversifying the type of technologies available and increasing the number and rate of flow of technologies to resource-poor farmers, thereby improving adoption, farm incomes, and welfare. New technology selected with farmer participation is better adapted locally than that recommended by researchers working on their own (Sperling et al 1993, Worede and Mekbib 1993).

The project was initiated in a pilot area in Cauca Department, southern Colombia. Cauca is one of the poorest departments in the country. The pilot area is characterized by hilly terrain, poor infrastructure , and small farms averaging 5 ha in size (average cultivated area is less than 3 ha). All farms engage in a mix of commercial and subsistence production. The area is a marginal coffee production area, with infertile acid soils, often badly eroded. Most farmers cultivate coffee, together with cassava, as a cash crop; some maize and climbing beans are grown traditionally for subsistence. Livestock are scarce (only 13% of farms have cattle).

The project began with the formation of CIALs in five communities (veredas) in 1990; the number increased to 18 in late 1991, to 32 in 1992-93, and to 55 in 1994. A further 30 CIALs, formed in Bolivia, Ecuador, Peru, and Honduras by international trainees in the method, brought the total to 85. This paper reports information obtained from monitoring the 48 CIALs formed between 1990 and mid-1994 in the pilot area in Colombia. These CIALs cover an area of approximately 1,605 km^2, involving an estimated 50,000 families and direct contact with more than 4,000 farmers, of whom 220 participated in training as members of the CIAL.

Each CIAL is formed with four farmers elected at a community meeting. The CIAL meets regularly during the first training cycle (or experimental period, usually equivalent to a cropping season of about 6 mo). The first training cycle involves up to 10 training visits by a support-farmer who has had at least 1 yr of prior experience in CIAL. Over the next cycle or cropping season, the number of visits is progressively reduced, as the CIAL gains experience and conduct experiments with increasing autonomy (Table 1).

The support farmer is backed up by an agronomist who provides input to the statistical design of CIAL experiments and analysis of data taken by the CIAL members. At present, the 48 CIALs are attended by three support farmers, backed up principally by one trainer-agronomist.

Results and discussion

The procedure in forming CIALs (Table 1) was developed in a pilot phase in 1990-91. Five farmers' research committees were established and trained in techniques of partici-

Table 1. Procedures in forming a new CIAL.

1.	On-farm research and extension staff of the host institution (the state agency, an NGO, or a farmer cooperative) receive training in the CIAL methodology and select communities or respond to requests from communities to form a CIAL.
2.	The host institution calls a community meeting in which farmers make a group analysis of what it means to experiment with new agricultural practices, discuss local experience with experimentation and its results, and explain the purpose of a local research committee.
3.	If the community decides to establish a CIAL, it forms a four-member committee composed of farmers recognized locally as experimenters (before they are elected to this committee, the leadership qualities of these farmers are assessed by members of the community).
4.	The CIAL conducts a diagnosis during the community meetings. In these meetings, possible topics for the CIAL experiment are prioritized.
5.	In a planning meeting with their host institution's agronomist, the CIAL defines the objective of their experiment, decides on the treatments and the check, determines criteria for site selection, specifies the timing, inputs, and data needed to draw conclusions from the trial, and outlines responsibilities for different tasks. In the first training cycle, a support farmer visits the CIAL regularly as these tasks are being implemented.
6.	After making the plans, the CIAL carries out the activities from planting to harvest and manages the community's CIAL fund. The CIAL fund is a collective rotating fund, in which each CIAL has a share. In Colombia, the CIAL fund amounts to less than 50% of the value of a head of livestock in the pilot area (US$375 per CIAL at current exchange rates).
7.	Once the crop has been harvested, the CIAL meets with the agronomist to draw conclusions from the data they have taken; the CIAL then presents the results in a community meeting.
8.	The community meets to hear CIAL's oral report of its activities, results, and financial status. If appropriate, another diagnosis is done to plan CIAL's activities for the next season.
9.	In the second and subsequent cycles of experimentation, the support farmer conducts two or three monitoring visits.

patory diagnosis, planning and establishing replicated on-farm trials, participatory evaluation of technology, analysis and interpretation of results, and budget analysis of the total cost of the trial and of individual treatments. Part of the process was planning and presenting a short oral report on the results to the five CIALs which met as a group, and to each community. After the first training cycle, the agronomists in the IPRA project team began to gradually hand over each operation in the process to the farmers. A sociologist made regular monitoring visits to assess how the farmers managed each operation and to detect when follow-up training was required.

On the basis of this experience, twelve CIAL handbooks were prepared for training using discussions with the farmers involved. At the end of 1991 (second phase), these training materials were used in a course with NGOs in the pilot area to prepare agronomists and their staff to establish new CIALs. As a result, 13 CIALs were established. Monitoring of the second phase now covered 18 CIALs; the training handbooks were revised and finalized. In 1992, the project began a regular course on the CIAL method for NGO trainees (who were university students doing a 6 mo agricultural extension practicum in the rural areas with the NGO), state extension agents, and local community leaders.

In the third phase, 28 more CIALs were formed in response to requests from communities and farmer associations. In 1993, trainees in the course on the CIAL method included three farmers who were members of CIALs formed in the second phase. These support farmers were contracted (one by an NGO, one by a farmer cooperative, one by the project) to form new CIALs in the third phase.

Devolution of responsibility for adaptive testing

One of the most important questions is "What types of responsibility for location-specific technology testing can be successfully taken on by experimenting farmers organized in a CIAL?" The 48 CIALs in the pilot area have been established successively by trainees after learning the methodology. This has allowed us to evaluate the training requirements for setting up new CIALs, the rate at which CIAL can be progressively "detached" from their trainer, and the rate at which they can take over the responsibility for carrying out experiments in the absence of a trainer.

Table 2 summarizes the type of institutional support such as training and monitoring visits required by a fully trained CIAL to conduct a crop-related on-farm experiment. When farmers' committees working on their own are interested in the results, they confidently and accurately recorded results for separate treatments without confusing them. In the planning meeting, when farmers have defined the data they want to take in measurement units that make sense to them, they are able to analyze the data to compare treatments and to assess germination rates and crop development. For example, farmers evaluated height of maize plants as too low (dogs can reach and steal cobs), medium (will withstand wind and resist lodging, (desired), and too tall (susceptible to lodging). Farmers commonly process yield data in terms of yield per unit of seed because they do not customarily use measures of area, although experimental plots are measured and staked out.

Table 2. Activities carried out by a CIAL in a crop-related experiment and institutional support required.

Type of activity	Number of activities	Training and assistance by agronomist	Para-professional	Monitoring by para-professional farmer
Group diagnosis	1	–	–	(1)[a]
Planning	1	1	–	–
Plot selection	1	–	1	–
Land preparation	1	–	–	–
Obtaining inputs	1	–	–	–
Establish experiment	2	–	–	–
Check germination	2	–	–	–
Crop management	Variable	–	–	–
Midterm evaluation	2	–	1	–
Harvest evaluation	1	–	–	–
Analysis of results	1	1	–	–
Community report	2	–	–	1
Total	15	2	2	2

[a]Infrequent.

Monitoring of the 48 CIALs showed that a fully trained CIAL (with two cycles of experience) required training support in 4 of the 15 activities listed in Table 2. In phases I and II, the presence of a trained agronomist was required in two activities: planning, including statistical design, and analysis of results. Two visits by a support-farmer were desirable to check that plot selection is consistent with the experimental objectives and to ensure that the data are taken accurately at midterm or harvest evaluation. Monitoring visits routinely involve a visit by a support farmer to the community meetings where diagnosis and reports are given. By phase III, the support farmers were beginning to take over the responsibility of planning the CIAL trials and analyzing the results, in consultation with one of the host institution's agronomists.

Most of the CIALs' research questions can be addressed by single-factor experiments (e.g., 6-10 varieties superimposed on local cultural practices, or 3-4 fertilizer or pest control treatments). This makes it possible for paraprofessional farmers to plan trials and analyze data, with an agronomist checking the design and helping in the interpretation of results. (In practice, as much training in design and analysis of farm trials is required of trainee agronomists as that required of farmers.)

The research agenda defined by the CIALs are evolving from principally germplasm-based strategies (new crops and varieties) at the beginning to an interest in cultural practices once a viable new crop or locally adapted variety has been selected. Paraprofessional farmers with adequate training can support CIALs in the design of a two-factor experiment: planting density x fertilizer dosage

Table 3 shows the rate of increase in activities carried out by CIALs independently of institutional support such as training. The data for the first cycle show how the number of training visits required has gone down from 17 needed to develop the method and the training materials in Phase I to 10 in Phase III. The number of training visits has been reduced because some activities listed in Table 3, such as obtaining inputs, and repeated activities like observation or evaluation of the experiment, can be carried out independently by the farmers even during the first training cycle.

Table 3. Rate of increase in independence of CIALs from institutional support, 1991-94.

Phase of CIAL formation	No. of CIALs	Mean number of visits per training cycle				
		Cycle 1	Cycle 2	Cycle 3	Cycle 4	Cycle 5
Phase I (pilot)	5	17	10	4	2	2
Phase II	13	12	7	3	3	–
Phase III	30	10	4	4	3	–

The newest CIALs formed in Phase III are operating, by cycle 2, with an average of 4 visits for training and support (Table 3). The five pilot CIALs formed in Phase I have continued increasing their autonomy: by the last cycle, pilot CIALs were operating with only two support visits (one for planning the experiment, one for analysis of results) by the paraprofessional farmer. In conclusion, the training of farmer research committees can be accomplished in two cycles (i.e., during two experiments); and a fully trained CIAL can take over the responsibility for the majority of the activities required for farmers to implement on-farm experiments.

Experience in 1993-94 (phase III) indicates that support farmers (with two cycles of experience as a CIAL member) who are contracted as support farmers can provide almost all of the training and monitoring support required in CIAL's simple on-farm trials. This permits the agronomists involved in the project to delegate the job of planning and analysis needed to routinely support fully trained CIALs and to form more CIALs. This, however, places heavier requirements on the paraprofessionals' skills, particularly with regard to reinforcing the CIAL's management capacities.

Quality of research conducted by CIALs

The next issue is the quality of research carried out by the CIALs. Evaluating the quality of farmers' research is related to the issue discussed earlier--the usefulness to farmers of the scientific method as compared with folk experimentation. The project's strategy is to combine both approaches: a formal experiment is planned and established, but if farmers decide to make changes in treatments or alter the experiment along the way in the style of folk experimentation, then the only requirement is that this is a decision made in the committee by the collaborating farmers.

CIAL experiments are established with a minimum of three replications (farmers), and on occasions are replicated within each site, if recommended by an agronomist. Trials are established on land belonging to members of the CIAL, on communal land, or on land belonging to other farmers, and have included rental or sharecropping arrangements common in the community. CIAL selects the site and conducts a follow-up visit to check if the proposed sites are consistent with the planned experimental objectives. Consequently, the trials involve land rental, share cropping, labor donation, and others carried out under a variety of collaborative arrangements. Observations and evaluation of the trial's progress may involve several experimenting farmers, already identified by the CIAL committee members as knowledgeable experts on the topic chosen during the community's diagnostic meeting. These farmers then take part in planning and implementing the replications. There exists, therefore, scope for farmers to intervene and to combine folk experimentation with the formal experimental design.

The project has monitored the quality of CIAL's research based on three criteria: (1) Was the experiment interpretable by farmers and also statistically analyzable? (2) Were farmers still satisfied that they could draw useful conclusions from the experiment, even if statistically not analyzable? (3) Did farmers conclude that they could not draw useful information from the experiment? The evaluation asks, therefore, if farmers perceive the experiments as useful for generating information, and if the experiments have the potential to supply useful information to formal research and extension systems.

Of the 273 CIAL-managed trial plots during 1991-94, an average 75% could be used for statistical analysis, while 90% were successful by farmers' criteria. In phase I, 91% of the plots were judged interpretable by farmers, although fewer (84%) were statistically analyzable. In Phase II, only 62% were statistically analyzable, although farmers still judged 89% to be interpretable for their purposes. The reasons for this drop in percentage of statistically analyzable plots were detected in the self-evaluation exercise conducted with each CIAL. In the second phase, delayed supply of inputs for the CIAL experiments together with those for the NGO's credit program caused the problem. When CIALs were allowed to manage the petty cash fund to buy inputs, their capacity to implement their trials in a timely fashion improved significantly. In phase III, the number of plots lost to analysis due to late planting decreased to three (managed by one CIAL); the remainder were lost due to other factors.

Reasons for success or failure of devolution

Why is this degree of responsibility and accuracy in conducting adaptive trials achieved by resource-poor farmers who are themselves struggling with their own plots and farms? This is especially puzzling in view of the huge resources devoted in the past to training and equipping teams of on-farm (i.e., farming systems) researchers for whom obtaining farmer collaboration or participation in formal experiments was a major source of frustration (Lightfoot and Barker 1988). An analysis indicates that five CIALs became inactive in 1991-94 (11% of the total number formed), and that several factors determine the degree to which a CIAL makes a commitment to managing its experiments with minimum institutional support.

First, the CIAL's training must successfully impart the principle that the committee's objective is to experiment, to generate knowledge, and to disprove or discredit unreliable recommendations. If this objective is not clear, the CIAL members experience loss of purpose if an experiment shows that the local practice is in fact the best available alternative to the innovation being tested. Our experience shows a truly impressive persistence of some CIALs in the face of several experiments which did not identify a promising innovation compared with local practice. In this situation, contact among CIALs is an important ingredient of success; one CIAL benefits from the others' experimentation and is motivated by it.

Second, it is useful for a CIAL to include one literate member who can read the CIAL handbooks aloud to the other members and who can keep records and tally accounts because this facilitates management. However, written records have proved important mainly to the host institution which is collating data from several CIALs. Illiterate farmers recall, with amazing facility, complicated varietal code names, layout of treatments, and differences among treatments if they perceive the information to be important and useful to them. Thus, literacy may not be a prerequisite for farmers to carry out the CIAL's adaptive research responsibilities; support by the paraprofessional may have to be more intensive over an extended period. Nor does functional illiteracy prevent farmers from exchanging results with one another, since the oral tradition is strong in these communities.

More critical to success is identification, in the group diagnosis, of a problem or question for the CIAL's experiment which the farmers want to answer and which is of interest to the community. This is why monitoring visits at the diagnostic stage and during reporting time are important: they ensure that the committee feels accountable to its community and, at the same time, gets encouragement from the interest shown in its results. The sense of community service and responsibility to the group welfare created and reinforced in these meetings is possibly the single most important determinant of successful completion of the experiment by a CIAL. Monitoring shows that conflict in the community and/or in the CIAL is conversely most likely to result in an inactive CIAL. For this reason, the approach includes use of a technique for CIAL's periodic evaluation of how its members feel about each other and their relationship to their community, which the committee can use with or without the support farmer's presence.

Managing a CIAL and experiments that command the respect of "outsiders" has proved to be a useful tool for "pulling in" the attention and resources of institutions external to the community. One CIAL has, for example, successfully negotiated a grant of land for a communal farm from the state, on the strength of demonstrated management capacity and teamwork. Others have attracted marketing arrangements with middlemen who previously would not journey to a distant village but are now attracted by the quality and quantity of produce resulting from experimentation. Yet others have persuaded the NGOs to introduce results of CIAL experiments into their credit programs. The motivation to run an experiment autonomously is as much related to its organizational function as an interface with external organizations as its usefulness to generate knowledge.

Impact of CIAL trials

The 48 CIALs in the pilot area have, with a decreasing amount of institutional support, carried out a large and increasing number of on-farm trials which farmers consider useful for knowledge generation and which are, to a very large extent, statistically analyzable. A rapid appraisal of the CIALs' impact showed that in 75% of the participating communities, there was a perceived benefit from their CIAL in the form of new seeds, new cultural practices, or information about which recommendations to follow. Of those CIALs with no perceived benefit, all but two were formed in phase III and are, therefore, newcomers with less likelihood of impact so far.

State institutions in the pilot area set research and extension priorities on the basis of area devoted to different crops in the municipality: thus, in the pilot area, the priorities are cassava, pastures, sugarcane, and coffee. Small farmers participating in the CIALs' diagnostic meetings had different priorities, as shown by the crops selected for CIAL experiments listed in Table 4. It is apparent that the communities identified a much more diverse research agenda than the institutions. For example, not one community prioritized cassava in their group diagnosis, although more than 4,000 farmers have participated in community meetings for this purpose. Diversity in the CIAL agenda reflects the farmers' objectives to identify alternatives to traditional cash crops (coffee and cassava) and to increase their food sufficiency by growing staples, such as potato, beans, and maize (for feeding chickens, an important source of locally produced protein, and an important ingredient of traditional staple dishes). Cauca Department imports these staples from other parts of the country to meet its food requirements (SAG 1989), so the local food self-sufficiency agenda reflects a regional problem.

At present, the CIALs are having some success in exerting demand pull and improving the diversity of technologies on offer via state and NGO programs. For example, in the NGO credit and technical assistance programs, maize and peas were introduced as a result of CIAL's experiments, and beans were given more importance. The municipal credit and technical assistance agencies (UMATA) also began to respond to farmers' priorities, especially when some of the CIALs started producing seed of varieties they had selected in their trials, which the UMATA recommended and distributed to other farmers. One UMATA recently began to use the CIALs' results to formulate recommendations to

Table 4. Experiments of local agricultural research committees.

Research topic	Number of experiments		
	Phase I	Phase II	Phase III
Peas and related cultural practices	1	3	5
Potato	6	5	3
Maize and related practices	7	8	7
Peanut	1	3	4
Fruits and related fertilizer dosages, pest control	3	6	11
Beans	6	4	0
Snap beans	0	1	1
Tomato	1	1	1
Soya bean	1	2	1
Sugarcane	1	1	1
Vegetables	4	7	6
Chicken feed mixes	1	1	0
Forage grasses	1	1	1
Cover crops (green manure)	0	0	1
Guinea pigs	0	0	1
Total	33	43	43

farmers participating in its credit program. Another responded to a CIAL's request for help in identifying peanut varieties for testing by obtaining new varieties from ICRISAT.

An important development was the evolution of some CIALs that had successfully selected new locally adapted crop varieties into small seed-production enterprises delivering seed of improved and local varieties to farmers in the area. By mid-1996, six CIALs had begun to produce seed from six varietal trials conducted over 3 yr, which have progressed to commercial-scale plots. These CIALs receive additional training in simple seed production, processing, and quality control techniques from the agronomist-trainer. When the national seed certification agency has visited these CIAL seed plots and given its approval, the seed can be sold under the category "farmer-improved seed."

Table 5 shows the quantity of seed produced by the six CIAL seed enterprises. The seed is distributed locally in the village stores and weekend markets. An estimated 281 ha of maize, 3064 ha of beans, and 3.5 ha of field peas (an entirely new crop introduced by CIAL experimentation into the pilot area) have been planted with CIAL seed. More than 10,000 farmers have purchased CIAL seed, which over one planting season is estimated to have produced grain with a gross value of more than US$2 million.

The differential yield between locally available and improved varieties selected by CIALs represents an additional gross income of US$765,000 to local farmers from maize and beans, and a newly introduced income source worth more than US$8,000 from peas. On a per capita basis, this represents an increment worth about a month's wage income in one planting season to farmers who purchased CIAL seed.

The seed enterprises also generate employment since they must hire additional labor to plant, harvest, sort, clean, and pack the seed in 1- to 5- kg sacks, also made locally by women. The bean seed enterprises have, for example, generated an average of 20,000 labor days of employment locally over five seasons, worth an estimated US$85,000 at current wage rates, over the 5 yr of operation.

This impact has been achieved by six CIALs formed early in the project. There is no guarantee, of course, that the newer CIALs will repeat this experience by identifying new practices or germplasm with comparable impact. The six CIALs that have developed into seed enterprises may have already captured the best opportunity and the windfall profits from participatory breeding and seed production. The impact of the newer CIALs may be more difficult to realize especially because, as Table 5 indicates, their research agenda are shifting emphasis from grains to perishables. On the other hand, the recent

Table 5. Seed production by six CIALs and its estimated impact over one planting season.

Crop	CIALs (no.)	Total seed production (kg)	Estimated area planted (ha)	Estimated production (t)	Farmgate price (US$ t^{-1})	Gross value (US$ 000)	Increment in production (%)	Value of increment (US$ 000)
Beans	2	147,080	3064	3,064	683	2,093	30	628
Maize	2	8,430	281	1,124	488	549	25	137
Peas	2	136	7	3.5	2,439	8	100	8
Total	6	155,646	3352	4,191.5	–	2,650	–	773

introduction of field peas, via CIAL experimentation, suggests that there could be scope for a significant increase in impact from CIAL experimentation with high-value crops.

Scaling up and the costs of the CIAL program

One important question is to what extent, and at what cost, this type of farmer participation in research could be scaled up to achieve broad coverage. Before participatory research became fashionable, critics often asked whether this approach was an expensive luxury, attractive on a case-to-case basis and when supported by highly skilled professionals, but not affordable for working with large number of farmers (Farrington and Martin 1988).

In this section we present information on the potential of the CIAL method to increase the efficiency of salaried personnel working in on-farm adaptive research and extension, by decreasing the amount of time required for them to carry out field trials. We also examine the operating costs of the CIAL corporation, a second-order organization formed by the CIALs in Cauca, to provide some insights into the feasibility and costs of creating self-sustaining CIALs.

We compare the amount of time required to conduct an on-farm trial with and without a CIAL to assess its contribution to public sector efficiency. Table 6 presents the estimated staff requirement and cost for an on-farm trial run by an extension agent, by a new CIAL in the first cycle of training, and by a fully trained CIAL. The analysis is based on the activities in Table 2, for which we estimate that an extension agent would require 8 person-days for a trial with up to three replicates (sites). A new CIAL requires an average of 10 training visits by the support farmer plus 1 person-day of extension agent input to do the same job. A fully trained CIAL can carry out a trial with 4 person-days of support training from the support farmer and a fraction of the input from the extensionist, conservatively valued here at 1 d. Estimates of management costs show that even training a new CIAL to conduct an on-farm trial is less costly than running a trial with a salaried professional, given the pay differentials for the pilot area. More important, devolving an on-farm trial to a fully trained CIAL costs 60% less than running a trial using an extension agent. Thus, adaptive research programs run by CIALs could potentially reduce

Table 6. Comparison of staff requirement and cost of an on-farm trial managed by CIAL and formal extension.

Trial management by	Person-days required (no.)[a]	Relative cost of salaried staff time (%)[b]
Extension/research	8	100[c]
New CIAL (cycle 1)	11	74
Fully trained CIAL	5	37

[a]Excludes crop management after trial establishment which is variable depending on the crop; and initial diagnosis.
[b]Support farmer's time valued at minimum wage; extension agent valued at twice the minimum wage; agronomist time valued at current average salary in the pilot area.
[c]100% works out to US$62.00 per on-farm trial.

management costs by up to 60%. Alternatively, a given amount of professional staff time can be expected to at least double its coverage by working with CIALs.

Important variables that affect the efficiency and coverage of adaptive research are the variability of micro-agroecological regimes, the population density, and the type of terrain, which affect the amount of time required for site visits. Before going on to examine the current operating costs of the CIAL corporation, the socio-geographical context of the project area needs to be described.

As of early 1995, there were a total of 55 CIALs scattered in nine *municipios* in the Department of Cauca (Fig. 1), which together composed an area of 6648 km^2. The participating communities presented an area of influence of approximately 2605 km^2 with a much higher mean density of 132 persons km^{-2} when the extensive cattle and forest holdings were excluded. The 55 CIALs were not distributed geographically across this area on the basis of, for example, the requirements of particular agroecological niches or client groups for location-specific adaptive testing. The project's primary objective was methodology development; the second was assessment of how different institutional linkages affect farmers' capacity to do participatory research in agriculture, so that CIALs were located geographically to facilitate comparison among institutional settings. Mean per capita crop land of 0.25 ha in the study area was comparable with estimates for Bolivia (0.33 ha), Ecuador (0.25 ha), and Peru (0.17 ha) (Pachico et al 1994).

Since 1991, the CIALs in the project area have decided to meet annually to exchange results. In this meeting, financed by raising money in their communities for transportation and lodging, CIALs report on their experiments, exchange seed, swap notes about their host institutions, and formulate recommendations on how to achieve the goals which they themselves establish in each meeting. This experience prompted the election of a central coordinating committee (*junta*) in 1993 and 1994, leading to the CIALs' decision to incorporate legally at the recommendation of the junta. Donations were obtained, which enabled the CIAL corporation to establish an investment fund from which the corporation could draw up to 70% of the interest (the remainder going back into the capital) for operating expenses. This allowed the CIALs to be self-sustaining. In addition, the support farmers began to give courses to the municipal extension services (UMATAS), which contracted them to form a number of pilot CIALs elsewhere, paying up to 50% of their salary, and so generating additional income for the corporation.

Very little data is published on the costs of doing adaptive research with groups of farmers, making comparison of figures shown in Table 7 difficult. Total operating cost per CIAL per year is US$502, while annual per capita cost ranges from $125 (if we consider only the 220 CIAL committee members) to under $1 (if we look at the total population in the area of influence). It would be $6.50 if we assume that only a third of the population in the CIAL communities has any contact with the CIAL trials. Based on the estimated number of purchasers of CIAL seed, the cost per capita would be approximately $3. The total annual operating budget of the CIAL corporation currently amounts to an equivalent of about two agronomists' salaries. These figures compare favorably with annual costs cited by Nimlos and Savage (1991) of $36 per capita and $2664 per community for an extension program using village-level support farmers in Ecuador. Also in Ecuador, Romanoff (1993, as cited in Bebbington et al 1994) reports

Table 7. Annual costs of the CIAL corporation for operating 55 CIALs.

Item	Annual costs (US$)
Annual cost per CIAL	
Personnel costs[a]	290.00
Cost of experiments[b]	90.00
Other operational costs[c]	122.00
Total	502.00
Annual per capita cost center	
Cost per capita	*$ per capita per year*
Total population (50,000)	0.55
CIAL communities (12,900)	2.10
33% of CIAL communities (4,260)	6.50
Seed purchasers (10,500)	2.60
CIAL Committee members (220)	125.50

[a]Includes agronomist (0.33), farmer coordinator (1.0), and support farmers (2.0).
[b]Average of costs per CIAL charged against CIAL funds in 1994.
[c]Average of transportation, supplies, and capital depreciation on four motorcycles.

that the cost of forming farmer groups of 10-30 members each for processing and marketing cassava is much larger and more complex than that incurred in forming CIALs.

With regard to coverage, Schwartz (1994) cites figures ranging from 100 to 300 farmers per private sector extension agent to 3,000 per public sector extension agent from case studies in Nigeria, Kenya, and Thailand. Comparable figures for the CIAL corporation can be estimated between 66 (direct contact with committee members) and around 3,000 (population in CIAL communities or number of seed purchasers, for example) per salaried paraprofessional/agronomist. However, because the CIALs do not at this time have a complete extension function, but rather a partial research/extension function—that of facilitating adaptation of technology—this comparison is not completely equivalent.

One reason for the relatively low cost of forming and running CIALs may be that the procedures for creating these groups were formalized fairly early on in the process in the form of easy-to-use training materials. Use of these materials means that support farmers with practical experience in the procedures can form and run new CIALs without previous training in the method (Soria, 1995, pers. commun.).

Nonetheless, the cost data presented here should not be viewed as conclusive, to the extent that further testing of the approach without the intervention of the originators (the IPRA project team) is under way, and will permit assessment of how robust and replicable the method is in different environments, with variant cost structures.

Equity

An issue related to assessing the effectiveness of the CIAL method in broadening the coverage of adaptive research is the question "How equitable is the distribution of benefits?" We have yet to conduct a comprehensive analysis to address this question, but survey data on a subpopulation of 11 communities provide some insights. As discussed earlier, selection of CIAL members is predicated on the assumption that experimenting farmers are likely to be the relatively better-off members of the rural community. Moreover, the CIALs are not designed to involve a large population in research; the committee mobilizes a capacity to test technology within the community, on the basis of a limited participation in the actual research. Therefore, distribution of knowledge about a CIAL's activities rather than participation in it is a more important test of the extent and nature of the coverage achieved.

A comparison of three social strata differentiated on an index of well-being (Ravnborg 1994) shows that of the 64 farmers actively participating in 11 CIALs, 39% came from the upper stratum and 22% were from the lower stratum (chi square $P = 0.046$). However, the community population was essentially one of small farmers and in this subsample, there was no significant difference in farm size between those who participated in the committees (average farm size is 4.4 ha) and those who did not (average farm size 3.5; probability of $T = 0.1484$). Similarly the knowledge about CIALs was evenly distributed between the poor (49%) on one side and the two better-off strata (51%) on the other side (chi square: $P = 0.491$).

Another issue is to what extent special interest groups in the community are able to get their priorities onto the CIAL research agenda defined in the community diagnosis. The project observed a marked tendency for few women to attend these meetings, and those who do attend often propose research problems that are not prioritized. To address this need, the project established a separate fund for two communities to set up two exclusive women research committees, groups formed CIALs; and four others added women members to the committees. Women still represent only 7% of committee members. The main reason appears to be the difficulty women face in devoting time to regular meetings that take them out of the home. For special interest groups like women or the landless poor, the research committee may not be an appropriate instrument for addressing their special research agenda. Several options have yet to be explored: for example, separate diagnosis with special interest groups to identify priorities, which are then included as treatments in trials carried out by CIAL members. This raises the question, however, of the degree of motivation of CIAL members to carry out trials on topics of secondary importance to them, and of the more powerful members of the community. Another option is to have special interest groups evaluate the trials so that their criteria for what is desirable innovation are included in the analysis and recommendations drawn from CIALs' research. Increasing the equity of coverage by adaptive research may have to be achieved by targeting the very poor, with the "slack" research/extension capacity of intermediary organizations created by devolving part of the research agenda to CIALs. These issues are topics for further empirical research, now starting in the project.

Conclusions

Forming and monitoring the evolution of the CIALs is an ongoing experiment to assess the feasibility and the implications of devolving the responsibilities for adaptive agricultural research to farmers. The CIALs were formed to investigate to what extent the methods for participatory diagnosis and problem definition, planning and evaluation, and, ultimately, monitoring of adaptive technology testing could be handed over to community-level organizations to generate "demand-pull" on formal research and extension systems, and to improve the access of resource-poor farmers to an adaptive technology testing service at a reasonable cost. Our experience so far suggests that it is possible to institutionalize this responsibility with farmers, that it is not unrealistic to expect "hard-data" from farmer-managed trials, and that this demonstration of farmers' capability wins respect for farmers, which is catalyzing a gradual reorientation of bureaucratic institutions' priorities. Current costs and coverage compare favorably with some state or private sector systems, although the basis for comparison is very limited. A favorable cost structure is clearly related to the demonstrated effectiveness of paraprofessionals, such as support farmers, for scaling up and for achieving devolution.

The project is entering a new phase with an international training program and monitoring of new CIALs, which are spreading in widely contrasting sociocultural environments as distant as Brazil and Honduras. Many questions remain about the long-run viability of the CIALs as an approach to institutionalizing farmer participation in agricultural research. But there are already some signals--such as the CIAL Miske in Bolivia which is reaching out to serve 22 communities, or the CIALs (called CALITS) in Peru which organized as a group to campaign for support from their regional state experiment station--that a quiet revolution may be under way to bring to full recognition the farmers' capacity as researchers.

References

Ashby J A, Sperling L (1994) Institutionalizing participatory client driven research and technology development in agriculture. Agricultural Administration (Research and Extension) Network Paper No. 49. Overseas Development Institute, London.

Ashby J, Quiros C A, Rivera Y (1987) Farmer participation in on-farm varietal trials. Agricultural Administration Network Discussion Paper No. 21. Overseas Development Institute, London.

Axinn G (1994) Responses to Newsletter 29. In Agricultural Research and Extension Network Newsletter 30. Overseas Development Institute, London. 18 p.

Bebbington A J, Merrill-Sands D, Farrington. J (1994) Farmer and community organizations in agricultural research and extension: functions, impacts and questions. Agricultural Administration (Research and Extension) Network Paper No. 47. Overseas Development Institute, London.

Berg T (1993) The science of plant breeding--support of alternative to traditional practices. In de Boef W, Amanor K, Wellard K (Eds.) Cultivating knowledge, genetic diversity, farmer experimentation and crop research. Intermediate Technology Publications, London. p 72-77.

Chambers R (1994) Foreword to Beyond farmer first. Scoones I, Thompson J (Eds.) Intermediate Technology Publications Ltd., London.

Drinkwater M (1994a) Knowledge, consciousness and prejudice. In Scoones I, Thompson J (Eds) Beyond farmer first. Intermediate Technology Publications, London. p 32-41.

Drinkwater M (1994b) Developing interaction and understanding: RRA and farmer research groups in Zambia. In Scoones I, Thompson J (Eds.) Beyond farmer first. Intermediate Technology Publications, London. p 133-139.

Farrington J, Martin A (1988) Farmer participation in agricultural research: a review of concepts and practices. Agricultural Administration Unit Occasional Paper No. 9. Overseas Development Institute, London.

Hardon J J, de Boef W S (1993) Linking farmers and breeders in local crop development. In de Boef W, Amanor K, Wellard K (Eds.) Cultivating knowledge, genetic diversity, farmer experimentation and crop research. Intermediate Technology Publications Ltd., London. p 64-71.

Heinrich G, Worman F, Koketso C (1991) Integrating FPR with conventional on farm research programs: an example from Botswana. J. Farming Syst. Res. Ext. 2:1-15.

Lightfoot C (1987) Indigenous research and on-farm trials. Agricultural Administration and Extension. p 2479-2489.

Lightfoot C, Barker R (1988) On farm trials: a survey of methods. Agricultural Administration and Extension. p 3015-3023.

Mattee AZ, Lasalle T (1994) Diverse and limited: farmers' organizations in Tanzania. Agricultural Administration (Research and Extension) Network Paper No. 50. Overseas Development Institute, London.

Merrill-Sands D, Biggs S D, Bingen R J Ewell P T, McAllister J L, Poats S R (1991) Integrating on-farm research into national agricultural research systems: lessons for research policy. Organization and management. In Tripp R (Ed.) Planned change in farming systems: progress in on-farm research. John Wiley & Sons, Chichester, UK. p 287-316.

Muchagata M G, de Reynal V, Verga Jr. I P (1994) Building a dialogue between researchers and small farmers: the Tocantins AgroEcology Centre (CAT) in Brazil. Agricultural Administration (Research and Extension) Paper No. 50 d. Overseas Development Institute, London.

Mushita T A (1993) Strengthening the informal seed system in communal areas of Zimbabwe. In de Boef W, Amanor K, Wellard K (Eds.). Cultivating knowledge, genetic diversity, farmer experimentation and crop research. Intermediate Technology Publications, London. p 85-88.

Nimlos T J, Savage R F (1991) Successful soil conservation in the Ecuadorian Highlands. J. Soil Water Conserv. September-October. p. 341-343.

Okali C, Sumberg V, Farrington. J (1994) Battlefields and trial plots: rhetoric and reality of farmer participatory research. Intermediate Technology Publications, London

Pachico D, Ashby J A, Sanint L R (1994) Natural resource and agricultural prospects for the hillsides of Latin America. Annual report, 1993-1994. CIAT Hillside Program, Centro Internacional de Agricultura Tropical, Palmira. p 283-321.

Ravnborg H (1994) Methodology for identifying different levels of well-being, and their distribution. Internal Discussion Paper. CIAT Hillside Program, Centro Internacional de Agricultura Tropical, Palmira.

Richards P (1989) Agriculture as a performance. In Chambers Pacey R A, Thrupp L A (Eds.) Farmer First: Farmer innovations and agricultural research. Intermediate Technology Publications, London.

SAG (1989)_El Cauca: tierras de progreso. SAG, Popayan, Cauca.

Salas M A (1994) The technicians only believe in science and cannot read the sky: the cultural dimension of the knowledge conflict in the Andes. In Scoones I, Thompson J (Eds.). Beyond farmer first. Intermediate Technology Publications, London. p 57-69.

Schwartz L A (1994) The role of the private sector in agricultural extension: economic analysis and case studies. Agricultural Administration (Research and Extension) Network Paper No. 48. Overseas Development Institute, London.

Sperling L, Loevinhson M, Ntabomvura B (1993) Rethinking the farmers' role in plant breeding: local bean experts and extension selection in Rwanda. Exp. Agric.

Worede M, Mekbib H (1993) Linking genetic resource conservation to farmers in Ethiopia. In de Boef W, Amanor K, Wellard K (Eds) Cultivating knowledge, genetic diversity, farmer experimentation and crop research. Intermediate Technology Publications, London. p 78-84.

Developments in Plant and Soil Sciences

1. J. Monteith and C. Webb (eds.): *Soil Water and Nitrogen in Mediterranean-type Environments.* 1981 ISBN 90-247-2406-6
2. J.C. Brogan (ed.): *Nitrogen Losses and Surface Run-off from Landspreading of Manures.* 1981 ISBN 90-247-2471-6
3. J.D. Bewley (ed.): *Nitrogen and Carbon Metabolism.* 1981 ISBN 90-247-2472-4
4. R. Brouwer, I. Gašparíková, J. Kolek and B.C. Loughman (eds.): *Structure and Function of Plant Roots.* 1981 ISBN 90-247-2510-0
5. Y.R. Dommergues and H.G. Diem (eds.): *Microbiology of Tropical Soils and Plant Productivity.* 1982 ISBN 90-247-2624-7
6. G.P. Robertson, R. Herrara and T. Rosswall (eds.): *Nitrogen Cycling in Ecosystems of Latin America and the Caribbean.* 1982 ISBN 90-247-2719-7
7. D. Atkinson, K.K.S. Bhat, M.P. Coutts, P.A. Mason and D.J. Read (eds.): *Tree Root Systems and Their Mycorrhizas.* 1983 ISBN 90-247-2821-5
8. M.R. Sarić and B.C. Loughman (eds.): *Genetic Aspects of Plant Nutrition.* 1983 ISBN 90-247-2822-3
9. J.R. Freney and J.R. Simpson (eds.): *Gaseous Loss of Nitrogen from Plant-Soil Systems.* 1983 ISBN 90-247-2820-7
10. United Nations Economic Commission for Europe (ed.): *Efficient Use of Fertilizers in Agriculture.* 1983 ISBN 90-247-2866-5
11. J. Tinsley and J.F. Darbyshire (eds.): *Biological Processes and Soil Fertility.* 1984 ISBN 90-247-2902-5
12. A.D.L. Akkermans, D. Baker, K. Huss-Danell and J.D. Tjepkema (eds.): Frankia *Symbioses.* 1984 ISBN 90-247-2967-X
13. W.S. Silver and E.C. Schröder (eds.): *Practical Application of* Azolla *for Rice Production.* 1984 ISBN 90-247-3068-6
14. P.G.L. Vlek (ed.): *Micronutrients in Tropical Food Crop Production.* 1985 ISBN 90-247-3085-6
15. T.P. Hignett (ed.): *Fertilizer Manual.* 1985 ISBN 90-247-3122-4
16. D. Vaughan and R.E. Malcolm (eds.): *Soil Organic Matter and Biological Activity.* 1985 ISBN 90-247-3154-2
17. D. Pasternak and A. San Pietro (eds.): *Biosalinity in Action.* Bioproduction with Saline Water. 1985 ISBN 90-247-3159-3
18. M. Lalonde, C. Camiré and J.O. Dawson (eds.): Frankia *and Actinorhizal Plants.* 1985 ISBN 90-247-3214-X
19. H. Lambers, J.J. Neeteson and I. Stulen (eds.): *Fundamental, Ecological and Agricultural Aspects of Nitrogen Metabolism in Higher Plants.* 1986 ISBN 90-247-3258-1
20. M.B. Jackson (ed.): *New Root Formation in Plants and Cuttings.* 1986 ISBN 90-247-3260-3
21. F.A. Skinner and P. Uomala (eds.): *Nitrogen Fixation with Non-Legumes* (Proceedings of the 3rd Symposium, Helsinki, 1984). 1986 ISBN 90-247-3283-2
22. A. Alexander (ed.): *Foliar Fertilization.* 1986 ISBN 90-247-3288-3
23. H.G. v.d. Meer, J.C. Ryden and G.C. Ennik (eds.): *Nitrogen Fluxes in Intensive Grassland Systems.* 1986 ISBN 90-247-3309-X
24. A.U. Mokwunye and P.L.G. Vlek (eds.): *Management of Nitrogen and Phosphorus Fertilizers in Sub-Saharan Africa.* 1986 ISBN 90-247-3312-X

Developments in Plant and Soil Sciences

25. Y. Chen and Y. Avnimelech (eds.): *The Role of Organic Matter in Modern Agriculture.* 1986 ISBN 90-247-3360-X
26. S.K. De Datta and W.H. Patrick Jr. (eds.): *Nitrogen Economy of Flooded Rice Soils.* 1986 ISBN 90-247-3361-8
27. W.H. Gabelman and B.C. Loughman (eds.): *Genetic Aspects of Plant Mineral Nutrition.* 1987 ISBN 90-247-3494-0
28. A. van Diest (ed.): *Plant and Soil: Interfaces and Interactions.* 1987 ISBN 90-247-3535-1
29. United Nations Economic Commission for Europe and FAO (eds.): *The Utilization of Secondary and Trace Elements in Agriculture.* 1987 ISBN 90-247-3546-7
30. H.G. v.d. Meer, R.J. Unwin, T.A. van Dijk and G.C. Ennik (eds.): *Animal Manure on Grassland and Fodder Crops.* Fertilizer or Waste? 1987 ISBN 90-247-3568-8
31. N.J. Barrow: *Reactions with Variable-Charge Soils.* 1987 ISBN 90-247-3589-0
32. D.P. Beck and L.A. Materon (eds.): *Nitrogen Fixation by Legumes in Mediterranean Agriculture.* 1988 ISBN 90-247-3624-2
33. R.D. Graham, R.J. Hannam and N.C. Uren (eds.): *Manganese in Soils and Plants.* 1988 ISBN 90-247-3758-3
34. J.G. Torrey and J.L. Winship (eds.): *Applications of Continuous and Steady-State Methods to Root Biology.* 1989 ISBN 0-7923-0024-6
35. F.A. Skinner, R.M. Boddey and I. Fendrik (eds.): *Nitrogen Fixation with Non-Legumes* (Proceedings of the 4th Symposium, Rio de Janeiro, 1987). 1989 ISBN 0-7923-0059-9
36. B.C. Loughman, O. Gašparíková and J. Kolek (eds.): *Structural and Functional Aspects of Transport in Roots.* 1989 ISBN 0-7923-0060-2; Pb 0-7923-0061-0
37. P. Plancquaert and R. Haggar (eds.): *Legumes in Farming Systems.* 1990 ISBN 0-7923-0134-X
38. A.E. Osman, M.M. Ibrahim and M.A. Jones (eds.): *The Role of Legumes in the Farming Systems of the Mediterranean Areas.* 1990 ISBN 0-7923-0419-5
39. M. Clarholm and L. Bergström (eds.): *Ecology of Arable Land – Perspectives and Challenges.* 1989 ISBN 0-7923-0424-1
40. J. Vos, C.D. van Loon and G.J. Bollen (eds.): *Effects of Crop Rotation on Potato Production in the Temperate Zones.* 1989 ISBN 0-7923-0495-0
41. M.L. van Beusichem (ed.): *Plant Nutrition – Physiology and Applications.* 1990 ISBN 0-7923-0740-2
42. N. El Bassam, M. Dambroth and B.C. Loughman (eds.): *Genetic Aspects of Plant Mineral Nutrition.* 1990 ISBN 0-7923-0785-2
43. Y. Chen and Y. Hadar (eds.): *Iron Nutrition and Interactions in Plants.* 1991 ISBN 0-7923-1095-0
44. J.J.R. Groot, P. de Willigen and E.L.J. Verberne (eds.): *Nitrogen Turnover in the Soil-Crop System.* 1991 ISBN 0-7923-1107-8
45. R.J. Wright, V.C. Baligar and R.P. Murrmann (eds.): *Plant-Soil Interactions at Low pH.* 1991 ISBN 0-7923-1105-1
46. J. Kolek and V. Kozinka (eds.): *Physiology of the Plant Root System.* 1992 ISBN 0-7923-1205-8
47. A.U. Mokwunye (ed.): *Alleviating Soil Fertility Constraints to Increased Crop Production in West Africa.* 1991 ISBN 0-7923-1221-X; Pb 0-7923-1222-8

Developments in Plant and Soil Sciences

48. M. Polsinelli, R. Materassi and M. Vincenzini (eds.): *Nitrogen Fixation* (Proceedings of the 5th Symposium, Florence, 1990). 1991 ISBN 0-7923-1410-7
49. J.K. Ladha, T. George and B.B. Bohlool (eds.): *Biological Nitrogen Fixation for Sustainable Agriculture.* 1992 ISBN 0-7923-1774-2
50. P.J. Randall, E. Delhaze, R.A. Richards and R. Munns (eds.): *Genetic Aspects of Plant Mineral Nutrition.* 1993 ISBN 0-7923-2118-9
51. K.S. Kumarasinghe and D.L. Eskew (eds.): *Isotopic Studies of Azolla and Nitrogen Fertilization of Rice.* 1993 ISBN 0-7923-2274-6
52. F.A. Bliss and G. Hardarson (eds.): *Enhancement of Biological Nitrogen Fixation of Common Baen in Latin America.* 1993 ISBN 0-7923-2451-X
53. M.A.C. Fragoso and M.L. van Beusichem (eds.): *Optimization of Plant Nutrition.* 1993 ISBN 0-7923-2519-2
54. N.J. Barrow (ed.) *Plant Nutrition – From Genetic Engineering to Field Practice.* 1993 ISBN 0-7923-2540-0
55. A.D. Robson (ed.): *Zinc in Soils and Plants.* 1993 ISBN 0-7923-2631-8
56. A.D. Robson, L.K. Abbott and N. Malajczuk (eds.): *Management of Mycorrhizas in Agriculture, Horticulture and Forestry.* 1994 ISBN 0-7923-2700-4
57. P.H. Graham, M.J. Sadowsky and C.P. Vance (eds.): *Symbiotic Nitrogen Fixation.* 1994 ISBN 0-7923-2781-0
58. F. Baluška, M. Čiamporová, O. Gašparíkova and P.W. Barlow (eds.): *Structure and Function of Roots.* 1995 ISBN 0-7923-2832-9
59. J. Abadía (ed.): *Iron Nutrition in Soils and Plants.* 1995 ISBN 0-7923-2900-7
60. P.S. Curtis, E.G. O'Neill, J.A. Teeri, D.R. Zak and K.S. Pregitzer (eds.): *Belowground Responses to Rising Atmospheric* CO_2. Implications for Plants, Soil Biota, and Ecosystem Processes. 1994 ISBN 0-7923-2901-5
61. P.C. Struik, W.J. Vredenberg, J.A. Renkema and J.E. Parlevliet (eds.): *Plant Production on the Threshold of a New Century.* 1994 ISBN 0-7923-2903-1
62. L.O. Nilsson, R.F. Hüttl and U.T. Johansson (eds.): *Nutrient Uptake and Cycling in Forest Ecosystems* (CEC/IUFRO Symposium, Halmstad, Sweden, 1993). 1995 ISBN 0-7923-3030-7
63. H.P. Collins, G.P. Robertson and M.J. Klug (eds.): *The Significance and Regulation of Soil Biodiversity* (International Symposium, East Lansing, 1993). 1995 ISBN 0-7923-3138-9
64. R.A. Date, N.J. Grundon, G.E. Rayment and M.E. Probert (eds.): *Plant-Soil Interactions at Low pH: Principles and Management* (3rd International Symposium, Brisbane, 1993). 1995 ISBN 0-7923-3198-2
65. J.K. Ladha and M.B. Peoples (eds.): *Management of Biological Nitrogen Fixation for the Development of More Productive and Sustainable Agricultural Systems.* (15th Congress of Soil Science, Acapulco, Mexico, 1994) 1995 ISBN 0-7923-3413-2; Pb 0-7923-3414-0
66. C. Rodriguez-Barrueco (ed.): *Fertilizers and Environment.* 1996 ISBN 0-7923-3729-8
67. B. Buerkert, B.E. Allison and M. von Oppen (eds.): *Wind Erosion in Niger.* Implications and Control Measures in a Millet-based Farming System. 1996 ISBN 0-7923-3885-5

Developments in Plant and Soil Sciences

68. O. van Cleemput, G. Hofman and A. Vermoesen (eds.): *Progress in Nitrogen Cycling.* 1996 ISBN 0-7923-3962-2
69. N. Ahmad (ed.): *Nitrogen Economy in Tropical Soils.* 1996 ISBN 0-7923-4094-9
70. M. Rahman, A. Kumar Podder, C. van Hove, Z.N. Tahmida Begum, T. Heulin and A. Hartmann (eds.): *Biological Nitrogen Fixation Associated with Rice Production.* 1996 ISBN 0-7923-4197-X
71. D. Rosen, E. Tel-Or, Y. Hadar and Y. Chen (eds.): *Modern Agriculture and the Environment.* 1997 ISBN 0-7923-4295-X
72. G.H. Elkan and R.G. Upchurch (eds.): *Current Issues in Symbiotic Nitrogen Fixation.* 1997 ISBN 0-7923-4367-0
73. H.M. Anderson, P.W. Barlow, D.T. Clarkson, M.B. Jackson and P.R. Shewry (eds.): *Plant Roots – From Cells to Systems.* 1997 ISBN 0-7923-4369-7
74. Zhu Zhao-liang, Wen Qi-xiao and J.R. Freney (eds.): *Nitrogen in Soils of China.* 1997 ISBN 0-7923-4372-7
75. J.K. Ladha, F.J. de Bruijn and K.A. Malik (eds.): *Opportunities for Biological Nitrogen Fixation in Rice and Other Non-Legumes* (2nd Working Group Meeting, Faisalabad, Pakistan, 1996). 1997 ISBN 0-7923-4514-2
76. R.W. Bell and B. Rerkasem (eds.): *Boron in Soils and Plants* (International Symposium, Chiang Mai, Thailand, 1997). 1997 ISBN 0-7923-4705-6
77. B. Dell, P.H. Brown and R. Bell (eds.): *Boron in Soils and Plants: Reviews.* 1997 ISBN 0-7923-4729-3
78. T. Ando, K. Fujita, T. Mae, H. Matsumoto, S. Mori and J. Sekiya (eds.): *Plant Nutrition for Sustainable Food Production and Environment.* 1997 ISBN 0-7923-4796-X
79. K.A. Malik, M.S. Mirza and J.K. Ladha (eds.): *Nitrogen Fixation with Non-Legumes.* 1998 ISBN 0-7923-4873-7
80. P.A. Finke, J. Bouma and M.R. Hoosbeek (eds.): *Soil and Water Quality at Different Scales.* 1998 ISBN 0-7923-4994-6
81. V. Balasubramanian, J.K. Ladha and G.L. Denning (eds.): *Resource Management in Rice Systems: Nutrients.* 1999 ISBN 0-7923-5199-1
82. J.E. Box, Jr. (ed.): *Root Demographics and Their Efficiencies in Sustainable Agriculture, Grasslands and Forest Ecosystems.* 1998 ISBN 0-7923-5230-0
83. G. Hardarson and W. Broughton (eds.): *Molecular Microbial Ecology of the Soil.* Results from an FAO/IAEA Co-ordinated Research Programme, 1992–1996. 1998 ISBN 0-7923-5252-1
84. C. Kirda, P. Moutonnet, C. Hera and D.R. Nielsen (eds.): *Crop Yield Response to Deficit Irrigation.* Report of an FAO/IAEA Co-ordinated Research Program by using Nuclear Techniques. 1999 ISBN 0-7923-5299-8

Kluwer Academic Publishers – Dordrecht / Boston / London

GPSR Compliance
The European Union's (EU) General Product Safety Regulation (GPSR) is a set of rules that requires consumer products to be safe and our obligations to ensure this.

If you have any concerns about our products, you can contact us on

ProductSafety@springernature.com

In case Publisher is established outside the EU, the EU authorized representative is:

Springer Nature Customer Service Center GmbH
Europaplatz 3
69115 Heidelberg, Germany

www.ingramcontent.com/pod-product-compliance
Ingram Content Group UK Ltd.
Pitfield, Milton Keynes, MK11 3LW, UK
UKHW051008190926
475607UK00009BA/117
9780792351993